MATHEMATICAL AND COMPUTATIONAL MODELING

MATHEMATICAL AND COMPUTATIONAL MODELING

With Applications in Natural and Social Sciences, Engineering, and the Arts

Edited by

RODERICK MELNIK
Wilfrid Laurier University
Waterloo, Ontario, Canada

Published by John Wiley & Sons, Inc., Hoboken, New Jersey
Published simultaneously in Canada

For general information on our other products and services or for technical support, please contact our Customer Care Department within the United States at (800) 762-2974, outside the United States at (317) 572-3993 or fax (317) 572-4002.

Wiley also publishes its books in a variety of electronic formats. Some content that appears in print may not be available in electronic formats. For more information about Wiley products, visit our web site at www.wiley.com.

Library of Congress Cataloging-in-Publication Data:

Mathematical and computational modeling : with applications in natural and social sciences, engineering, and the arts / edited by Roderick Melnik, Wilfrid Laurier University, Waterloo, Ontario, Canada.
    pages   cm
  Includes bibliographical references and index.
  ISBN 978-1-118-85398-6 (cloth)
1. Mathematical models.   I. Melnik, Roderick, editor.
  QA401.M3926 2015
  511′.8—dc23
                                        2014043267

Set in 10/12pts Times Lt Std by SPi Publisher Services, Pondicherry, India

Printed in the United States of America

10 9 8 7 6 5 4 3 2 1

1   2015

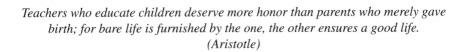

Teachers who educate children deserve more honor than parents who merely gave birth; for bare life is furnished by the one, the other ensures a good life.
(Aristotle)

To my parents who were my first and most devoted teachers.

CONTENTS

3   Numerical Challenges in a Cholesky-Decomposed Local Correlation Quantum Chemistry Framework   59

David B. Krisiloff, Johannes M. Dieterich, Florian Libisch, and Emily A. Carter

LIST OF CONTRIBUTORS

Julien Arino, Department of Mathematics, University of Manitoba, Winnipeg, Canada

Steven J. Brams, Department of Politics, New York University, New York, NY, USA

Emily A. Carter, Department of Mechanical and Aerospace Engineering, Princeton University, Princeton, NJ, USA; and Program in Applied and Computational Mathematics, and Andlinger Center for Energy and the Environment, Princeton University, Princeton, NJ, USA

Julia Chelen, Center for Advanced Modeling in the Social, Behavioral, and Health Sciences, Department of Emergency Medicine, Johns Hopkins University, Baltimore, MD, USA

Ronald R. Coifman, Mathematics Department, Yale University, New Haven, CT, USA

Johannes M. Dieterich, Department of Mechanical and Aerospace Engineering, Princeton University, Princeton, NJ, USA

Matan Gavish, Statistics Department, Stanford University, Palo Alto, CA, USA

Dimitrios Giannakis, Courant Institute of Mathematical Sciences, New York University, New York, NY, USA

Antonios Gonis, Physics and Life Sciences, Lawrence Livermore National Laboratory, Livermore, CA, USA

Ali Haddad, Mathematics Department, Yale University, New Haven, CT, USA

Bernd Hofmann, Faculty of Mathematics, Technische Universität Chemnitz, Chemnitz, Germany

D. Marc Kilgour, Department of Mathematics, Wilfrid Laurier University, Waterloo, Ontario, Canada

Eili Y. Klein, Center for Advanced Modeling in the Social, Behavioral, and Health Sciences, Department of Emergency Medicine, Johns Hopkins University, Baltimore, MD, USA; and Center for Disease Dynamics, Economics & Policy, Washington, DC, USA

David B. Krisiloff, Department of Chemistry, Princeton University, Princeton, NJ, USA

Mel Levy, Department of Chemistry, Duke University, Durham, NC, USA; Department of Physics, North Carolina A&T State University, Greensboro, NC, USA; and Department of Chemistry and Quantum Theory Group, School of Science and Engineering, Tulane University, New Orleans, LO, USA

Florian Libisch, Department of Mechanical and Aerospace Engineering, Princeton University, Princeton, NJ, USA

Andrew J. Majda, Courant Institute of Mathematical Sciences, New York University, New York, NY, USA

Michael D. Makowsky, Center for Advanced Modeling in the Social, Behavioral, and Health Sciences, Department of Emergency Medicine, Johns Hopkins University, Baltimore, MD, USA

Roderick Melnik, The MS2Discovery Interdisciplinary Research Institute, M^2NeT Laboratory and Department of Mathematics, Wilfrid Laurier University, Waterloo, Ontario, Canada

Israel Michael Sigal, Department of Mathematics, University of Toronto, Ontario, Canada

Paul E. Smaldino, Center for Advanced Modeling in the Social, Behavioral, and Health Sciences, Department of Emergency Medicine, Johns Hopkins University, Baltimore, MD, USA

Iman A. Soliman, Department of Mathematics, Cairo University, Giza, Egypt

Ronen Talmon, Mathematics Department, Yale University, New Haven, CT, USA

Nicolae Tarfulea, Department of Mathematics, Computer Science, and Statistics, Purdue University Calumet, Hammond, IN, USA

Godfried T. Toussaint, Department of Computer Science, New York University Abu Dhabi, Abu Dhabi, United Arab Emirates

PREFACE

Mathematical and computational modeling has become a major driving force in scientific discovery and innovation, covering an increasing range of diverse application areas in the natural and social sciences, engineering, and the arts. Mathematical models, methods, and algorithms have been ubiquitous in human activities from the ancient times till now. The fundamental role they play in human knowledge, as well as in our well-being, is indisputable, and it continues to grow in its importance.

Significant sources of some of the most urgent and challenging problems the humanity faces today are coming not only from traditional areas of mathematics applications in natural and engineering sciences, but also from life, behavioral, and social sciences. We are witnessing an unprecedented growth of model-based approaches in practically every domain of human activities. This expands further interdisciplinary horizons of mathematical and computational modeling, providing new and strengthening existing links between different disciplines and human activities. Integrative, holistic approaches and systems–science methodologies are required in an increasing number of areas of human endeavor. In its turn, such approaches and methodologies require the development of new state-of-the-art mathematical models and methods.

Given this wide spectrum of applications of mathematical and computational modeling, we have selected five representative areas, grouped in this book into sections. These sections contain 12 selective chapters, written by 25 experts in their respective fields. They open to the reader a broad range of methods and tools important in many applications across different disciplines. The book provides details on state-of-the-art achievements in the development of these methods and tools, as well as their applications. Original results are presented on both fundamental theoretical and applied developments, with many examples emphasizing interdisciplinary nature of mathematical and computational modeling and universality of models in our better understanding nature, society, and the man-made world.

Aimed at researchers in academia, practitioners, and graduate students, the book promotes interdisciplinary collaborations required to meet the challenges at the interface of different disciplines on the one hand and mathematical and computational modeling on the other. It can serve as a reference on theory and applications of mathematical and computational modeling in diverse areas within the natural and social sciences, engineering, and the arts.

I am thankful to many of my colleagues in North America, Europe, Asia, and Australia whose encouragements were vital for the completion of this project. Special thanks go to the referees of this volume. Their help and suggestions were invaluable. Finally, I am very grateful to the John Wiley & Sons editorial team, and in particular, Susanne Steitz-Filler and Sari Friedman for their highly professional support.

Waterloo, ON, Canada   RODERICK MELNIK
August 2014–2015

SECTION 1

INTRODUCTION

1

UNIVERSALITY OF MATHEMATICAL MODELS IN UNDERSTANDING NATURE, SOCIETY, AND MAN-MADE WORLD

RODERICK MELNIK

The MS2Discovery Interdisciplinary Research Institute, M^2NeT Laboratory and Department of Mathematics, Wilfrid Laurier University, Waterloo, Ontario, Canada

1.1 HUMAN KNOWLEDGE, MODELS, AND ALGORITHMS

There are various statistical and mathematical models of the accumulation of human knowledge. Taking one of them as a starting point, the Anderla model, we would learn that the amount of human knowledge about 40 years ago was 128 times greater than in the year A.D. 1. We also know that this has increased drastically over the last four decades. However, most such models are economics-based and account for technological developments only, while there is much more in human knowledge to account for. Human knowledge has always been linked to models. Such models cover a variety of fields of human endeavor, from the arts to agriculture, from the description of natural phenomena to the development of new technologies and to the attempts of better understanding societal issues. From the dawn of human civilization, the development of these models, in one way or another, has always been connected with the development of mathematics. These two processes, the development of models representing the core of human knowledge and the development of mathematics, have always gone hand in hand with each other. From our knowledge

Mathematical and Computational Modeling: With Applications in Natural and Social Sciences, Engineering, and the Arts, First Edition. Roderick Melnik.
© 2015 John Wiley & Sons, Inc. Published 2015 by John Wiley & Sons, Inc.

in particle physics and spin glasses [4, 6] to life sciences and neuron stars [1, 5, 16], universality of mathematical models has to be seen from this perspective.

Of course, the history of mathematics goes back much deeper in the dawn of civilizations than A.D. 1 as mentioned earlier. We know, for example, that as early as in the 6th–5th millennium B.C., people of the Ancient World, including predynastic Sumerians and Egyptians, reflected their geometric-design-based models on their artifacts. People at that time started obtaining insights into the phenomena observed in nature by using quantitative representations, schemes, and figures. Geometry played a fundamental role in the Ancient World. With civilization settlements and the development of agriculture, the role of mathematics in general, and quantitative approaches in particular, has substantially increased. From the early times of measurements of plots of lands and of the creation of the lunar calendar, the Sumerians and Babylonians, among others, were greatly contributing to the development of mathematics. We know that from those times onward, mathematics has never been developed in isolation from other disciplines. The cross-fertilization between mathematical sciences and other disciplines is what produces one of the most valuable parts of human knowledge. Indeed, mathematics has a universal language that allows other disciplines to significantly advance their own fields of knowledge, hence contributing to human knowledge as a whole. Among other disciplines, the architecture and the arts have been playing an important role in this process from as far in our history as we can see. Recall that the summation series was the origin of harmonic design. This technique was known in the Ancient Egypt at least since the construction of the Chephren Pyramid of Giza in 2500 BCE (the earliest known is the Pyramid of Djoser, likely constructed between 2630 BCE and 2611 BCE). The golden ratio and Fibonacci sequence have deep roots in the arts, including music, as well as in the natural sciences. Speaking of mathematics, H. Poincare once mentioned that "it is the unexpected bringing together of diverse parts of our science which brings progress" [11]. However, this is largely true with respect to other sciences as well and, more generally, to all branches of human endeavor. Back to Poincare's time, it was believed that mathematics "confines itself at the same time to philosophy and to physics, and it is for these two neighbors that we work" [11]. Today, the quantitative analysis as an essential tool in the mathematics arsenal, along with associated mathematical, statistical, and computational models, advances knowledge in pretty much every domain of human endeavor. The quantitative-analysis-based models are now rooted firmly in the application areas that were only recently (by historical account) considered as non-traditional for conventional mathematics. This includes, but not limited to, life sciences and medicine, user-centered design and soft engineering, new branches of arts, business and economics, social, behavioral, and political sciences.

Recognition of universality of mathematical models in understanding nature, society, and man-made world is of ancient origin too. Already Pythagoras taught that in its deepest sense the reality is mathematical in nature. The origin of quantification of science goes back at least to the time of Pythagoras' teaching that numbers provide a key to the ultimate reality. The Pythagorean tradition is well reflected in the Galileo statement that "the *Book of Nature* is written in the language of mathematics." Today, we are witnessing the areas of mathematics applications not only growing rapidly in more traditional natural and engineering sciences but also in social and behavioral sciences as well. It should be noted that the term "universality" is also used in the literature in different, more specific and narrow contexts. For example, in statistical mechanics, universality is the observation that there are properties for a large class of systems that are independent of the dynamical details of the system. A pure mathematical definition of a universal property is usually given based on representations of category theory. Another example is provided by computer science and computability theory where the word "universal" is usually applied to a system which is Turing complete. There is also a universality principle, a system property often modeled by random matrices. These concepts are useful for corresponding mathematical or statistical models and are subject of many articles (see, e.g., [2–7, 14, 16] and references therein). For example, the authors of Ref. [2] discuss universality classes for complex networks with possible applications in social and biological dynamic systems. A universal scaling limit for a class of Ising-type mathematical models is discussed in Ref. [6]. The concept of universality of predictions is discussed in Ref. [14] within the Bayesian framework. Computing universality is a subject of discussions in Ref. [3], while universality in physical and life sciences are discussed in Refs. [7] and [5], respectively. Given a brief historical account demonstrating the intrinsic presence of models in human knowledge from the dawn of civilizations, "universality" here is understood in a more general, Aristotle's sense: "To say of what is, that it is not, or of what is not, that it is, is false; while to say of what is, that it is, and of what is not, that it is not, is true." The underlying reason for this universality lies with the fact that models are inherently linked to algorithms. From the ancient times till now, human activities and practical applications have stimulated the development of model-based algorithms. If we note that abstract areas of mathematics are also based on models, it can be concluded that mathematical algorithms have been at the heart of the development of mathematics itself. The word "algorithm" was derived from Al-Khwarizmi (c. 780 – c. 850), a mathematician, astronomer and geographer, whose name was given to him by the place of his birth (Khwarezm or Chorasmia). The word indicated a technique with numerals. Such techniques were present in human activities well before the ninth century, while specific algorithms, mainly stimulated by geometric considerations at that time, were also known. Examples include algorithms for approximating the area of a given circle (known to Babylonians and Indians), an algorithm for calculating π by inscribing and then circumscribing a polygon around a circle (known to Antiphon and Bryson already in the fifth century B.C.), Euclid's algorithm to determine the greatest common divisor of two integers, and many others. Further development of the subject was closely interwoven with applications and other disciplines. It led to what in the second part of the twentieth century was called by E. Wigner as "the unreasonable effectiveness of mathematics in the natural sciences." In addition to traditional areas of natural sciences and engineering, the twentieth century saw an ever increasing role of mathematical models in the life and environmental sciences too. This development was based on earlier achievements. Indeed, already during the 300 B.C., Aristotle studied the manner in which species evolve to fit their environment. His works served as an important stepping stone in the development of modern evolutionary theories, and his holistic views and teaching that "the whole is more than the sum of its parts" helped the progress of systems science in general and systems biology in particular. A strong growth of genetics and population biology in the twentieth century effectively started from the rediscovery of G. Mendel's laws in 1900 (originally published in 1865–1866), and a paramount impetus for this growth to be linked with mathematical models was given by R. A. Fisher's *Fundamental Theorem of Natural Selection* in 1930. This result was based on a partial differential equation (PDE), expressing the rate of fitness increase for any living organism. Mathematical models in other areas of life sciences were also developing and included A. J. Lotka and V. Volterra's predator–prey systems (1925–1931), A. A. Malinovsky's models for evolutionary genetics and systems analysis (1935), R. Fisher and A. Kolmogorov equation for gene propagation (1937), A. L. Hodgkin and A. F. Huxley's equations for neural axon membrane potential (1952), to name just a few. New theories, such as self-organization and biological pattern formation, have appeared, demonstrating the powerful cross-fertilization between mathematics and the life sciences (see additional details in Ref. [1]). More recently, the ready availability of detailed molecular, functional, and genomic data has led to the unprecedented development of new data-driven mathematical models. As a result, the tools of mathematical modeling and computational experiment are becoming largely important in today's life sciences. The same conclusion applies to environmental, earth, and climate sciences as well. Based on the data since 1880, by now we know that global warming has been mostly caused by the man-made world with its emission from the burning of fossil fuels, environmental pollution, and other factors. In moving forward, we will need to solve many challenging environmental problems, and the role of mathematical and computational modeling in environmental, earth, and climate sciences will continue to increase [13].

Mathematical models and algorithms have become essential for many professionals in other areas, including sociologists, financial analysts, political scientists, public administration workers, and the governments [12], with this list continuing to grow. Our discussion would be incomplete if we do not mention here a deep connection between mathematics and the arts. Ancient civilizations, including Egyptians, Mesopotamians, and Chinese, studied the mathematics of sound, and the Ancient Greeks investigated the expression of musical scales in terms of the ratios of small integers. They considered harmony as a branch of science, known now as musical acoustics. They worked to demonstrate that the mathematical laws of harmonics and rhythms have a fundamental character not only to the understanding of the world but also to human happiness and prosperity. While a myriad of examples of the intrinsic connection between mathematics and the arts are found in the Ancient World, undoubtedly the Renaissance brought an enriched rebirth of classical ancient world cultures and mathematical ideas not only for better understanding of nature but also for the arts. Indeed, painting three-dimensional scenes on a two-dimensional canvas presents just one example where such a connection was shown to be critical. Not only philosophers, but artists too, were convinced that the whole universe, including the arts, could be explained with geometric and numerical techniques. There are many examples from the Renaissance period where painters were also mathematicians, with Piero della Francesca (c.1415–1492) and Leonardo da Vinci (1452–1519) among them. Nowadays, the arts and mathematics are closely interconnected, continuing to enrich each other. There are many architectural masterpieces and paintings that have been preserved based on the implementation of sophisticated mathematical models. Efficient computer graphics algorithms have brought a new dimension to many branches of the modern arts, while a number of composers have incorporated mathematical ideas into their works (and the golden ratio and Fibonacci numbers are among them). Musical applications of number theory, algebra, and set theory, among other areas of mathematics, are well known.

While algorithms and models have always been central in the development of mathematical sciences, providing an essential links to the applications, their importance has been drastically amplified in the computer age, where the role of mathematical modeling and computational experiment in understanding nature and our world becomes paramount.

1.2 LOOKING INTO THE FUTURE FROM A MODELING PERSPECTIVE

Although on a historical scale electronic computers belong to a very recent invention of humans, the first computing operations were performed from ancient times by people themselves. From abacus to Napier's Bones, from the Pascaline to the Leibnitz's Stepped Reckoner, from the Babbage's Difference (and then Analytic) Engine to the Hollerith's Desk invention as a precursor to IBM, step by step, we have drastically improved our ability to compute. Today, modern computers allow us to increase productivity in intellectual performance and information processing to a level not seen in the human history before. In its turn, this process leads to a rapid development of new mathematics-based algorithms that are changing the entire landscape of human activities, penetrating to new and unexpected areas. As a result, mathematical modeling expands its interdisciplinary horizons, providing links between different disciplines and human activities. It becomes pervasive across more and more disciplines, while practical needs of human activities and applications, as well as the interface between these disciplines, human activities, mathematics and its applications, stimulate the development of state-of-the-art new methods, approaches, and tools. A special mention in this context deserves such areas as social, behavioral, and life sciences. The ever expanding range of the two-way interaction between mathematical modeling and these disciplines indicates that this interaction is virtually unlimited. Indeed, taking life sciences as an example, the applications of mathematical algorithms, methods, and tools in drug design and delivery, genetic mapping and cell dynamics, neuroscience, and bionanotechnology have become ubiquitous. In the meantime, new challenges in these disciplines, such as sequencing macromolecules (including those already present in biological databases), provide an important catalyst for the development of new mathematics, new efficient algorithms, and methods [1]. Euclidian, non-Euclidian, and fractal geometries, as well as an intrinsic link between geometry and algebra highlighted by R. Descartes through his coordinate system, have all proved to be very important in these disciplines, while the discovery of what is now known as the Brownian motion by Scottish botanist R. Brown has revolutionized many branches of mathematics. Game theory and the developments in control and cybernetics were influenced by the developments in social, behavioral, and life sciences, while the growth of systems science has provided one of the fundamentals for the development of systems biology where biological systems are considered in a holistic way [1]. There is a growing understanding that the interactions between different components of a biological system at different scales (e.g., from the molecular to the systemic level) are critical. Biological systems provide an excellent example of coupled systems and multiscale dynamics. A multiscale spatiotemporal character of most systems in nature, science, and engineering is intrinsic, demonstrating complex interplay of its components, well elucidated in the literature (e.g., [8,9,13] and references therein). In life sciences, the number of such examples of multiscale coupled systems and associated problems is growing rapidly in many different, albeit often interconnected, areas. Some examples are as follows:

- Complex biological networks, genomics, cellular systems biology, and systems biological approaches in other areas, studies of various organs, their systems, and functions;
- Brain dynamics, neuroscience and physiology, developmental biology, evolution and evolutionary dynamics of biological games;
- Immunology problems, epidemiology and infectious diseases, drug development, delivery, and resistance;
- Properties, dynamics, and interactions at various length and time scales in bio-macromolecules, including DNA, RNA, proteins, self-assembly and spatiotemporal pattern formation in biological systems, phase transitions, and so on.

Many mathematical and computational modeling tools are ubiquitous. They are universal in a sense that they can be applied in many other areas of human endeavors. Life sciences have a special place when we look into the future developments of mathematical and computational modeling. Indeed, compared to other areas, for example, those where we study physical or engineering systems, our knowledge of biological systems is quite limited. One of the reasons behind this is biological system complexity, characterized by the fact that most biological systems require dealing with multiscale interactions of their highly heterogeneous parts on different time scales.

In these cases in particular, the process of mathematical and computational modeling becomes frequently a driving source for the development of hierarchies of mathematical models. This helps determine the range of applicability of models. What is especially important is that based on such hierarchies, mathematical models can assist in explaining the behavior of a system under different conditions and the interaction of different system components. Clearly, different models for the same system can involve a range of mathematical structures and can be formalized with various mathematical tools such as equation- or inequality-based models, graphs, and logical and game theoretic models. We know by now that the class of the models amenable to analytical treatments, while keeping assumptions realistic, is strikingly small, when compared to the general class of mathematical models that are at the forefront of modern science and engineering [10]. As a result, most modern problems are treated numerically, in which case the development of efficient algorithms becomes critical. As soon as such algorithms are implemented on a computer, we can run the model multiple times under varying conditions, helping us to answer outstanding questions quicker and more efficiently, providing us an option to improve the model when necessary. Model-algorithm-implementation is a triad which is at the heart of mathematical modeling and computational experiment. It is a pervasive, powerful, theoretical, and practical tool, covering the entire landscape of mathematical applications [10]. This tool will play an increasingly fundamental role in the future as we can carry out mathematical modeling and computational experiment even in those cases when natural experiments are impossible. At the same time, given appropriate validation and verification procedures, we can provide reliable information more quickly and with less expense compared to natural experiments. The two-way interactions between new developments in information technology and mathematical modeling and computational experiment are continuously increasing predictive capabilities and research power of mathematical models.

Looking into the future from a modeling perspective, we should also point out that such predictive capabilities and research power allow us to deal with complex systems that have intrinsically interconnected (coupled) parts, interacting in nontrivial dynamic manner. In addition to life, behavioral, and social sciences, mentioned earlier, such systems arise in many other areas, including, but not limited to, fusion and energy problems, materials science and chemistry, high energy and nuclear physics, cosmology and astrophysics, earth, climate, environmental, and sustainability sciences.

In addition to the development of new models and efficient algorithms, the success of predictive mathematical modeling in applications is dependent also on further advances in information sciences and the development of statistical, probabilistic, and uncertainty quantification methods. Uncertainty comes from many different sources, among which we will mention parameters with uncertain values, uncertainty in the model as a representation of the underlying phenomenon, process, or system, and uncertainty in collecting/processing/measurements of data for model calibration. The task of quantifying and mitigating these uncertainties in mathematical models leads to the development of new statistical/stochastic methods, along with methods for efficient integration of data and simulation.

Further to supporting theories and increasing our predictive capabilities, mathematical and computational modeling can often suggest sharper natural experiments and more focused observations, providing in their turn a check to the model accuracy. Natural experiments and results of observations may produce large amounts of data sets that can intelligently be processed only with efficient mathematical data mining algorithms, and powerful statistical and visualization tools [15]. The application of these algorithms and tools requires a close collaboration between different disciplines. As a result, observations and experiments, theory and modeling reinforce each other, leading together to our better understanding of phenomena, processes, and systems we study, as well as to the necessity of even more close interactions between mathematical modeling, computational analyses, and experimental approaches.

1.3   WHAT THIS BOOK IS ABOUT

The rest of the book consists of 4 main sections, containing 11 state-of-the-art chapters on applications of mathematical and computational modeling in natural and social sciences, engineering, and the arts. These chapters are based on selected invited contributions from leading specialists from all over the world. Inevitably, given the vast range of research areas within the field of mathematical and computational modeling, the book such as this can present only selective topics. At the same time, these selective topics open to the reader a broad spectrum of methods and tools important in these applications, and ranging from infectious disease dynamics and epidemic modeling to superconductivity and quantum mechanical challenges, from the models for voting systems to the modeling of musical rhythms. The book provides both theoretical advances in these areas of applications, as well as some representative examples of modern problems from these applications. Following this introductory section, each remaining section with its chapters stands alone as an in-depth research or a survey within a specific area of application of mathematical and computational modeling. We highlight the main features of each such chapter within four main remaining sections of this book.

- **Advanced Mathematical and Computational Models in Physics and Chemistry.** This section consists of three chapters.

  - This section is opened by a chapter written by I. M. Sigal who addresses the macroscopic theory of superconductivity. Superconducting vortex states provide a rich area of research. In the 1950s A. Abrikosov solved the Ginzburg–Landau (GL) equations in an applied magnetic field for certain values of GL parameter (later A. Abrikosov received a Nobel Prize for this work). This led to what is now known as the famous vortex solution, characterized by the fact that the superconducting order parameter contains a periodic lattice of zeros. In its turn, this led to studies of a new mixed Abrikosov vortex phase between the Meissner state and the normal state. The area keeps generating new interesting results in both theory and application. For example, unconventional vortex pattern formations (e.g., vortex clustering) were recently discovered in multiband superconductors (e.g., [17] and references therein). Such phenomena, which are of both fundamental and practical significance, present a subject of many experimental and theoretical works. Recently, it was shown that at low temperatures the vortices form an ordered Abrikosov lattice both in low and in high fields. The vortices demonstrate distinctive modulated structures at intermediate fields depending on the effective intervortex attraction. These and other discoveries generate an increasing interest to magnetic vortices and Abrikosov lattices. Chapter by I. M. Sigal reminds us that the celebrated GL equations form an integral part, namely the Abelian-Higgs component, of the standard model of particle physics, having fundamental consequences for many areas of physics, including those beyond the original designation area of the model. Not only this chapter reviews earlier works on key solutions of the GL model, but it presents some interesting recent results. Vortex lattices, their existence, stability, and dynamics are discussed, demonstrating also that automorphic functions appear naturally in this context and play an important role.

– A prominent role in physics and chemistry is played by the Hartree-Fock method which is based on an approximation allowing to determine the wave function and the energy of a quantum many-body system in a stationary state. More precisely, the Hartree-Fock theoretical framework is based on the variational molecular orbital theory, allowing to solve Schrödinger's equation in such a way that each electron spatial distribution is described by a single, one-electron wave function, known as molecular orbital. While in the classical Hartree-Fock theory the motion of electrons is uncorrelated, correlated wavefunction methods remedy this drawback. The second chapter in this section is devoted to a multireference local correlation framework in quantum chemistry, focusing on numerical challenges in the Cholesky decomposition context. The starting point of the discussion, presented by D. K. Krisiloff, J. M. Dieterich, F. Libisch, and E. A. Carter, is based on the fact that local correlation methods developed for solving Schrödinger's equation for molecules have a reduced computational cost compared to their canonical counterparts. Hence, the authors point out that these methods can be used to model notably larger chemical systems compared to the canonical algorithms. The authors analyze in detail local algorithmic blocks of these methods.

– Variational methods are in the center of the last chapter of this section, written by M. Levy and A. Gonis. The basic premises here lie with the Rayleigh-Ritz variational principle which, in the context of quantum mechanical applications, reduces the problem of determining the ground-state energy of a Hamiltonian system consisting of N interacting electrons to the minimization of the energy functional. The authors then move to the main part of their results, closely connected to a fundamental element of quantum mechanics. In particular, they provide two alternative proofs of the generalization of the variational theorem for Hamiltonians of N-electron systems to wavefunctions of dimensions higher than N. They also discuss possible applications of their main result.

- **Mathematical and Statistical Models in Life Science Applications.** This section consists of two chapters.

  – The first chapter deals with mathematical modeling of infectious disease dynamics, control, and treatment, focusing on a model for the spread of tuberculosis (TB). TB is considered to be the second highest cause of infectious disease-induced mortality after HIV/AIDS. Written by J. Arino and I. A. Soliman, this chapter provides a detailed account of a model that incorporates three strains, namely (1) drug sensitive, (2) emerging multidrug resistant, and (3) extensively drug-resistant. The authors provide an excellent introduction to the subject area, followed by the model analysis. In studying the dynamics of the model, they characterize parameter regions where backward bifurcation may occur. They demonstrate the global stability of the disease-free equilibrium in regions with no backward bifurcation. In conclusion, the authors discuss possible options for their model improvement and how mathematical epidemiology contributes to our better understanding of disease transmission processes and their control.

– Epidemiological modeling requires the development and application of an integrated approach. The second chapter of this section focuses on these issues with emphasis on antibiotic resistance. The chapter is written by E. Y. Klein, J. Chelen, M. D. Makowsky, and P. E. Smaldino. They stress the importance of integrating human behavior, social networks, and space into infectious disease modeling. The field of antibiotic resistance is a prime example where this is particularly critical. The authors point out that the annual economic cost to the US health care system of antibiotic-resistant infections is estimated to be $21–$34 billion, and given human health and economics reasons, they set a task of better understanding how resistant bacterial pathogens evolve and persist in human populations. They provide a selection of historical achievements and limitations in mathematical modeling of infectious diseases. This is followed by a discussion of the integrated approach, the authors advocate for, in addressing the multifaceted problem of designing innovative public health strategies against bacterial pathogens. The interaction of epidemiological, evolutionary, and behavioral factors, along with cross-disciplinary collaboration in developing new models and strategies, is becoming crucial for our success in this important field.

• **Mathematical Models and Analysis for Science and Engineering.** This section consists of four chapters.

– The first chapter is devoted to mathematical models in climate modeling, with a major focus given to examples from climate atmosphere-ocean science (CAOS). However, it covers potentially a much larger area of applications in science and engineering. Indeed, as pointed out by the authors of this chapter, D. Giannakis and A. J. Majda, large-scale data sets generated by dynamical systems arise in a vast range of disciplines in science and engineering, for example, fluid dynamics, materials science, astrophysics, earth sciences, to name just a few. Therefore, the main emphasis of this chapter is on data-driven methods for dynamical systems, aiming at quantifying predictability and extracting spatiotemporal patterns. In the context of CAOS, we are dealing with a system of time-dependent coupled nonlinear PDEs. The dynamics of this system takes place in an infinite-dimensional phase space, where the corresponding equations of fluid flow and thermodynamics are defined. In this case, the observed data usually correspond to well-defined physical functions of that phase space, for example, temperature, pressure, or circulation, measured over a set of spatial points. Data-driven methods appear to be critical in our better understanding of many important phenomena intrinsic to the dynamics of climate system, including El Nino Southern Oscillation in the ocean and the Madden–Julian oscillation in the atmosphere. The authors provide a comprehensive review on data-driven methods and illustrate their techniques with applications, most of which are pertinent to CAOS. They conclude with a discussion of open problems and possible connections between the developed techniques.

– Inverse problems lie at the heart of many scientific and engineering inquiries. Broadly speaking, they provide a framework that is used to convert observed measurements (or desired effects) into information about a physical object or system (or causes). This framework covers an extremely diverse range of applications, from imaging science, computer graphics, and computer vision, to earth science, and to astrophysics. Many problems from life sciences, discussed in the previous section, can also be formulated as inverse. Some specific examples from science and engineering include the development of underwater detection devices, location of oil and mineral deposits, creation of astrophysical images from telescope data, finding cracks and interfaces within materials, shape optimization, and so on. Regularization techniques are fundamental in solving inverse problems efficiently, and the Tikhonov regularization technique plays a particularly prominent role in this. In this chapter, written by B. Hofmann, the author provides an overview of a number of new aspects and recent developments in Tikhonov's regularization for nonlinear inverse problems. The author formulates such problems as operator equations in Banach spaces. In order to construct stable and convergent approximate solutions to these problems, stabilizing penalty functionals are necessary. The author discusses in detail the interplay of convergence properties and approximate choices of the regularization parameters, as well as solution smoothness and the nonlinearity structure. In order to express and characterize the latter properties, a number of variational inequalities were presented. Examples on how to construct such inequalities were given, and their significance for obtaining convergence rates was also discussed.

– It is well known that many mathematical models, which play a fundamental role in science and engineering, can be formulated in the form of first-order symmetric hyperbolic (FOSH) systems of differential equations, supplemented by constraints. Examples include Maxwell's equations, as well Einstein's field equations, to name just a few. If we consider the initial value (Cauchy) problem for either Maxwell's or Einstein's equations, it is known that the constraints will be preserved by their evolution. In other words, whenever the initial data satisfy the constraints, the solution will satisfy them too for all times. This is a starting point of the discussion in the chapter written by N. Tarfulea. The author explains that for bounded computational domains, for example, artificial space cut offs may be needed for such evolution systems. The task then is to impose appropriate boundary conditions so that the numerical solution of the reduced (or cut off) system would approximate (in the best possible way) the original model defined on infinite space. The author provides a survey of known techniques for finding constraint preserving boundary conditions for some constrained FOSH systems, followed by a number of new ideas for such constructions. Theory is exemplified by a constrained FOSH system originating from a system of wave equations with constraints, as well as by more complex systems. In particular, Einstein's equations in the Einstein–Christoffel and Alekseenko–Arnold formulations are also analyzed with the proposed methodology.

– An overview of recent developments in methodologies for empirical organization of data is given in the chapter written by R. R. Coifman, R. Talmon, M. Cavish, and A. Haddad. Through these methodologies, this chapter provides a link between various applications of mathematics, ranging from natural sciences to engineering and to social sciences and the arts. Based on geometric and analytic ideas, the authors present a general mathematical framework for learning. These ideas are centered around building a network or a graph whose nodes are observations. In this framework, the authors propose to maintain connections between observations by constantly reconfiguring them and calibrating in order to achieve learning characteristics for specific tasks. The developed methods are then related to ideas from Harmonic Analysis. The intrinsic connection between harmonic analysis and other disciplines is well known. Indeed, this area of mathematics is central to a wide range of applications, from signal and image processing to machine learning, quantum mechanics, neuroscience, and biomedical engineering. It enriches the rapidly growing field of data representation and analysis, and stimulate interdisciplinary research. The authors of this chapter illustrate their ideas on examples taken from both natural and social sciences. They show how such different things as text documents, psychological questionnaires, medical profiles, physically measured engineering data, or financial data can be organized and map out in an automatic and purely data-driven manner.

• **Mathematical Methods in Social Sciences and Arts.** This section consists of two chapters.

– This section is opened by chapter written by S. J. Brams and D. M. Kilgour. It provides an important example of application of mathematical models in social and behavioral sciences. The range and diversity of such applications and developed models are growing continuously, from mathematical demography to models of contagion in finance, social dynamics and networks, arms races, social mobility, coalitions and consensus formation, quantification of power, experimental games, reduction of structural complexity, and decision theory. A notable example presented in this chapter deals with approval voting (AV). It is a voting system in which voters can vote for, or approve, as many candidates as they like. The authors proposed a new voting system for multiwinner election, called satisfaction approval voting (SAV). They considered the use of this system in different types of elections. For example, they considered a case where there are no political parties, as well as a number of other possible cases. Their system elects the set of candidates that maximizes the satisfaction of all voters, where a candidate's satisfaction score is the sum of the satisfactions that her/his election would give to all voters, while a voter's satisfaction is the fraction of her/his approved candidates who are elected. The authors demonstrated that SAV and AV may elect disjoint sets of candidates. In this context, an example of a recent election of the Game Theory Society was given. In conclusion, the authors explained why the most compelling application of their SAV is to party-list systems. This observation has important social implications because SAV is likely to lead to more informed voting and more responsive government in parliamentary systems.

– The concluding chapter of this section and the book provides an example of application of mathematical methods to arts, focusing on music, an art form whose medium is sound and silence. Ancient civilizations, including Egyptians, Chinese, Indian, Mesopotamians, and Greek, studied mathematics of sound. The expression of musical scales in terms of the ratios of small integers goes deep into the human history. Harmony arising out of numbers was sought in all natural phenomena by the Ancient Greeks, starting from Pythagoras. The word "harmonikos" was reserved in that time for those skilled in music. Nowadays, we use the word "harmonics" indicating waves with frequencies that are integer multiples of one another. The applications of mathematical methods from number theory, algebra, and geometry in music are well known, as well as the incorporation of Fibonacci numbers and the golden ratio in musical compositions. The concluding chapter, written by G. T. Toussaint, is devoted to the field of evolutionary musicology where one concerns with characterizing what music is, determining its origin and cross-cultural universals. The author notes that a phylogeny of music may sometimes be correctly constructed from rhythmic features alone. Then, a phylogenetic analysis of a family of rhythms can be carried out based on dissimilarity matrix that is calculated from all pairs of rhythms in the family. How do we define musical rhythms? How do we analyze them? Asking these questions, the author provides a comprehensive account to what is known in this field, focusing on the mathematical analysis of musical rhythms. The working horse of the discussion is the well-known *clave son* rhythm popular in many cultures around the world. The main methodology developed for the analysis is based on geometric quantization. Different types of models are considered and compared, highlighting most important musicological properties.

1.4   CONCLUDING REMARKS

Mathematical and computational modeling, their methods, and tools are rapidly becoming a major driving force in scientific discovery and innovation, providing us with increasingly more reliable predictive capabilities in many areas of human endeavor. In this section, we have presented a brief historical account and an overview of new trends in this field, demonstrating universality of mathematical models. We highlighted a unique selection of topics, representing part of a vast spectrum of the interface between mathematics and its applications, that are discussed in detail in subsequent sections of the book. These topics cover mathematical and computational models from natural and social sciences, engineering, and the arts.

REFERENCES

1. Antoniouk, A. and Melnik, R. (Eds), Mathematics and life sciences, Berlin, De Gruyter, 2012.

2. Barzel, B. and Barabasi, A. L., Universality in network dynamics, Nature Physics, 9 (10), 673–681 (2013).

3. Burgin, M. and Eberbach, E., Universality for turing machines, inductive turing machines and evolutionary algorithms, Fundamenta Informaticae, 91 (1), 53–77 (2009).

4. Carmona, P. and Hu, Y. Y., Universality in Sherrington-Kirkpatrick's spin glass model, Annales de L'Institut Henri Poincare - Probabilities et Statistiques, 42 (2), 215–222 (2006).

5. Chay, T. R., Fan, Y. S., and Lee, Y. S., Bursting, spiking, chaos, fractals, and universality in biological rhythms, International Journal of Bifurcation and Chaos, 5 (3) 595–635 (1995).

6. Chelkak, D. and Smirnov, S., Universality in the 2D Ising model and conformal invariance of fermionic observables, Inventiones Mathematicae, 189 (3), 515–580 (2012).

7. Dolan, B. P., Modular invariance, universality and crossover in the quantum Hall effect, Nuclear Physics B, 554 (3), 487–513 (1999).

8. Engquist, B., Lötstedt, P., and Runborg, O. (Eds.), Multiscale modeling and simulation in science, Berlin; Springer, 2009.

9. Foster, D. H., Costa, T., Peszynska, M., and Schneider, G., Multiscale modeling of solar cells with interface phenomena, Journal of Coupled Systems and Multiscale Dynamics, 1, 179–204 (2013).

10. Melnik, R. and Kotsireas, I. (Eds), Advances in applied mathematics, modeling, and computational science, Fields Institute Communications, 66, New York, Springer, 2013.

11. Poincare, H., L'avenir des mathématiques ("The Future of Mathematics"). Revue générale des sciences pures et appliquées, 19 (23), 930–939 (1908).

12. Rapoport, A., Mathematical models in the social and behavioral sciences, New York, John Wiley & Sons, Inc., 1983.

13. Regenauer-Lieb, K., Veveakis, M., Poulet, T., et al., Multiscale coupling and multiphysics approaches in earth sciences: Theory, Journal of Coupled Systems and Multiscale Dynamics, 1, 49–73 (2013).

14. Sancetta, A., Universality of Bayesian predictions, Bayesian Analysis, 7 (1), 1–36 (2012).

15. Sokolowski, J. A. and Banks, C. M., Modeling and simulation fundamentals: theoretical underpinnings and practical domains, Hoboken, NJ, John Wiley & Sons, Inc., 2010.

16. Tsui, L. K. and Leung, P. T., Universality in quasi-normal modes of neutron stars, Monthly Notices of the Royal Astronomical Society, 357 (3), 1029–1037 (2005).

17. Xu, X. B., Fangohr, H., Gu, M., et al., Simulation of the phase diagram of magnetic vortices in two-dimensional superconductors: evidence for vortex chain formation, Journal of Physics - Condensed Matter, 26 (11), 115702 (2014).

SECTION 2

ADVANCED MATHEMATICAL AND COMPUTATIONAL MODELS IN PHYSICS AND CHEMISTRY

2

MAGNETIC VORTICES, ABRIKOSOV LATTICES, AND AUTOMORPHIC FUNCTIONS

ISRAEL MICHAEL SIGAL

Department of Mathematics, University of Toronto, Ontario, Canada

2.1 INTRODUCTION

In this chapter, we present some recent results on the Ginzburg–Landau equations of superconductivity and review appropriate background. The Ginzburg–Landau equations describe the key mesoscopic and macroscopic properties of superconductors and form the basis of the phenomenological theory of superconductivity. They are thought of to be the result of coarse-graining the Bardeen–Cooper–Schrieffer microscopic model, and were derived from that model by Gorkov [36]. (Recently, the rigorous derivation in the case of nondynamic magnetic fields was achieved by Frank et al. [34].)

These equations appear also in particle physics, as the Abelean-Higgs model, which is the simplest, and arguably most important, ingredient of the standard model [93]. Geometrically, they are the simplest equations describing the interaction of the electromagnetic field and a complex field, and can be thought of as the 'Dirichlet' problem for a connection of $U(1)-$principal bundle and a section of associated vector bundle.

One of the most interesting mathematical and physical phenomena connected with Ginzburg–Landau equations is the presence of *vortices* in their solutions. Roughly speaking, a vortex is a spatially localized structure in the solution, characterized by

Mathematical and Computational Modeling: With Applications in Natural and Social Sciences, Engineering, and the Arts, First Edition. Roderick Melnik.
© 2015 John Wiley & Sons, Inc. Published 2015 by John Wiley & Sons, Inc.

a nontrivial topological degree (winding number). It represents a localized defect where the normal state intrudes and magnetic flux penetrates. It is called the magnetic vortex.

Vortices exist on their own, or, as predicted by A. Abrikosov [1] in 1957, they can be arrayed in a lattice pattern. (In 2003, Abrikosov received the Nobel Prize for this discovery.)

Individual vortices and vortex lattices are the subjects of this chapter. In this we present already classical results on the former and recent results on the latter. It can be considered as an update of the review [45], from which for convenience of the reader, we reproduce some material. Some additional discussions can be found in the reviews [21, 69] and in [18, 59, 60].

Like the latter review, we do not discuss the two important areas, the $\kappa \to \infty$ regime (the quasiclassical limit of the theory) and the linear eigenvalue problem related to the second critical magnetic field. Fairly extensive reviews of these problems are given in Refs. [71] and [31], respectively (see [3–5, 8] for more recent work). We also mention the book [12] that inspired much of the activity in this area. Some additional material and physics background can be found in the reviews [21, 69] and in the papers [18, 59, 60]. Related equations are treated in Refs. [4, 6, 10, 22, 52, 56].

2.2   THE GINZBURG–LANDAU EQUATIONS

The Ginzburg–Landau theory [35] gives a macroscopic description of superconducting materials in terms of a pair $(\Psi, A) : \mathbb{R}^d \to \mathbb{C} \times \mathbb{R}^d$, $d = 1, 2, 3$, a complex-valued function $\Psi(x)$, called an *order parameter*, so that $|\Psi(x)|^2$ gives the local density of (Cooper pairs) superconducting electrons, and the vector field $A(x)$, so that $B(x) := \operatorname{curl} A(x)$ is the magnetic field. In equilibrium, they satisfy the system of nonlinear partial differential equations (PDE) called the *Ginzburg–Landau equations*:

$$-\Delta_A \Psi = \kappa^2 (1 - |\Psi|^2)\Psi$$
$$\operatorname{curl}^2 A = \operatorname{Im}(\bar{\Psi} \nabla_A \Psi)$$
(2.1)

where $\nabla_A = \nabla - iA$ and $\Delta_A = \nabla_A^2$, the covariant derivative and covariant Laplacian, respectively, and $\kappa > 0$ is a parameter, called the Ginzburg–Landau parameter, which depends on the material properties of the superconductor. For $d = 2$, $\operatorname{curl} A := \partial_1 A_2 - \partial_2 A_1$ is a scalar, and for scalar $B(x) \in \mathbb{R}$, $\operatorname{curl} B = (\partial_2 B, -\partial_1 B)$ is a vector. The vector quantity $J(x) := \operatorname{Im}(\bar{\Psi} \nabla_A \Psi)$ is the superconducting current. (for example, see Refs. [86, 88]).

Particle physics. In the Abelian-Higgs model, ψ and A are the Higgs and $U(1)$ gauge (electromagnetic) fields, respectively. Geometrically, one can think of A as a connection on the principal $U(1)$−bundle $\mathbb{R}^d \times U(1)$, $d = 2, 3$.

Cylindrical geometry. In the commonly considered idealized situation of a superconductor occupying all space and homogeneous in one direction, one is led to a problem on $\mathbb{R}^2$ and so may consider $\Psi : \mathbb{R}^2 \to \mathbb{C}$ and $A : \mathbb{R}^2 \to \mathbb{R}^2$. This is the case we deal within this chapter.

2.2.1   Ginzburg–Landau energy

The Ginzburg–Landau equations (2.1) are the Euler–Lagrange equations for critical points of the *'Ginzburg–Landau energy functional* (written here for a domain $Q \in \mathbb{R}^2$)

$$\mathcal{E}_Q(\Psi, A) := \frac{1}{2} \int_Q \left\{ |\nabla_A \Psi|^2 + (\mathrm{curl}A)^2 + \frac{\kappa^2}{2}(|\Psi|^2 - 1)^2 \right\}. \tag{2.2}$$

Superconductivity. In the case of superconductors, the functional $\mathcal{E}(\psi, A)$ gives the difference in (Helmholtz) free energy (per unit length in the third direction) between the superconducting and normal states, near the transition temperature.

This energy depends on the temperature (through κ) and the average magnetic field, $b = \lim_{Q \to \mathbb{R}^2} \frac{1}{|Q|} \int_Q \mathrm{curl}A$, in the sample, as thermodynamic parameters. Alternatively, one can consider the free energy depending on the temperature and an applied magnetic field, h. For a sample occupying a finite domain Q, this leads (through the Legendre transform) to the Ginzburg–Landau Gibbs free energy $G_Q(\Psi, A) := \mathcal{E}_Q(\Psi, A) - \Phi_Q h$, where $\Phi_Q = b|Q| = \int_Q \mathrm{curl}A$ is the total magnetic flux through the sample.

$$G_Q(\Psi, A) := \frac{1}{2} \int_Q \left\{ |\nabla_A \Psi|^2 + \frac{\kappa^2}{2}(|\Psi|^2 - 1)^2 + (\mathrm{curl}A - h)^2 \right\}. \tag{2.3}$$

The parameters b or h do not enter the equations (2.1) explicitly, but they determine the density of vortices, which we describe below.

In what follows, we write $\mathcal{E}(\Psi, A) = \mathcal{E}_{\mathbb{R}^2}(\Psi, A)$ and $G(\Psi, A) = G_{\mathbb{R}^2}(\Psi, A)$.

Particle physics. In the particle physics case, the functional $\mathcal{E}(\Psi, A)$ gives the energy of a static configuration in the $U(1)$ Yang–Mills–Higgs classical gauge theory.

2.2.2   Symmetries of the equations

The Ginzburg–Landau equations (2.1) admit several symmetries, that is, transformations that map solutions to solutions.

Gauge symmetry: for any sufficiently regular function $\gamma : \mathbb{R}^2 \to \mathbb{R}$,

$$T_\gamma^{\mathrm{gauge}} : (\Psi(x), A(x)) \mapsto (e^{i\gamma(x)}\Psi(x), A(x) + \nabla\gamma(x)); \tag{2.4}$$

Translation symmetry: for any $h \in \mathbb{R}^2$,

$$T_h^{\mathrm{trans}} : (\Psi(x), A(x)) \mapsto (\Psi(x+h), A(x+h)); \tag{2.5}$$

Rotation symmetry: for any $\rho \in SO(2)$,

$$T_\rho^{\mathrm{rot}} : (\Psi(x), A(x)) \mapsto (\Psi(\rho^{-1}x), \rho^{-1}A((\rho^{-1})^T x)), \tag{2.6}$$

One of the analytically interesting aspects of the Ginzburg–Landau theory is the fact that, because of the gauge transformations, the symmetry group is infinite-dimensional.

2.2.3   Quantization of flux

Consider first finite energy states (Ψ, A) that have the regularity $H^2_{loc}(\mathbb{R}^2)$ of solutions to (2.1) (see Ref. [15] for the regularity results). Such states are classified by their topological degree (the winding number of ψ at infinity):

$$\deg(\Psi) := \text{degree}\left(\left.\frac{\Psi}{|\Psi|}\right|_{|x|=R} : \mathbb{S}^1 \to \mathbb{S}^1\right),$$

for $R \gg 1$, s.t. $|\Psi(x)| \geq \delta > 0$ for $x : |x| = R$. (Since $\Psi \in H^2_{loc}(\mathbb{R}^2)$ and $\int (1 - |\Psi|^2)^2 dx < \infty$, one can show that such an R exists.) For more on the degree on Sobolev spaces see Ref. [16]. For each such state, we have the quantization of magnetic flux:

$$\int_{\mathbb{R}^2} B(x)dx = 2\pi \deg(\Psi) \in 2\pi\mathbb{Z},$$

which follows from integration by parts (Stokes theorem) and the requirement that $|\Psi(x)| \to 1$ and $|\nabla_A \Psi(x)| \to 0$ as $|x| \to \infty$.

For vortex lattices (see below) the energy is infinite, but the flux quantization still holds for each lattice cell because of gauge-periodic boundary conditions (see below for details).

2.2.4   Homogeneous solutions

The simplest solutions to the Ginzburg–Landau equations (2.1) are the trivial ones corresponding to physically homogeneous states:

1. the perfect superconductor solution, (Ψ_s, A_s), where $\Psi_s \equiv 1$ and $A_s \equiv 0$ (so the magnetic field $\equiv 0$),
2. the normal metal solution, (Ψ_n, A_n), where $\Psi_n \equiv 0$ and A_n corresponds to a constant magnetic field.

(Of course, any gauge transformation of one of these solutions has the same properties.)

We see that the perfect superconductor is a solution only when the magnetic field $B(x)$ is zero. On the other hand, there is a normal solution for any constant magnetic field (to be thought of as determined by applied external magnetic field).

2.2.5   Type I and Type II superconductors

Assuming magnetic fields are weak and consequently neglecting variations of $|\Psi|^2$, we see from the second equation in (2.1) that (in our units) the magnetic field varies on the length scale 1, the *penetration depth*. Furthermore, if the magnetic field in the first equation in (2.1) vanishes, then the order parameter varies on the length scale $\frac{1}{\kappa}$, the *coherence length*.

The two length scales $1/\kappa$ and 1 coincide at $\kappa = 1/\sqrt{2}$. This critical value $\kappa = 1/\sqrt{2}$ separates superconductors into two classes with different properties:

$\kappa < 1/\sqrt{2}$: Type I superconductors—exhibit first-order (discontinuous, finite size nucleation) phase transitions from the non-superconducting state to the superconducting state (essentially, all pure metals);

$\kappa > 1/\sqrt{2}$: Type II superconductors—exhibit second-order (continuous) phase transitions and the formation of vortex lattices (dirty metals and alloys).

An important quantifier of the difference between type I and type II superconductors is the surface tension. As was observed first in Ref. [35], the surface tension at the interface between the normal and superconducting phases changes sign from positive for $\kappa < 1/\sqrt{2}$ to negative for $\kappa > 1/\sqrt{2}$. In detail, consider a flat interface. Assuming the material is uniform in the directions orthogonal to the x_1-axis, becoming normal as $x_1 \to -\infty$ and superconducting as $x_1 \to \infty$. The interface between these phases is the plane $\{x_1 = 0\}$. (By a translation and a rotation, we can always reduce to this case.) Thus we look for a solution depending only on x_1, $(\Psi(x), A(x)) = (\psi(x_1), a(x_1))$, with the magnetic field in the direction of x_3, the vector potential, a, in the direction of x_2, and with the boundary conditions, $\psi(x_1) \to 0$ and $\operatorname{curl} a(x_1) \to h$ as $x_1 \to -\infty$ and $\psi(x_1) \to 1$ and $\operatorname{curl} a(x_1) \to 0$, as $x_1 \to \infty$.

The boundary conditions at $x_1 = -\infty$ and $x_1 = \infty$ are consistent with the equations, if the applied field h satisfies $h = \kappa/\sqrt{2}$. However, in our units, $h_c := \kappa/\sqrt{2}$ is the thermodynamic critical magnetic field, at which the Gibbs free energies of the superconducting and normal phases are equal. (As the problem is one-dimensional, the integration in the functional (2.2) or (2.3) should be taken in the variable $x = x_1$ only, with the energy being interpreted as the energy per unit area of the interface $\{x_1 = 0\}$.) Then, by the definition [35], the surface tension is the surplus of the Gibbs free energy of such a solution compared to the normal (or superconducting) phase at the applied magnetic field h_c,

$$\sigma := \int_{-\infty}^{\infty} \{g_{h_c}(\psi, a) - g_{h_c}(0, a_c)\}\, dx, \tag{2.7}$$

where $g_h(\psi, a) := \frac{1}{2}[|\nabla_a \psi|^2 + \frac{\kappa^2}{2}(|\psi|^2 - 1)^2 + (\operatorname{curl} a - h)^2]$ (see (2.3)) and $\operatorname{curl} a_c = h_c$. It is shown in Ref. [19] that $\sigma > 0$, if $\kappa < 1/\sqrt{2}$ and $\sigma_{\text{surftens}} > 0$, if $\kappa > 1/\sqrt{2}$, with $\sigma = 0$, if $\kappa = 1/\sqrt{2}$.

2.2.6   Self-dual case $\kappa = 1/\sqrt{2}$

In the *self-dual* case $\kappa = 1/\sqrt{2}$ of (2.1), vortices effectively become noninteracting, and there is a rich multi-vortex solution family. Bogomolnyi [14] found the topological energy lower bound

$$\mathcal{E}(\Psi, A)\big|_{\kappa=1/\sqrt{2}} \geq \pi \,|\deg(\Psi)| \tag{2.8}$$

and showed that this bound is saturated (and hence the Ginzburg–Landau equations are solved) when certain *first-order* equations are satisfied.

2.2.7   Critical magnetic fields

In superconductivity, there are several critical magnetic fields, two of which (the first and the second critical magnetic fields) are of special importance:

h_{c1} is the field at which the first vortex enters the superconducting sample and

h_{c2} is the field at which a mixed state bifurcates from the normal one.

(The critical field h_{c1} is defined by the condition $G(\Psi_s, A_s) = G(\Psi^{(1)}, A^{(1)})$, where (Ψ_s, A_s) is the perfect superconductor solution, defined above, and $(\Psi^{(1)}, A^{(1)})$ is the 1-vortex solution, defined below, while h_{c2}, by the condition that the linearization of the l.h.s. of (2.1) on the normal state (Ψ_n, A_n) has zero eigenvalue. The field h_{c1} depends on Q and is 0 for $Q = \mathbb{R}^2$. Its asymptotics, as $\kappa \to \infty$, was found rigorously in [4,8]. One can show that $h_{c2} = \kappa^2$.)

For type I superconductors $h_{c1} > h_{c2}$ and for type II superconductors $h_{c1} < h_{c2}$. In the former case, the vortex states have relatively large energies, that is, are metastable, and therefore are of little importance.

For type II superconductors, there are two important regimes to consider: (1) average magnetic fields per unit area, b, are less than but sufficiently close to h_{c2},

$$0 < h_{c2} - b \ll h_{c2} \tag{2.9}$$

and (2) the external (applied) constant magnetic fields, h, are greater than but sufficiently close to h_{c1},

$$0 < h - h_{c1} \ll h_{c1}. \tag{2.10}$$

The reason the first condition involves b, while the second h, is that the first condition comes from the Ginzburg–Landau equations (which do not involve h), while the second from the Ginzburg–Landau Gibbs free energy.

One of the differences between the regimes (2.9) and (2.10) is that $|\Psi|^2$ is small in the first regime (the bifurcation problem) and large in the second one. If a superconductor fills in the entire $\mathbb{R}^2$, then in the second regime, the average magnetic field per unit area, $b \to 0$, as $h \to h_{c1}$.

2.2.8 Time-dependent equations

A number of dynamical versions of the Ginzburg–Landau equations appear in the literature. Here we list the most commonly studied and physically relevant.

Superconductivity. In the leading approximation, the evolution of a superconductor is described by the gradient-flow-type equations for the Ginzburg–Landau energy

$$\begin{cases} \gamma \partial_{t,\Phi}\Psi = \Delta_A \Psi + \kappa^2(1 - |\Psi|^2)\Psi, \\ \sigma \partial_{t,\Phi} A = -\operatorname{curl}^* \operatorname{curl} A + \operatorname{Im}(\bar{\Psi}\nabla_A \Psi). \end{cases} \quad (2.11)$$

Here Φ is the scalar (electric) potential, γ a complex number, and σ a two-tensor, and $\partial_{t,\Phi}$ is the covariant time derivative $\partial_{t,\Phi}(\Psi, A) = ((\partial_t + i\Phi)\Psi, \partial_t A + \nabla\Phi)$. The second equation is Ampère's law, $\operatorname{curl} B = J$, with $J + J_N + J_S$, where $J_N = -\sigma(\partial_t A + \nabla\Phi)$ (using Ohm's law) is the normal current associated to the electrons not having formed Cooper pairs, and $J_S = \operatorname{Im}(\bar{\Psi}\nabla_A \Psi)$, the supercurrent.

These equations are called the *time-dependent Ginzburg–Landau equations* or the *Gorkov–Eliashberg–Schmidt equations* proposed by Schmid [74] and Gorkov and Eliashberg [37] (earlier versions are proposed by Bardeen and Stephen and Anderson, Luttinger and Werthamer).

Particle physics. The time-dependent $U(1)$ Higgs model is described by

$$\begin{aligned} \partial_{t\Phi}^2 \Psi &= \Delta_A \Psi + \kappa^2(1 - |\Psi|^2)\Psi \\ \partial_t \partial_{t\Phi} A &= -\operatorname{curl}^* \operatorname{curl} A + \operatorname{Im}(\bar{\Psi}\nabla_A \Psi), \end{aligned} \quad (2.12)$$

coupled (covariant) wave equations describing the $U(1)$-gauge Higgs model of elementary particle physics (written here in the *temporal gauge*). Equations (2.12) are sometimes also called the *Maxwell-Higgs equations*.

For the existence results for these two sets of equations see [17, 25].

In what follows, we concentrate on the Gorkov–Eliashberg–Schmidt equations, (2.11) and, for simplicity of notation, we use the gauge, in which the scalar potential, Φ, vanishes, $\Phi = 0$.

2.3 VORTICES

2.3.1 *n*-vortex solutions

A model for a vortex is given, for each degree $n \in \mathbb{Z}$, by a "radially symmetric" (more precisely *equivariant*) solution of the Ginzburg–Landau equations (2.1) of the form

$$\Psi^{(n)}(x) = f_n(r)e^{in\theta} \quad \text{and} \quad A^{(n)}(x) = a_n(r)\nabla(n\theta), \quad (2.13)$$

where (r, θ) are the polar coordinates of $x \in \mathbb{R}^2$. Note that $\deg(\Psi^{(n)}) = n$. The pair $(\Psi^{(n)}, A^{(n)})$ is called the *n-vortex* (*magnetic* or *Abrikosov* in the case of superconductors and *Nielsen–Olesen* or *Nambu string* in the particle physics case). For superconductors, this is a mixed state with the normal phase residing at the point where the vortex vanishes. The existence of such solutions of the Ginzburg–Landau equations was already noticed by Abrikosov [1] and proven in Ref. [11] (For results on symmetry breaking solutions with finite number of vortices and on pinning of vortices see Refs. [61, 89] and [75], respectively.).

Using self-duality, and consequent reduction to a first-order equations, Taubes [84, 85] has showed that for a given degree n, the family of solutions modulo gauge transformations (*moduli space*) is $2|n|$-dimensional, and the $2|n|$ parameters describe the locations of the zeros of the scalar field, that is, the vortex centers. A review of this theory can be found in the book of Jaffe-Taubes [47].

The n-vortex solution exhibits the length scales discussed above. Indeed, the following asymptotics for the field components of the n-vortex (2.13) were established in Ref. [68] (see also Ref. [47]):

$$
\begin{aligned}
J^{(n)}(x) &= n\beta_n K_1(r)[1 + o(e^{-m_\kappa r})]\hat{Jx} \\
B^{(n)}(r) &= n\beta_n K_1(r)\left[1 - \tfrac{1}{2r} + O\left(\tfrac{1}{r^2}\right)\right] \\
|1 - f_n(r)| &\le ce^{-m_\kappa r}, \ |f_n'(r)| \le ce^{-m_\kappa r},
\end{aligned}
\tag{2.14}
$$

as $r := |x| \to \infty$, where $J^{(n)} := \mathrm{Im}(\overline{\Psi^{(n)}}\nabla_{A^{(n)}}\Psi^{(n)})$ is the n-vortex supercurrent, $B^{(n)} := \mathrm{curl}A^{(n)}$ is the n-vortex magnetic field, $\beta_n > 0$ is a constant, and K_1 is the modified Bessel function of order 1 of the second kind. The length scale of $\Psi^{(n)}$ is $1/m_\kappa$. Since $K_1(r)$ behaves like $ce^{-r}/\sqrt{r}$ for large r, we see that the length scale for $J^{(n)}$ and $B^{(n)}$ is 1. (In fact, for $x \ne 0$, $\Psi^{(n)}$ vanishes as $\kappa \to \infty$.)

2.3.2 Stability

We say the n-vortex is *(orbitally) stable*, if for any initial data sufficiently close to the n-vortex (which includes initial momentum field in the (2.12) case), the solution remains, for all time, close to *an element of the orbit of the n-vortex under the symmetry group*. Here "close" can be taken to mean close in the "energy space" Sobolev norm H^1.

Similarly, for *asymptotic stability*, the solution converges, as $t \to \infty$, to an element of the symmetry orbit (i.e., to a spatially-translated, gauge-transformed n-vortex).

We spell out the definition of the latter. We define the manifold, obtained by action of the symmetry group $\mathbb{R}^2 \times H^2(\mathbb{R}^2, \mathbb{R})$ of the vortex on the vortex $u^{(n)} := (\Psi^{(n)}, A^{(n)})$,

$$
\mathcal{M}^{(n)} = \{T_h^{\mathrm{trans}} T_\gamma^{\mathrm{gauge}} u^{(n)} : h \in \mathbb{R}^2, \ \gamma \in H^2(\mathbb{R}^2, \mathbb{R})\}.
$$

Let dist_{H^1} denote the H^1-distance to this manifold. We say that the vortex $u^{(n)}$ is *asymptotically stable* under H^1-perturbations, if there is $\delta > 0$ s.t. for any initial condition u_0 satisfying $\mathrm{dist}_{H^1}(u_0, \mathcal{M}^{(n)}) \le \delta$ there exists $g(t) := (h(t), \gamma(t)) \in \mathbb{R}^2 \times H^2(\mathbb{R}^2, \mathbb{R})$, s.t. the solution $u(t)$ of the time-dependent equation ((2.11) or (2.12)) satisfies

$$\|u(t) - T_{h(t)}^{\text{trans}} T_{g(t)}^{\text{gauge}} u_\omega\|_{H^1} \to 0, \quad \text{as } t \to \infty.$$

The basic result on vortex stability is the following:

Theorem 2.1 [40,41]   *1. For type I superconductors, all n-vortices are asymptotically stable.*

   2. For type II superconductors, the ± 1-vortices are stable, while the n-vortices with $|n| \geq 2$ are unstable.

This stability behavior was long conjectured (see Ref. [47]), based on numerical computations (e.g., [46]) leading to a "vortex interaction" picture wherein intervortex interactions are always attractive in the type-I case but become repulsive for like-signed vortices in the type-II case.

This result agrees with the fact, mentioned earlier, that the surface tension is positive for $\kappa < 1/\sqrt{2}$ and negative for $\kappa > 1/\sqrt{2}$, so the vortices try to minimize their "surface" for $\kappa < 1/\sqrt{2}$ and maximize it for $\kappa > 1/\sqrt{2}$.

Stability for vortices for the Ginzburg–Landau equations with impurity potentials was proven in Ref. [44].

For the Maxwell-Higgs equations (2.12), the above result was proven for the orbital stability only (see Ref. [40]). The *asymptotic stability* of the *n*-vortex for these equations is not known.

Also see Refs. [24,73] for extensions of these results to domains other than $\mathbb{R}^2$.

To demonstrate the above theorem, we first prove the linearized/energetic stability or instability. To formulate the latter, we observe that the *n*-vortex is a critical point of the Ginzburg–Landau energy (2.2), and the second variation of the energy

$$L^{(n)} := \text{Hess } \mathcal{E}(\Psi^{(n)}, A^{(n)})$$

is the linearized operator for the Ginzburg–Landau equations (2.1) around the *n*-vortex, acting on the space $X = L^2(\mathbb{R}^2, \mathbb{C}) \oplus L^2(\mathbb{R}^2, \mathbb{R}^2)$. (Hess $\mathcal{E}(u)$ is the Gâteaux derivative of the L^2−gradient of the l.h.s. of (2.1) and $u = (\Psi, A)$. Since Hess $\mathcal{E}(u)$ is only real-linear, to apply the spectral theory, it is convenient to extend it to a complex-linear operator. However, in order not to introduce extra notation, we ignore this point here and *deal with $L^{(n)}$ as if it were a complex-linear operator*.)

The symmetry group of $\mathcal{E}(\Psi, A)$, which is infinite-dimensional due to gauge transformations, gives rise to an infinite-dimensional subspace of $\text{Null}(L^{(n)}) \subset X$, which we denote here by Z_{sym}. We say the *n*-vortex is *(linearly) stable* if for some $c > 0$,

$$L^{(n)}|_{Z_{\text{sym}}^\perp} \geq c > 0, \tag{2.15}$$

and *unstable* if $L^{(n)}$ has a negative eigenvalue. By this definition, a stable state is a local energy minimizer which is a *strict* minimizer in directions orthogonal to the infinitesimal symmetry transformations. An unstable state is an energy saddle point.

Once the linearized (spectral) stability is proven, the main task in proving the orbital stability is the construction of a path in the (infinite dimensional due to gauge symmetry) symmetry group orbit of the n-vortex, to which the solution remains close. For the gradient-flow equations (2.11), the orbital stability can be easily strengthened to and with little more work, the asymptotic stability can be accomplished.

A few brief remarks on the proof of the key step (2.15):

- Since the vortices are gauge equivalent under the action of rotation, that is,

$$\Psi(R_\alpha x) = e^{in\alpha}\Psi(x), \ R_{-\alpha}A(R_\alpha x) = A(x),$$

where R_α is counterclockwise rotation in $\mathbb{R}^2$ through the angle α, the linearized operator $L^{(n)}$ commutes with the representation $\rho_n : U(1) \to \mathcal{B}([L^2(\mathbb{R}^2; \mathbb{C})]^4)$, where $\mathcal{B}([L^2(\mathbb{R}^2; \mathbb{C})]^4)$ is the space of bounded operators on $[L^2(\mathbb{R}^2; \mathbb{C})]^4$, of the group $U(1)$, given by

$$\rho_n(e^{i\theta})(\xi, \alpha)(x) = (e^{in\theta}\xi, e^{-i\theta}\alpha)(R_{-\theta}x).$$

It follows that $L^{(n)}$ leaves invariant the eigenspaces of $d\rho_n(s)$ for any $s \in i\mathbb{R} = Lie(U(1))$. (The representation of $U(1)$ on each of these subspaces is multiple to an irreducible one.) According to a representation of the symmetry group, this results in (fiber) block decomposition of $L^{(n)}$,

$$X \approx \bigoplus_{m \in \mathbb{Z}} (L^2_{\mathrm{rad}})^4, \qquad L^{(n)} \approx \oplus_{m \in \mathbb{Z}} L^{(n)}_m, \tag{2.16}$$

where $L^2_{\mathrm{rad}} \equiv L^2(\mathbb{R}^+, rdr)$ and $\approx$ stands for the unitary equivalence, which is described below. One can then study each operator $L^{(n)}_m$, which acts on (vectors of) radially-symmetric functions.

- The gauge adjusted translational zero modes each lies within a single subspace of the decomposition (2.16) and correspond, after complexification and rotation, to the vector

$$T = (f'_n(r), \ b^{(n)}(r)f_n(r), \ na'_n(r)/r, \ na'_n(r)/r), \tag{2.17}$$

where we defined, for convenience, $b^{(n)}(r) = n(1 - a^{(n)}(r))/r$, in the $m = \pm 1$ sectors. The stability proof is built on Perron–Frobenius-type arguments ("positive zero mode $\iff$ the bottom of the spectrum"), adapted to the setting of systems. In particular, T lies in the "positivity cone" of vector functions with positive components, and we are able to conclude that $L^{(n)}_{\pm 1} \geq 0$ with non-degenerate zero-eigenvalue.

- A key component is the exploitation of the special structure, hinted at by the Bogomolnyi lower bound (2.8), of the linearized operator $L^{(n)}$ at the self-dual value $\kappa = 1/2$ of the Ginzburg–Landau parameter. In fact,

$$L^{(n)}_m|_{\kappa=1/2} = (F^{(n)}_m)^* F^{(n)}_m$$

for a first-order operator, F_m, having $2|n|$ zero–modes that can be calculated semi-explicitly. These modes can be thought of as arising from independent relative motions of vortices, and the fact that they are energy-neutral relates to the vanishing of the vortex interaction at $\kappa = 1/2$ [14,92]. Two of the modes arise from translational symmetry, while careful analysis shows that as κ moves above (respectively below) $1/2$, the $2|n| - 2$ "extra" modes become unstable (respectively stable) directions.

Technically, it is convenient, on the first step, to effectively remove the (infinite-dimensional subspace of) gauge-symmetry zero modes, by modifying $L^{(n)}$ to make it coercive in the gauge directions—this leaves only the two zero modes arising from translational invariance remaining.

Let $\mathcal{C}$ be the operation of taking the complex conjugate. The results in (fiber) block decomposition of $L^{(n)}$, mentioned above is given in the following.

Theorem 2.2 [41]   *1. Let $\mathcal{H}_m := [L^2_{rad}]^4$ and define $U : X \to \mathcal{H}$, where $\mathcal{H} = \bigoplus_{m \in \mathbf{Z}} \mathcal{H}_m$, so that on smooth compactly supported v it acts by the formula*

$$(Uv)_m(r) = J_m^{-1} \int_0^{2\pi} \chi_m^{-1}(\theta)\rho_n(e^{i\theta})v(x)d\theta.$$

where $\chi_m(\theta)$ are characters of $U(1)$, that is, all homomorphisms $U(1) \to U(1)$ (explicitly we have $\chi_m(\theta) = e^{im\theta}$) and

$$J_m : \mathcal{H}_m \to e^{i(m+n)\theta}L^2_{rad} \oplus e^{i(m-n)\theta}L^2_{rad} \oplus -ie^{i(m-1)\theta}L^2_{rad} \oplus ie^{i(m+1)\theta}L^2_{rad}$$

acting in the obvious way. Then U extends uniquely to a unitary operator.
2. Under U the linearized operator around the vortex, $K^{(n)}_\#$, decomposes as

$$UL^{(n)}U^{-1} = \bigoplus_{m \in \mathbf{Z}} L_m^{(n)}, \tag{2.18}$$

where the operators $L_m^{(n)}$ act on $\mathcal{H}_m$ as $J_m^{-1}L^{(n)}J_m$.
3. The operators $K_m^{(n)}$ have the following properties:

$$K_m^{(n)} = RK_{-m}^{(n)}R^T, \text{ where } R = \begin{pmatrix} Q & 0 \\ 0 & Q \end{pmatrix}, Q = \begin{pmatrix} 0 & \mathcal{C} \\ \mathcal{C} & 0 \end{pmatrix}, \tag{2.19}$$

$$\sigma_{ess}(K_m^{(n)}) = [\min(1,\lambda), \infty), \tag{2.20}$$

for $|n| = 1$ and $m \geq 2$, $L_m^{(n)} - L_1^{(n)} \geq 0$ with no zero-eigenvalue, $\quad$ (2.21)

$$L_0^{(n)} \geq c > 0 \quad \text{for all } \kappa, \tag{2.22}$$

$L_1^{(\pm 1)} \geq 0$ with non-degenerate zero-mode given by (2.17). $\quad$ (2.23)

Since by (2.20) and (2.23), $L_1^{(\pm 1)}|_{T^\perp} \geq \tilde{c} > 0$ and, by (2.22) and (2.23), $L_m^{(\pm 1)} \geq$ $c' > 0$ for $|m| \geq 2$, this theorem implies (2.15).

2.4 VORTEX LATTICES

In this section, we describe briefly recent results on vortex lattice solutions, that is solutions that display vortices arranged along vertices of a lattice in $\mathbb{R}^2$. Since their discovery by Abrikosov in 1957, solutions have been studied in numerous experimental and theoretical works (of the more mathematical studies, we mention the articles of Eilenberger [30] and Lasher [53]).

The rigorous investigation of Abrikosov solutions was carried out by Odeh [58], soon after their discovery. Odeh has given a detailed sketch of the proof of the bifurcation of Abrikosov solutions at the second critical magnetic field. Further details were provided by Barany, Golubitsky, and Tursky [9], using equivariant bifurcation theory, and by Takáč [83], who obtained results on the zeros of the bifurcating solutions. The proof of existence was completed by Tzneteas and Sigal [89] and extended further by Tzaneteas and Sigal [90] beyond the cases covered in the works above.

Existence of Abrikosov solutions at low magnetic fields near the first critical magnetic field was given in Ref. [76].

Moreover, Odeh has also given a detailed sketch of the proof, with details in Ref. [29], of the existence of Abrikosov solutions using the variational minimization of the Ginzburg–Landau energy functional reduced to a fundamental cell of the underlying lattice. However, this proof provides only very limited information about the solutions.

Chapman [20] and Almag [6] gave a detailed analysis of extension of Abrikosov solutions to higher magnetic fluxes per fundamental cells.

Moreover, important and fairly detailed results on asymptotic behavior of solutions, for $\kappa \to \infty$ and the applied magnetic fields, h, satisfying $h \leq \frac{1}{2}\log\kappa + $const (the London limit), were obtained by Aydi and Sandier [8] (see Ref. [71] for references to earlier works). Further extensions to the Ginzburg–Landau equations for anisotropic and high-temperature superconductors can be found in Refs. [4, 5].

Among related results, a relation of the Ginzburg–Landau minimization problem, for a fixed, finite domain and in the regime of the Ginzburg–Landau parameter $\kappa \to \infty$ and external magnetic field, to the Abrikosov lattice variational problem was obtained Ref. [3] (see also Ref. [7]). Dutour [28] (see also Ref. [29]) has found boundaries between superconducting, normal, and mixed phases. In Ref. [7, 8], the Ginzburg–Landau energy is connected to the thermodynamic limit of the Abrikosov energy. The complete proof of the thermodynamic limit of the Abrikosov energy is given in Ref. [33] and boundary effects on the Abrikosov energy are established in Ref. [32]. The connection between vortex lattice problems and the Ginzburg–Landau functional is established in the large kappa limit in Ref. [72].

The proof that the triangular lattices minimize the Ginzburg–Landau energy functional per the fundamental cell was obtained in [89]. The paper used original Abrikosov ideas and the results of [2, 56] on the Abrikosov "constant".

The stability of Abrikosov lattices is shown in Ref. [77] for gauge periodic perturbations, that is, perturbations having the same translational lattice symmetry as the
solutions themselves, and in Ref. [78] for local, more precisely, H^1, perturbations.

Here we describe briefly the existence and stability results and the main ideas
entering into their proofs.

2.4.1 Abrikosov lattices

In 1957, A. Abrikosov [1] discovered a class of solutions, (Ψ, A), to (2.1), presently
known as Abrikosov lattice vortex states (or just Abrikosov lattices), whose physi-
cal characteristics, density of Cooper pairs, $|\Psi|^2$, the magnetic field, $\mathrm{curl} A$, and the
supercurrent, $J_S = \mathrm{Im}(\bar{\Psi} \nabla_A \Psi)$, are double-periodic w.r.t a lattice $\mathcal{L}$. (This set of states
is invariant under the symmetries of the previous subsection.)

For Abrikosov states, for (Ψ, A), the magnetic flux, $\int_\Omega \mathrm{curl} A$, through a fundamental lattice cell, Ω, is quantized,

$$\frac{1}{2\pi} \int_\Omega \mathrm{curl} A = \deg \Psi = n, \tag{2.24}$$

for some integer n. Indeed, the periodicity of $n_s = |\Psi|^2$ and $J = \mathrm{Im}(\bar{\Psi} \nabla_A \Psi)$ implies
that $\nabla \varphi - A$, where $\Psi = |\Psi| e^{i\varphi}$, is periodic, provided $\Psi \neq 0$ on $\partial \Omega$. This, together
with Stokes's theorem, $\int_\Omega \mathrm{curl} A = \oint_{\partial \Omega} A = \oint_{\partial \Omega} \nabla \varphi$ and the single-valuedness of Ψ,
implies that $\int_\Omega \mathrm{curl} A = 2\pi n$ for some integer n. Using the reflection symmetry of the
problem, one can easily check that we can always assume $n \geq 0$.

Equation (2.24) implies the relation between the average magnetic flux, b, per
lattice cell, $b = 1/|\Omega| \int_\Omega \mathrm{curl} A$, and the area, $|\Omega|$, of a fundamental cell

$$b = \frac{2\pi n}{|\Omega|}. \tag{2.25}$$

Finally, it is clear that the gauge, translation, and rotation symmetries of the
Ginzburg–Landau equations map lattice states to lattice states. In the case of the
gauge and translation symmetries, the lattice with respect to which the solution is
gauge-periodic does not change, whereas with the rotation symmetry, the lattice is
rotated as well. The magnetic flux per cell of solutions is also preserved under the
action of these symmetries.

2.4.2 Existence of Abrikosov lattices

We assume always that the coordinate origin is placed at one of the vertices of the
lattice $\mathcal{L}$. Recall that we identify $\mathbb{R}^2$ with $\mathbb{C}$, via the map $(x_1, x_2) \to x_1 + i x_2$. We can
choose a basis in $\mathcal{L}$ so that $\mathcal{L} = r(\mathbb{Z} + \tau \mathbb{Z})$, where $\tau \in \mathbb{C}$, $\mathrm{Im}\,\tau > 0$, and $r > 0$, with
bases giving the same lattice related by elements of the modular group $SL(2, \mathbb{Z})$ (see
Appendix 2.A for details). Hence, it suffices to consider τ in the fundamental domain,

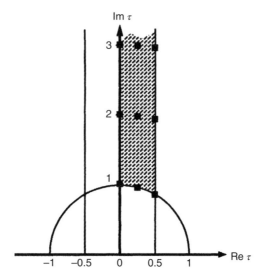

FIGURE 2.1   Half of the fundamental domain $\Pi^+/SL(2,\mathbb{Z})$.

$\Pi^+/SL(2,\mathbb{Z})$, of $SL(2,\mathbb{Z})$ acting on the Poincaré half-plane $\Pi^+ := \{\tau \in \mathbb{C} : \operatorname{Im}\tau > 0\}$ (see Fig. 2.1).

Due to the quantization relation (2.25), the parameters τ, b, and n determine the lattice $\mathcal{L}$ up to a rotation and a translation. As the equations (2.1) are invariant under rotations and translations, solutions corresponding to translated and rotated lattices are related by symmetry transformations and therefore can be considered equivalent, with equivalence classes determined by triples $\omega = (\tau, b, n)$, specifying the underlying lattice has shape τ, the average magnetic flux per lattice cell b, and the number n of quanta of magnetic flux per lattice cell. With this in mind, we will say that an Abrikosov lattice state (Ψ, A) is of type $\omega = (\tau, b, n)$, if it belongs to the equivalence class determined by $\omega = (\tau, b, n)$.

Let $\beta(\tau)$ be the Abrikosov "constant," defined in (2.33) below. The following critical value of the Ginzburg–Landau parameter κ plays an important role in what follows

$$\kappa_c(\tau) := \sqrt{\frac{1}{2}\left(1 - \frac{1}{\beta(\tau)}\right)}. \tag{2.26}$$

Recall that the value of the second critical magnetic field at which the normal material undergoes the transition to the superconducting state is that $h_{c2} = \kappa^2$.

For the case $n = 1$ of one quantum of flux per unit cell, the following result establishes the existence of nontrivial lattice solutions near the normal metal solution:

Theorem 2.3 [9, 29, 58, 90] *Fix a lattice shape τ and let b satisfy*

$$|\kappa^2 - b| \ll \kappa^2[(2\kappa^2 - 1)\beta(\tau) + 1] \qquad (2.27)$$

(uniformly in the parameters τ and b) and

$$\text{either } \kappa > \kappa_c(\tau), \ \kappa^2 > b \text{ or } \kappa < \kappa_c(\tau), \ \kappa^2 < b. \qquad (2.28)$$

Then for $\omega = (\tau, b, 1)$

- *there exists a smooth Abrikosov lattice solution $u_\omega = (\Psi_\omega, A_\omega)$ of type ω.*

Remark. For $\kappa > 1/\sqrt{2}$ and the triangular and square lattices the theorem was proven in Refs. [9, 29, 58, 89], and in the case stated in Refs. [89, 90].

Let $\mathcal{L}_\omega$ be the lattice specified by a triple $\omega = (\tau, b, n)$ and let Ω_ω denote its elementary cell. Define the average energy, $E_b(\tau) := \frac{1}{|\Omega_\omega|}\mathcal{E}_{\Omega_\omega}(u_\omega)$, per lattice cell, of the Abrikosov lattice solution, u_ω, $\omega = (\tau, b, 1)$, found in Theorem 2.3.

Theorem 2.4 [90] *Let $\kappa > 1/\sqrt{2}$ and let b satisfy $b < \kappa^2$ and (2.27). Then for a fixed b,*

- *$E_b(\tau)$ has the global minimum in τ at the hexagonal (equilateral triangular) lattice, $\tau = e^{i\pi/3}$.*

(Due to a calculation error, Abrikosov concluded that the lattice that gives the minimum energy is the square lattice. The error was corrected by Kleiner et al. [51], who showed that it is in fact the triangular lattice that minimizes the energy.)

Now, we formulate the existence result for low magnetic fields, those near the first critical magnetic field h_{c1}: Let $\mathcal{L}_\omega$ be a lattice specified by a triple $\omega = (\tau, b, n)$ and let Ω_ω denote its elementary cell. We have the following.

Theorem 2.5 [76] *Let $\kappa \neq 1/\sqrt{2}$ and fix a lattice shape λ and $n \neq 0$. Then there is $b_0 = b_0(\kappa) \ (\sim (\kappa - 1/\sqrt{2})^2) > 0$ such that for $b \leq b_0$, there exists an odd solution Abrikosov lattice solution $u_\omega \equiv (\Psi_\omega, A_\omega)$ of (2.1), s.t.*

$$u_\omega(x) = u^{(n)}(x - \alpha) + O(e^{-c\rho}) \text{ on } \Omega_\omega + \alpha, \ \forall \alpha \in \mathcal{L}_\omega, \qquad (2.29)$$

where $u^{(n)} := (\Psi^{(n)}, A^{(n)})$ is the n-vortex, $\rho = b^{-1/2}$, and $c > 0$, in the sense of the local Sobolev norm of any index.

In the next two subsections, we present a discussion of some key general notions. After this, we outline the proofs of the results above.

2.4.3   Abrikosov lattices as gauge-equivariant states

A key point in proving both theorems is to realize that a state (Ψ, A) is an Abrikosov lattice if and only if (Ψ, A) is gauge-periodic or gauge-equivariant (with respect to a lattice $\mathcal{L}$) in the sense that there exist (possibly multivalued) functions $g_s : \mathbb{R}^2 \to \mathbb{R}$, $s \in \mathcal{L}$, such that

$$T_s^{\text{trans}}(\Psi, A) = T_{g_s}^{\text{gauge}}(\Psi, A). \tag{2.30}$$

Indeed, if state (Ψ, A) satisfies (2.30), then all associated physical quantities are $\mathcal{L}$-periodic, that is, (Ψ, A) is an Abrikosov lattice. In the opposite direction, if (Ψ, A) is an Abrikosov lattice, then $\operatorname{curl} A(x)$ is periodic w.r.t. $\mathcal{L}$, and therefore $A(x + s) = A(x) + \nabla g_s(x)$, for some functions $g_s(x)$. Next, we write $\Psi(x) = |\Psi(x)| e^{i\phi(x)}$. Since $|\Psi(x)|$ and $J(x) = |\Psi(x)|^2 (\nabla \phi(x) - A(x))$ are periodic w.r.t. $\mathcal{L}$, we have that $\nabla \phi(x + s) = \nabla \phi(x) + \nabla \tilde{g}_s(x)$, which implies that $\phi(x + s) = \phi(x) + g_s(x)$, where $g_s(x) = \tilde{g}_s(x) + c_s$, for some constants c_s.

Since T_s^{trans} is a commutative group, we see that the family of functions g_s has the important cocycle property

$$g_{s+t}(x) - g_s(x + t) - g_t(x) \in 2\pi\mathbb{Z}. \tag{2.31}$$

This can be seen by evaluating the effect of translation by $s + t$ in two different ways. We call $g_s(x)$ the *gauge exponent*. It can be shown (see Appendix 2.B) that by a gauge transformation, we can pass from a exponential g_s satisfying the cocycle condition (2.31) is equivalent to the exponent

$$\frac{b}{2} s \wedge x + c_s, \text{ with } c_s \text{ satisfying } c_{s+t} - c_s - c_t - \frac{1}{2} b s \wedge t \in 2\pi\mathbb{Z}, \tag{2.32}$$

and b satisfying $b|\Omega| \in 2\pi\mathbb{Z}$. For more discussion of g_s, see Appendix 2.B.

Remark.   Relation (2.31) for Abrikosov lattices was isolated Sigal and Tzaneteas [77], and it played an important role. This condition is well known in algebraic geometry and number theory, where $e^{i g_s(x)}$ is called the automorphy factor (see for example Ref. [39]). However, there the associated vector potential (connection on the corresponding principal bundle) A is not considered.

2.4.4   Abrikosov function

We begin with some notation. Let $\langle f \rangle_\Omega$ denote the average, $\langle f \rangle_\Omega = 1/|\Omega| \int_\Omega f$, of a function f over $\Omega \subset \mathbb{R}^2$. Introduce the normalized lattice $\mathcal{L}_\tau = \sqrt{\frac{2\pi}{|\Omega_\omega|}} \mathcal{L}\omega$, for

$\omega = (\tau, b, n = 1)$. The key role in understanding the energetics of the Abrikosov lattice solution is played by the Abrikosov function

$$\beta(\tau) := \frac{\langle |\phi|^4 \rangle_{\Omega_\tau}}{\langle |\phi|^2 \rangle_{\Omega_\tau}^2}, \tag{2.33}$$

where Ω_τ is a fundamental cell of the lattice $\mathcal{L}_\tau$ and ϕ is the solution to the problem

$$(-\Delta_{A^n} - n)\phi = 0, \quad \phi(x + s) = e^{i\frac{n}{2}s \wedge x}\phi(x), \quad \forall s \in \mathcal{L}_\tau, \tag{2.34}$$

where $A^n(x) := -\frac{n}{2}Jx$ and $J := \begin{pmatrix} 0 & 1 \\ -1 & 0 \end{pmatrix}$, for $n = 1$. We will show below (Proposition 2.1) that, for $n = 1$, the problem (2.34) has a unique solution and therefore β is well-defined. It is not hard to see that β depends only on the equivalence class of $\mathcal{L}$.

Equations (2.33) and (2.34) imply that $\beta(\tau)$ is symmetric w.r.t. the imaginary axis, $\beta(-\bar{\tau}) = \beta(\tau)$. Hence it suffices to consider $\beta(\tau)$ on the $\operatorname{Re} \tau \geq 0$ half of the fundamental domain, $\Pi^+/SL(2, \mathbb{Z})$ (the heavily shaded area in Fig. 2.1).

Moreover, we can consider (2.33) and (2.34) on the entire Poincaré half-plane $\Pi^+ := \{\tau \in \mathbb{C} : \operatorname{Im} \tau > 0\}$, which allows us to define $\beta(\tau)$ as a modular function on Π^+,

- the function $\beta(\tau)$, defined on Π^+, is invariant under the action of $SL(2, \mathbb{Z})$.

This implies that it suffices to consider $\beta(\tau)$ on the fundamental domain, $\Pi^+/SL(2, \mathbb{Z})$, of the modular group $SL(2, \mathbb{Z})$ acting on the Poincaré half-plane Π^+ (see Fig. 2.1).

Remarks.

1) The term Abrikosov constant comes from the physics literature, where one often considers only equilateral triangular or square lattices.

2) The way we defined the Abrikosov constant $\beta(\tau)$, it is manifestly independent of b. Our definition differs from the standard one by rescaling: the standard definition uses the function $\phi_b(x) = \phi(\sqrt{b}x)$, instead of $\phi(x)$.

2.4.5 Comments on the proofs of existence results

Comments on the proof of Theorem 2.3. As mentioned in Section 2.4.3, we look for solutions of (2.11), satisfying condition (2.30), or explicitly as, for $s \in \mathcal{L}$,

$$\begin{cases} \Psi(x + s) = e^{ig_s(x)}\Psi(x), \\ A(x + s) = A(x) + \nabla g_s(x), \end{cases} \tag{2.35}$$

where g_s satisfies (2.31). By (2.32), it can be taken to be

$$g_s(x) = \frac{b}{2} s \wedge x + c_s, \tag{2.36}$$

where b is the average magnetic flux, $b = 1/|\Omega| \int_\Omega \operatorname{curl} A$ (satisfying (2.25) so that $bs \wedge t \in 2\pi\mathbb{Z}$), and c_s satisfies

$$c_{s+t} - c_s - c_t - \frac{1}{2} bs \wedge t \in 2\pi\mathbb{Z}. \tag{2.37}$$

The linearized problem. We expect that as the average flux b decreases below $h_{c2} = \kappa^2$, a vortex lattice solution emerges from the normal material solution (Ψ_n, A_n), where $\Psi_n = 0$ and A_n is a magnetic potential, with the constant magnetic field b. Note that $(\Psi_n, A_n) = (0, A^b)$ satisfies (2.35), if we take the gauge $A^b = (-b/2)Jx$. Linearizing (2.1) at $(0, A^b)$ leads to the linearized problem

$$(-\Delta_{A^b} - \kappa^2)\phi = 0, \tag{2.38}$$

with $\phi(x)$ satisfying

$$\phi(x+s) = e^{i\frac{b}{2} s \cdot Jx} \phi(x), \quad \forall s \in \mathcal{L}. \tag{2.39}$$

(The second equation in (2.1) leads to $\operatorname{curl} a = 0$ which gives, modulo gauge transformation, $a = 0$.) We show that this problem has n linearly independent solutions, provided $b|\Omega| = 2\pi n$ and $b = \kappa^2 = h_{c2}$.

Denote by L^b the operator $-\Delta_{A^b}$, defined on the lattice cell Ω with the lattice boundary conditions in (2.39), is self-adjoint, has a purely discrete spectrum, and evidently satisfies $L^b \geq 0$. We have the following well-known result

Proposition 2.1 *The operator L^b is self-adjoint, with the purely discrete spectrum given by the spectrum explicitly as*

$$\sigma(L^b) = \{ (2k+1)b : k = 0, 1, 2, \dots, \}, \tag{2.40}$$

and each eigenvalue is of the same multiplicity.
If $b|\Omega| = 2\pi n$, then this multiplicity is n and, in particular, we have

$$\dim_{\mathbb{C}} \operatorname{Null}(L^b - b) = n.$$

Proof: The self-adjointness is standard. Spectral information about L^b can be obtained by introducing the complexified covariant derivatives (harmonic oscillator annihilation and creation operators), $\bar{\partial}_{A^b}$ and $\bar{\partial}^*_{A^b} = -\partial_{A^b}$, with

$$\bar{\partial}_{A^b} := (\nabla_{A^b})_1 + i(\nabla_{A^b})_2 = \partial_{x_1} + i\partial_{x_2} + \frac{1}{2} bx_1 + \frac{1}{2} ibx_2. \tag{2.41}$$

One can verify that these operators satisfy the following relations:

1. $[\bar{\partial}_{A^b}, \bar{\partial}^*_{A^b}] = 2\,\mathrm{curl}A^b = 2b;$
2. $-\Delta_{A^b} - b = \bar{\partial}^*_{A^b}\bar{\partial}_{A^b}.$

As for the harmonic oscillator (see e.g., Ref. [42]), this gives the spectrum explicitly (2.40). This proves the first part of the theorem.

For the second part, a simple calculation gives the following operator equation

$$e^{\frac{b}{2}(ix_1 x_2 - x_2^2)}\bar{\partial}_{A^b}e^{-\frac{b}{2}(ix_1 x_2 - x_2^2)} = \partial_{x_1} + i\partial_{x_2}.$$

This immediately proves that $\Psi \in \mathrm{Null}\,\bar{\partial}_{A^b}$ if and only if $\xi(x) = e^{\frac{b}{2}(ix_1 x_2 - x_2^2)}\psi(x)$ satisfies $\partial_{x_1}\xi + i\partial_{x_2}\xi = 0$.

We identify $\mathbb{R}^2$ with $\mathbb{C}$, via the map $(x_1, x_2) \to x_1 + ix_2$. We can choose a basis in $\mathcal{L}$ so that $\mathcal{L} = r(\mathbb{Z} + \tau\mathbb{Z})$, where $\tau \in \mathbb{C}$, $\mathrm{Im}\,\tau > 0$, and $r > 0$. By the quantization condition (2.25), $r := \sqrt{\frac{2\pi n}{\mathrm{Im}\,\tau b}}$. Define $z = \frac{1}{r}(x_1 + ix_2)$ and

$$\theta(z) = e^{\frac{b}{2}(ix_1 x_2 - x_2^2)}\phi(x). \tag{2.42}$$

By the above, the function θ is entire and, due to the periodicity conditions on ϕ, satisfies

$$\theta(z+1) = \theta(z),$$

$$\theta(z+\tau) = e^{-2inz}e^{-in\tau z}\theta(z).$$

Hence θ is the theta function. By the first relation, θ has the absolutely convergent Fourier expansion

$$\theta(z) = \sum_{k=-\infty}^{\infty} c_k e^{2\pi k iz}. \tag{2.44}$$

with the coefficients satisfying $c_{k+n} = e^{in\pi\tau}e^{2ki\pi\tau}c_k$, which means such functions are determined by $c_0, \ldots, c_{n-1}$ and therefore form an n-dimensional vector space. This proves Proposition 2.1.

This also gives the form of the leading approximation (2.42)–(2.44) to the true solution.

The nonlinear problem.   Now let $n = 1$. Once the linearized map is well understood, it is possible to construct solutions, u_ω, $\omega = (\tau, b, 1)$, of the Ginzburg–Landau equations for a given lattice shape parameter τ, and the average magnetic flux b near h_{c2}, via a Lyapunov–Schmidt reduction.

Comments on the proof of Theorem 2.4   The relation between the Abrikosov function, $E_b(\tau) := \frac{1}{|\Omega_\omega|}\mathcal{E}_{\Omega_\omega}(u_\omega)$, of this solution is given by the following result (see Ref. [89]).

Proposition 2.2   *In the case* $\kappa > \frac{1}{\sqrt{2}}$, *the minimizers,* τ_b, *of* $\tau \mapsto E_b(\tau)$ *are related to the minimizer,* τ_*, *of* $\beta(\tau)$, *as* $\tau_b - \tau_* = O(\mu^{1/2})$. *In particular,* $\tau_b \to \tau_*$ *as* $b \to \kappa^2$.

This result was already found (non-rigorously) by Abrikosov [1]. Thus the problem of minimization of the energy per the lattice cell is reduced to finding the minima of $\beta(\tau)$ as a function of the lattice shape parameter τ.

Using symmetries of $\beta(\tau)$, one can also show (see Ref. [78] and remark after Theorem 2.8) that $\beta(\tau)$ has critical points at the points $\tau = e^{\pi i/3}$ and $\tau = e^{\pi i/2}$. However, to determine minimizers of $\beta(\tau)$ requires a rather delicate analysis, which gives the following.

Theorem 2.6 [2,57]   *The function* $\beta(\tau)$ *has exactly two critical points,* $\tau = e^{i\pi/3}$ *and* $\tau = e^{i\pi/2}$. *The first is minimum, while the second is a maximum.*

Hence the second part of Theorem 2.4 follows.

Comments on the proof of Theorem 2.5.   The idea here is to reduce solving (2.1) for (Ψ, A) on the space $\mathbb{R}^2$ to solving it for (ψ, a) on the fundamental cell Ω, satisfying the boundary conditions

$$
\begin{cases}
\psi(x+s) = e^{ig_s(x)}\psi(x), \\
a(x+s) = a(x) + \nabla g_s(x), \\
(\nu \cdot \nabla_a\psi)(x+s) = e^{ig_s(x)}(\nu \cdot \nabla_a\psi)(x), \\
\operatorname{curl}a(x+s) = \operatorname{curl}a(x), \\
x \in \partial_1\Omega/\partial_2\Omega \text{ and } s = \omega_1/\omega_2.
\end{cases}
\tag{2.45}
$$

induced by the periodicity condition (2.35). Here $\partial_1\Omega/\partial_2\Omega = $ the left/bottom boundary of Ω, $\{\omega_1,\omega_2\}$ is a basis in $\mathcal{L}$ and $\nu(x)$ is the normal to the boundary at x.

To this end we show that given a continuously differentiable function (ψ,a) on the fundamental cell Ω, satisfying the boundary conditions (2.45), with g_s satisfying (2.31), we can lift it to a continuous and continuously differentiable function (Ψ, A) on the space $\mathbb{R}^2$, satisfying the gauge-periodicity conditions (2.35). Indeed, we define for any $\alpha \in \mathcal{L}$,

$$
\Psi(x) = \psi(x-\alpha)e^{i\Phi_\alpha(x)}, \ A(x) = a(x-\alpha) + \nabla\Phi_\alpha(x), \ x \in \Omega+\alpha,
\tag{2.46}
$$

where $\Phi_\alpha(x)$ is a real, possibly multi-valued, function to be determined. (Of course, we can add to it any $\mathcal{L}-$periodic function.) We define

$$
\Phi_\alpha(x) := g_\alpha(x-\alpha), \text{ for } x \in \Omega+\alpha.
\tag{2.47}
$$

Lemma. Assume functions (ψ, a) on Ω are twice differentiable, up to the boundary, and obey the boundary conditions (2.45) and the Ginzburg–Landau equations (2.1). Then the functions (Ψ, A), constructed in (2.46) and (2.47), are smooth in $\mathbb{R}^2$ and satisfy the periodicity conditions (2.35) and the Ginzburg–Landau equations (2.1).

Proof: If (ψ, a) satisfies the Ginzburg–Landau equations (2.1) in Ω, then $U \equiv (\Psi, A)$, constructed in (2.46) and (2.47), has the following properties

  (1) (Ψ, A) is twice differentiable and satisfies (2.1) in $\mathbb{R}^2/(\cup_{t \in \mathcal{L}} S_t \partial \Omega)$, where S_t : $x \to x + t$;
  (2) (Ψ, A) is continuous with continuous derivatives ($\nabla_A \Psi$ and curl$A$) in $\mathbb{R}^2$ and satisfies the gauge-periodicity conditions (2.35) in $\mathbb{R}^2$.

Indeed, the periodicity condition (2.35) applied to the cells $\Omega + \alpha - \omega_i$ and $\Omega + \alpha$ and the continuity condition on the common boundary of the cells $\Omega + \alpha - \omega_i$ and $\Omega + \alpha$ imply that $\Phi_\alpha(x)$ should satisfy the following two conditions:

$$\Phi_\alpha(x) = \Phi_{\alpha - \omega_i}(x - \omega_i) + g_{\omega_i}(x - \omega_i), \text{ mod } 2\pi, \; x \in \Omega + \alpha, \qquad (2.48)$$

$$\Phi_\alpha(x) = \Phi_{\alpha - \omega_i}(x) + g_{\omega_i}(x - \alpha), \text{ mod } 2\pi, \; x \in \partial_i \Omega + \alpha, \qquad (2.49)$$

where $i = 1, 2$, and, recall, $\{\omega_1, \omega_2\}$ is a basis in $\mathcal{L}$ and $\partial_1 \Omega / \partial_2 \Omega$ is the left/bottom boundary of Ω.

To show that (2.47) satisfies the conditions (2.48) and (2.49), we note that, due to (2.31), we have $g_\alpha(x - \alpha) = g_{\alpha - \omega_i}(x - \alpha) + g_{\omega_i}(x - \omega_i)$, mod 2π, $x \in \Omega + \alpha$, and $g_\alpha(x - \alpha) = g_{\alpha - \omega_i}(x - \alpha + \omega_i) + g_{\omega_i}(x - \alpha)$, mod 2π, $x \in \partial_i \Omega + \alpha$, which are equivalent to (2.48) and (2.49), with (2.47).

The second pair of conditions in (2.45) implies that $\nabla_A \Psi$ and curl$A$ are continuous across the cell boundaries.

By (1) and (2), the derivatives $\Delta_A \Psi$ and curl$^2 A$ are continuous, up to the boundary, in $S_t \partial \Omega$, for every $t \in \mathcal{L}$. By (2.1), they are equal in $\mathbb{R}^2/(\cup_{t \in \mathcal{L}} S_t \partial \Omega)$ to functions continuous in $\mathbb{R}^2$ satisfying there the periodicity condition (2.35). Hence, they are also continuous and satisfy the periodicity condition (2.35) in $\mathbb{R}^2$. By iteration of the above argument (i.e., elliptic regularity), Ψ, A are smooth functions obeying (2.35) and (2.1).

Now, we use the $n-$vortex $(\Psi^{(n)}, A^{(n)})$, placed in the center of the fundamental cell Ω, to construct an approximate solution $(\psi^{\text{appr}}, a^{\text{appr}})$ to (2.1) in Ω, satisfying (2.45), and use it and the Lyapunov–Schmidt splitting technique to show that there is a true solution (ψ, a) nearby sharing the same properties. After that, we use Lemma 2.4.5 above to lift (ψ, a) to a solution (Ψ, A) on the space $\mathbb{R}^2$, satisfying the gauge-periodicity conditions (2.35).

2.4.6   Stability of Abrikosov lattices

The Abrikosov lattices are static solutions to (2.11) and their stability w.r.t. the dynamics induced by these equations is an important issue. In Ref. [77], we considered the stability of the Abrikosov lattices for magnetic fields close to the second critical magnetic field $h_{c2} = \kappa^2$, under the simplest perturbations, namely those having the same (gauge-) periodicity as the underlying Abrikosov lattices (we call such perturbations *gauge-periodic*) and proved for a lattice of arbitrary shape, $\tau \in \mathbb{C}$, $\mathrm{Im}\,\tau > 0$, that, under gauge-periodic perturbations, Abrikosov vortex lattice solutions are

(1) *asymptotically stable for $\kappa^2 > \kappa_c(\tau)$ and*

(2) *unstable for $\kappa^2 < \kappa_c(\tau)$.*

This result belies the common belief among physicists and mathematicians that Abrikosov-type vortex lattice solutions are stable only for triangular lattices and $\kappa > 1/\sqrt{2}$, and it seems this is the first time the threshold (2.26) has been isolated.

In Ref. [76], similar results are shown to hold also for low magnetic fields close to h_{c1}.

Gauge-periodic perturbations are not a common type of perturbations occurring in superconductivity. Now, we address the problem of the stability of Abrikosov lattices under local or finite-energy perturbations (defined precisely below). We consider Abrikosov lattices of arbitrary shape, not just triangular or rectangular lattices as usually considered, and for magnetic fields close to the second critical magnetic field $h_{c2} = \kappa^2$.

Finite-energy (H^1-) perturbations.   We now wish to study the stability of these Abrikosov lattice solutions under a class of perturbations that have finite-energy. More precisely, we fix an Abrikosov lattice solution u_ω and consider perturbations $v : \mathbb{R}^2 \to \mathbb{C} \times \mathbb{R}^2$ that satisfy

$$\Lambda_{u_\omega}(v) = \lim_{Q \to \mathbb{R}^2} \left(\mathcal{E}_Q(u_\omega + v) - \mathcal{E}_Q(u_\omega) \right) < \infty. \qquad (2.50)$$

Clearly, $\Lambda_{u_\omega}(v) < \infty$ for all vectors of the form $v = T_\gamma^{\mathrm{gauge}} u_\omega - u_\omega$, where $\gamma \in H^2(\mathbb{R}^2; \mathbb{R})$.

In fact, we will be dealing with the smaller class, H^1_{cov}, of perturbations, where H^1_{cov} is the Sobolev space of order 1 defined by the covariant derivatives, that is,

$$H^1_{\mathrm{cov}} := \{ v \in L^2(\mathbb{R}^2, \mathbb{C} \times \mathbb{R}^2) \mid \|v\|_{H^1} < \infty \},$$

where the norm $\|v\|_{H^1}$ is determined by the covariant inner product

$$\langle v, v' \rangle_{H^1} = \mathrm{Re} \int \bar{\xi}\xi' + \overline{\nabla_{A_\omega}\xi} \cdot \nabla_{A_\omega}\xi' + \alpha \cdot \alpha' + \sum_{k=1}^{2} \nabla\alpha_k \cdot \nabla\alpha'_k,$$

where $v = (\xi, \alpha)$, $v' = (\xi', \alpha')$, while the L^2-norm is given by

$$\langle v, v' \rangle_{L^2} = \mathrm{Re} \int \bar{\xi} \xi' + \alpha \cdot \alpha'. \tag{2.51}$$

An explicit representation for the functional $\Lambda_{u_\omega}(v)$, shows that $\Lambda_{u_\omega}(v) < \infty$ for all vectors $v \in H^1_{\mathrm{cov}}$.

To introduce the notions of stability and instability, we note that the hessian Hess $\mathcal{E}(u)$ is well-defined as a differential operator for say $u \in u_\omega + H^1_{\mathrm{cov}}$ and is a real-linear operator on H^1_{cov}. We define the manifold

$$\mathcal{M}_\omega = \{ T^{\mathrm{gauge}}_\gamma u_\omega : \gamma \in H^1(\mathbb{R}^2, \mathbb{R}) \}$$

of gauge equivalent Abrikosov lattices and the H^1-distance, dist_{H^1}, to this manifold.

Definition. We say that the Abrikosov lattice u_ω is *asymptotically stable* under $H^1_{\mathrm{cov}}-$ perturbations, if there is $\delta > 0$ s.t. for any initial condition u_0 satisfying $\mathrm{dist}_{H^1}(u_0, \mathcal{M}_\omega) \leq \delta$ there exists $g(t) \in H^1$, s.t. the solution $u(t)$ of (2.11) satisfies

$$\| u(t) - T^{\mathrm{gauge}}_{g(t)} u_\omega \|_{H^1} \to 0,$$

as $t \to \infty$. We say that u_ω is *energetically unstable* if the Hessian, $\mathcal{E}''(u_\omega)$, of $\mathcal{E}(u)$ at u_ω has a negative spectrum.

We restrict the initial conditions (Ψ_0, A_0) for (2.11) satisfying

$$T^{\mathrm{refl}}(\Psi_0, A_0) = (\Psi_0, A_0). \tag{2.52}$$

Note that, by uniqueness, the Abrikosov lattice solutions $u_\omega = (\Psi_\omega, A_\omega)$ satisfy $T^{\mathrm{refl}} u_\omega = u_\omega$ and therefore so are the perturbations, $v_0 := u_0 - u_\omega$, where $u_0 := (\Psi_0, A_0)$:

$$T^{\mathrm{refl}} v_0 = v_0. \tag{2.53}$$

Stability result. Recall that $\beta(\tau)$ is the Abrikosov "constant," introduced in (2.33).

Theorem 2.7 *There exists continuous modular functions $\gamma_\delta(\tau), \delta > 0$, depending on the lattice shape parameter τ, such that, for b sufficiently close to κ^2, in the sense of (2.27), and, under H^1-perturbations, satisfying (2.52), the Abrikosov lattice u_ω is*

- *asymptotically stable for all (τ, κ) s.t. $\kappa > 1/\sqrt{2}$ and $\gamma_\delta(\tau) > 0$, for all $\delta > 0$, and*
- *energetically unstable if either $\kappa < 1/\sqrt{2}$ or $\kappa > 1/\sqrt{2}$ and $\gamma_\delta(\tau) < 0$, for some $\delta > 0$.*

The functions $\gamma_\delta(\tau), \delta > 0$, $\mathrm{Im}\,\tau > 0$, $\delta > 0$, appearing in the theorem above are described below. Meantime we make the following important remark. Since we know that, for $\kappa > 1/\sqrt{2}$, the triangular lattice has the lowest energy (see Theorem 2.3), this seems to suggest that other lattices should be unstable. The reason that this energetics does not affect the stability under local perturbations can be gleaned from investigating the zero mode of the Hessian of the energy functional associated with different lattice shapes, τ. This mode is obtained by differentiating the Abrikosov lattice solutions w.r.t. τ, which shows that it grows linearly in $|x|$. To rearrange a non-triangular Abrikosov lattice into the triangular one, one would have to activate this mode and hence to apply a perturbation, growing at infinity (at the same rate).

This also explains why the Abrikosov "constant" $\beta(\tau)$ mentioned above, which plays a crucial role in understanding the energetics of the Abrikosov solutions, is not directly related to the stability under local perturbations, the latter is governed by $\gamma(\tau)$.

2.4.7   Functions $\gamma_\delta(\tau), \delta > 0$

For a lattice $\mathcal{L}$, considered as a group of lattice translations, $\hat{\mathcal{L}}$ denotes the dual group, i.e. the group of characters, $\chi : \mathcal{L} \to U(1)$. Furthermore, let $\mathcal{T} := \mathbb{R}^2/\mathcal{L}$.

Theorem 2.8   *The functions $\gamma_\delta(\tau), \delta > 0$, on lattice shapes τ, entering Theorem 2.7, are given by*

$$\gamma_\delta(\tau), \delta > 0 := \inf_{\chi \in \hat{\mathcal{L}}_{\tau,\delta}} \gamma_\chi(\tau), \text{ where } \gamma_\chi(\tau) := 2\langle |\phi_0|^2 |\phi_\chi|^2 \rangle_{\mathcal{T}_\tau}$$
$$- |\langle \phi_0^2 \bar{\phi}_\chi \bar{\phi}_{\chi^{-1}} \rangle_{\mathcal{T}_\tau}| - \langle |\phi_0|^4 \rangle_{\mathcal{T}_\tau}. \tag{2.54}$$

Here $\hat{\mathcal{L}}_{\tau,\delta}$ is a subset of $\hat{\mathcal{L}}_\tau$ defined below, $\phi_0 := \phi_{\chi=1}$ and the functions ϕ_χ, $\chi \in \hat{\mathcal{L}}_\tau$, are unique solutions of the equations

$$(-\Delta_{A^1} - 1)\phi = 0, \quad \phi(x + s) = e^{i\frac{1}{2}s \cdot Jx} \chi(s)\phi(x), \quad \forall s \in \mathcal{L}_\tau, \tag{2.55}$$

with $A^1(x) := -\frac{1}{2}Jx$ and $J := \begin{pmatrix} 0 & 1 \\ -1 & 0 \end{pmatrix}$, normalized as $\langle |\phi_\chi|^2 \rangle_{\mathcal{T}_\tau} = 1$.

The functions $\gamma_\delta(\tau)$, $\operatorname{Im}\tau > 0$, defined in (2.54), has the following properties

- $\gamma_\delta(\tau)$, $\operatorname{Im}\tau > 0$, are symmetric w.r.t. the imaginary axis, $\gamma(-\bar{\tau}) = \gamma(\tau)$;
- $\gamma_\delta(\tau)$ have critical points at $\tau = e^{i\pi/2}$ and $\tau = e^{i\pi/3}$, provided it is differentiable at these points.

We also see that the Abrikosov constant, $\beta(\tau)$, is related to $\gamma_\chi(\tau)$ as $\beta(\tau) = \frac{1}{2}\gamma_{\chi=1}(\tau)$.

The functions $\gamma_\delta(\tau)$ are studied numerically in Ref. [78], where the above conjecture is confirmed and is shown that they are positive for $\tau = e^{i\pi/3}$ and $\delta > 0$ sufficiently small.

We fix $\delta > 0$ sufficiently small and write $\gamma(\tau)$ for $\gamma_\delta(\tau)$. The definition of $\gamma(\tau)$ implies that it is symmetric w.r.t. the imaginary axis, $\gamma(-\bar{\tau}) = \gamma(\tau)$. Hence it suffices to consider $\gamma(\tau)$ on the $\text{Re}\,\tau \geq 0$ half of the fundamental domain, $\Pi^+/SL(2,\mathbb{Z})$, of the modular group $SL(2,\mathbb{Z})$ (the heavily shaded area in Fig. 2.1). Using the symmetries of $\gamma(\tau)$, we show in Ref. [78] that the points $\tau = e^{i\pi/2}$ and $\tau = e^{i\pi/3}$ are critical points of the function $\gamma(\tau)$, provided it is differentiable at these points. (While the function $\gamma_\chi(\tau)$ is obviously smooth, derivatives of $\gamma(\tau)$ might jump. In fact, the numerical computations described in Ref. [78] show that $\gamma(\tau)$ is likely to have the line of cusps at $\text{Re}\,\tau = 0$.) We *conjecture*:

- For fixed $\text{Re}\,\tau \in [0, 1/2]$, $\gamma(\tau)$ is a decreasing function of $\text{Im}\,\tau$.
- $\gamma(\tau)$ has a unique global maximum at $\tau = e^{\frac{i\pi}{3}}$ and a saddle point at $\tau = e^{\frac{i\pi}{2}}$.

Remark. The points $\tau = e^{i\pi/2}$ and $\tau = e^{i\pi/3}$ are distinguished by the fact that they are the only points in $\Pi^+/SL(2,\mathbb{Z})$, which are fixed points under the maps from $SL(2,\mathbb{Z})$, other than identity, namely, under $\tau \to -\bar{\tau}$, $\tau \to -\tau^{-1}$ and $\tau \to 1 - \bar{\tau}$, $\tau \to 1 - \tau^{-1}$, respectively. This fact is used to prove that the points $\tau = e^{i\pi/2}$ and $\tau = e^{i\pi/3}$ are critical points of the function $\gamma(\tau)$ (see the last statement of Theorem 2.8) and can be used similarly for the function $\beta(\tau)$.

In Ref. [78], we confirm this conjecture numerically (numerics is due to Dan Ginsberg, see Figure 2.2 for the result of computing $\gamma(\tau)$ in Matlab, on a uniform grid with step size 0.01, using the default Nelder–Mead algorithm. The second plot is the function plotted only on the Poincaré strip.).

Calculations of Ref. [78] show that $\gamma(\tau) > 0$ for all equilateral lattices, $|\tau| = 1$, and is negative for $|\tau| \geq 1.3$. Though Abrikosov lattices are not as rigid under finite energy perturbations, as for gauge-periodic ones, they are still surprisingly stable.

It is convenient to consider $\gamma(\tau)$ as a modular function on Π^+. To this end, recall that we identify $\mathbb{R}^2$ with $\mathbb{C}$, via the map $(x_1, x_2) \to x_1 + ix_2$ and choose a basis in $\mathcal{L}$ so that $\mathcal{L} = r(\mathbb{Z} + \tau\mathbb{Z})$, where $\tau \in \mathbb{C}$, $\text{Im}\,\tau > 0$, and $r > 0$. By the quantization condition (2.25) with $n = 1$, $r := \sqrt{2\pi/\text{Im}\,\tau b}$. Denote

$$\mathcal{L}_\tau = \sqrt{2\pi/\text{Im}\,\tau}(\mathbb{Z} + \tau\mathbb{Z}).$$

The dual to it is $\mathcal{L}_\tau^* = \sqrt{2\pi/\text{Im}\,\tau}\,i(\mathbb{Z} - \tau\mathbb{Z})$. (The dual, or reciprocal, lattice, $\mathcal{L}^*$, of $\mathcal{L}$ consists of all vectors $s^* \in \mathbb{R}^2$ such that $s^* \cdot s \in 2\pi\mathbb{Z}$, for all $s \in \mathcal{L}$.) We identify the dual group, $\hat{\mathcal{L}}_\tau$, with a fundamental cell, Ω_τ^*, of the dual lattice $\mathcal{L}_\tau^*$, chosen so that Ω_τ^* is invariant under reflections, $k \to -k$. This identification given explicitly by $\chi(s) \to \chi_k(s) = e^{ik\cdot s} \leftrightarrow k$. The subset $\hat{\mathcal{L}}_{\tau,\delta}$ used in Theorem 2.8, is given by $\hat{\mathcal{L}}_{\tau,\delta} := \{\chi(s) = e^{ik\cdot x} : |k| \geq \delta\}$.

Then $\gamma_k(\tau) = \gamma_{\chi_k}(\tau)$, $\chi_k(s) = e^{ik\cdot s}$, $k \in \Omega_\tau^*$, where $\gamma_\chi(\tau)$ are defined in (2.54) for all τ, $\text{Im}\,\tau > 0$. Since the functions $\gamma(\tau)$ are independent of the choice of a basis in $\mathcal{L}_\tau$, they are invariant under action of the modular group $SL(2,\mathbb{Z})$, $\gamma(g\tau) = \gamma(\tau)$, $\forall g \in SL(2,\mathbb{Z})$, that is, they are modular functions on Π^+.

The numerics mentioned above are based on the following explicit representation of the functions $\gamma_k(\tau)$:

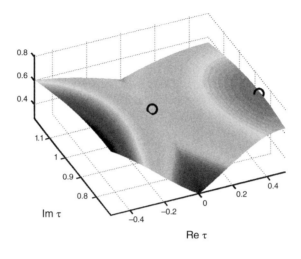

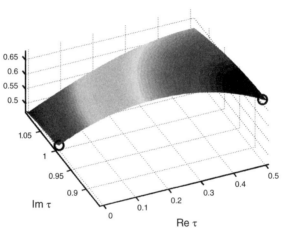

FIGURE 2.2 Plots of the function $\gamma^{\mathrm{approx}}(\tau)$. The circled points are $\tau = e^{i\pi/2}$ and $\tau = e^{i\pi/3}$.

Theorem 2.9 *The functions $\gamma_k(\tau)$ admit the explicit representation*

$$\gamma_k(\tau) = 2\sum_{t\in\mathcal{L}_\tau^*} e^{-\frac{1}{2}|t|^2}\cos[\mathrm{Im}(\bar{k}t)] - |\sum_{t\in\mathcal{L}_\tau^*} e^{-\frac{1}{2}|t+k|^2+i\,\mathrm{Im}(\bar{k}t)}| - \sum_{t\in\mathcal{L}_\tau^*} e^{-\frac{1}{2}|t|^2}. \quad (2.56)$$

Our computations also show that

- $\gamma_k(\tau)$ is minimized at $k \approx \sqrt{\frac{2\pi}{\mathrm{Im}\,\tau b}}(\frac{1}{2} - \frac{1}{2\sqrt{3}}i)$ at the point $\tau = e^{i\pi/3}$, and a value of $k \approx \sqrt{\frac{2\pi}{\mathrm{Im}\,\tau b}}(\frac{1}{2} + i\frac{1}{2})$ for $\tau = e^{i\pi/2}$, which corresponds to vertices of the corresponding Wigner–Seitz cells.

Interestingly, in Ref. [78], we show that the points $k \in \frac{1}{2}\mathcal{L}^{*}_{\tau}$ are critical points of the function $\gamma_k(\tau)$ in k. It is easy to see that $k = 0$ is a point of maximum of $\gamma_k(\tau)$ in $k \in \Omega^{*}_{\tau}$.

Remark. We think of $\gamma_k(\tau)$ as the "Abrikosov beta function with characteristic." While $\beta(\tau)$ is defined in terms of the standard theta function, $\gamma_k(\tau)$ is defined in terms of theta functions with finite characteristics, see (2.54) and the equations (2.66)–(2.68) below.

2.4.8   Key ideas of approach to stability

Let $u_\omega = (\Psi_\omega, A_\omega)$, $\omega := (\tau, b, 1)$, be an Abrikosov lattice solution. As usual, one begins with the Hessian $L_\omega := \mathcal{E}''(u_\omega)$ of the energy functional $\mathcal{E}$ at u_ω. To begin with, due to the fact that the solution u_ω of (2.1) breaks the gauge invariance, the operator L_ω has the gauge zero modes, $L_\omega G_{\gamma'} = 0$, where $G_{\gamma'} := (i\gamma' \Psi_\omega, \nabla \gamma')$. Then the stability of the static solution u_ω is decided by the sign of the infimum $\mu(\omega, \kappa) := \inf_{v \in H^1_\perp} \langle v, L_\omega v \rangle_{L^2} / \|v\|^2_{L^2}$, on the subspace $H^1_\perp$ of the Sobolev space H^1_{cov}, which is the orthogonal complement of these zero modes $G_{\gamma'}$.

As usual, the stability of the static solution $u_\omega = (\Psi_\omega, A_\omega)$, $\omega = (\tau, r)$, is decided by the location of the spectrum of the linearized operator, L_ω, for the map on the r.h.s. of (2.11) (L_ω is the (real) Hessian, $\mathcal{E}''(u_\omega)$, of the Ginzburg–Landau energy functional (2.2) at u_ω). Namely, whether L_ω is strictly positive or has some negative spectrum.

To address the question above, we first observe that since the solution u_ω of (2.1) breaks the gauge, translational and rotational invariance, the operator L_ω has the gauge, translational and rotational zero modes

$$G_{\gamma'} := (i\gamma' \Psi_\omega, \nabla \gamma'), \tag{2.57}$$

$$S_{h'} = ((h' \cdot \nabla_{A_\omega}) \Psi_\omega, (\text{curl} A_\omega) J h'), \quad h' \in \mathbb{R}^2, \tag{2.58}$$

$$R_{\varphi'} = \varphi'(Jx \cdot \nabla \Psi_\omega, -JA_\omega + Jx \cdot \nabla A_\omega), \tag{2.59}$$

that is, $L_\omega G_{\gamma'} = 0$, $L_\omega S_{h'} = 0$ and $L_\omega R_{\varphi'} = 0$. The gauge zero modes are square integrable, while the translational and rotational ones are not. However, the translational modes are bounded, which shows the presence of the essential spectrum at 0. The rotational zero modes grow at infinity and are not used in our analysis. (However, both, the translational and rotational modes would be important in its extensions to bounded and growing at infinity perturbations.)

In consequence, we should check whether $\inf_{v \in H^1_\perp, \|v\|_{L^2}=1} \langle v, L_\omega v \rangle_{L^2}$, where $H^1_\perp$ is the orthogonal complement of the gauge symmetry zero modes of L_ω, is strictly positive or negative. In our case, $\inf_{v \in H^1_\perp, \|v\|_{L^2}=1} \langle v, L_\omega v \rangle_{L^2} = 0$ and we have to dig deeper into the spectral structure of L_ω. The reason that the spectrum of L_ω has no gap lies in the spontaneous breaking of the translational symmetry by u_ω. This phenomenon is well known in particle physics and goes under the name of the Goldstone theorem, and the resulting gapless branch of the spectrum is called the Goldstone excitations and Goldstone particles. Thus we proceed to the symmetry analysis of L_ω.

Before proceeding, we recall that for $\mathcal{L}$, considered as a group of lattice translation, the dual group $\hat{\mathcal{L}}$ is the group of all continuous homomorphisms from $\mathcal{L}$ to $U(1)$, i.e. the group of characters, $\chi_\kappa(s) : \mathcal{L} \to U(1)$. ($\hat{\mathcal{L}}$ can be identified with the fundamental cell Ω^* of the dual lattice $\mathcal{L}^*$, with the identification given explicitly by identifying $\kappa \in \Omega^*$ with the character $\chi_\kappa : \mathcal{L} \to U(1)$ given by $\chi_\kappa(s) = e^{i\kappa \cdot s}$.)

The key idea of the proof of the first part of Theorem 2.7 stems from the observation that since the Abrikosov lattice solution $u_\omega = (\Psi_\omega, A_\omega)$ is gauge periodic (or equivariant) w.r.t. the lattice $\mathcal{L}_\omega$, that is, satisfies (2.30) and (2.31), the linearized map L_ω commutes with magnetic translations,

$$\rho_s = T_s^{\mathrm{mag-trans}} \oplus T_s^{\mathrm{trans}}, \qquad \forall s \in \mathcal{L}_\omega, \tag{2.60}$$

where $T_s^{\mathrm{mag-trans}} = (T_{g_s}^{\mathrm{gauge}})^{-1} T_s^{\mathrm{trans}}$ is the group of magnetic translations and, recall, T_s^{trans} denotes translation by s, which, due to (2.31), give a unitary group representation of $\mathcal{L}_\omega$. (Note that (2.30) implies that u_ω is invariant under the magnetic translations, $T_s^{\mathrm{mag-trans}} u_\omega = u_\omega$.) Therefore L_ω is unitary equivalent to a fiber integral over the dual group, $\hat{\mathcal{L}}_\omega$, of the group of lattice translations,

$$L_\omega \approx \int_{\hat{\mathcal{L}}_\omega}^{\oplus} L_{\omega\chi} d\hat{\chi} \quad \text{acting on} \quad \int_{\hat{\mathcal{L}}_\omega}^{\oplus} \mathcal{H}_\chi d\hat{\chi}. \tag{2.61}$$

Here $d\hat{\chi}$ is the usual Lebesgue measure normalized so that $\int_{\hat{\mathcal{L}}_\omega} d\hat{\chi} = 1$, $L_{\omega\chi}$ is the restriction of L_ω to $\mathcal{H}_\chi$ and $\mathcal{H}_\chi$ is the set of all functions, v_k, from $L_{\mathrm{loc}}^2(\mathbb{R}^2; \mathbb{C} \times \mathbb{R}^2)$, satisfying the gauge-periodicity conditions

$$\rho_s v_\chi(x) = \chi(s) v_\chi(x), \forall s \in \mathcal{L}_\omega, \tag{2.62}$$

where $\chi : \mathcal{L}_\omega \to U(1)$ are the characters acting on $v = (\xi, \alpha)$ as the multiplication operators

$$\chi(s)v = (\chi(s)\xi, \chi(s)\alpha).$$

Furthermore, $\mathcal{H}_\chi$ is endowed with the inner product $\langle v, v' \rangle_{L^2} = \frac{1}{|\mathcal{T}_\omega|} \int_{\mathcal{T}_\omega} \mathrm{Re}\, \bar{\xi}\xi' + \bar{\alpha}\alpha'$, where $\mathcal{T}_\omega := \mathbb{R}^2/\mathcal{L}_\omega$ is a $2-$torus and $v = (\xi, \alpha)$, $v' = (\xi', \alpha')$. The inner product in $\int_{\hat{\mathcal{L}}_\omega}^{\oplus} \mathcal{H}_\chi d\hat{\chi}$ is given by $\langle v, w \rangle_\mathcal{H} := \frac{1}{|\hat{\mathcal{L}}_\omega|} \int_{\hat{\mathcal{L}}_\omega} \langle v_\chi, w_\chi \rangle_{\mathcal{H}_\chi} d\hat{\chi}$.

As χ varies in $\hat{\mathcal{L}}_\omega$, the eigenvalues of $L_{\omega\chi}$ sweep the spectral bands of L_ω. Using the formula above and using $\chi_k(x) := e^{ik \cdot x}$ to pass from $\chi \in \hat{\mathcal{L}}_\omega$ to $k \in \Omega_\omega^*$, we show that the lowest spectral branch (band) of L_ω is of the form

$$\nu_{\omega\kappa}(k) = \gamma_k(\tau, \kappa)|k|^2 + O(|k|^2 \epsilon^4), \quad k \in \Omega_\tau^*, \tag{2.63}$$

where $\gamma_k(\tau, \kappa) \geq \gamma_\delta(\tau)$, for some $\delta > 0$, with $\gamma_\delta(\tau)$ given in (2.54), and ϵ is the natural perturbation parameter defined as

$$\epsilon = \sqrt{\frac{\kappa^2 - b}{\kappa^2[(2\kappa^2 - 1)\beta(\tau) + 1]}}. \tag{2.64}$$

There is another, higher for $k \neq 0$, gapless branch and the remaining spectral branches are separated from 0 by a gap of order $O(\epsilon^2)$. These gapless spectral branches are due to the breaking of the translational symmetry (by u_ω) and represents, what is known in particle physics as, the Goldstone excitation spectrum.

The linear result above gives the weak linearized (energetic) stability of u_ω, if $\gamma(\tau, \kappa) := \inf_{k \in \Omega_\tau^*} \gamma_k(\tau, \kappa) > 0$, and the instability, if $\gamma(\tau, \kappa) < 0$. To lift the stability part to the (nonlinear) asymptotic stability, we split the fluctuation v as $v = v' + v''$, where v' is the projection of v onto the subspace of the lowest spectral branch of L_ω and v'' is the orthogonal complement. For v'', we use the fact that L_ω has a gap of order ϵ^2 on v''s. We exploit this in order to estimate the Lyapunov functional $\Lambda_\omega(v) = \frac{1}{2}\langle v'', L_\omega v'' \rangle + \cdots$ using an appropriate differential inequality for $\Lambda_\omega(v)$. (The resulting estimate of v'' will, of course, depend on some information about v'!)

For v', we use the evolution equation, which follows from the equation for v, and study this equation in the spectral representation of the corresponding spectral branch L_ω ($v' \leftrightarrow f$, with $f : \Omega^* \to \mathbb{C}$). This equation is roughly of the form of a nonlinear heat equation in two dimensions, with the nonlinearity of order two. To the authors' knowledge, the long-time behavior of such equations for small initial data is not understood presently. However, under the condition (2.52), the function f is odd which allows us to eke out an extra decay, provided we can control derivatives of f in k (and assuming some information about v''!). Bootstrapping the estimates on v' and v'', we arrive at the long-time estimates of v and consequently, the stability result.

Remark. $\mathscr{H}_\chi$ can be thought of as the space of L^2−sections of the vector bundle $[\mathbb{R}^2 \times (\mathbb{C} \times \mathbb{R}^2)]/\mathcal{L}$, with the group $\mathcal{L}$, which acts on $\mathbb{R}^2 \times (\mathbb{C} \times \mathbb{R}^2)$ as $s(x, v) = (x + s, \chi_k(s)\tau_{g_s(x)}^{-1}v)$, where $\tau_\alpha = e^{i\alpha} \oplus e^{-i\alpha} \oplus 1 \oplus 1$, for $\alpha \in \mathbb{R}$.

Finally we derive the explicit representation (2.56) and the uniqueness for (2.55). As in (2.42), we introduce the new function $\theta_q(z, \tau)$, by the equation

$$\phi_k(x) = c_0 e^{\frac{\pi}{2\operatorname{Im}\tau}(z^2 - |z|^2)}\theta_q(z, \tau), \tag{2.65}$$

where c_0 is such that $\langle |\phi_k|^2 \rangle_{\Omega_\tau} = 1$, and $x_1 + ix_2 = \sqrt{\frac{2\pi}{\operatorname{Im}\tau}}z$ and $k = \sqrt{\frac{2\pi}{\operatorname{Im}\tau}}iq$. Then, again as above, we can show that the functions $\theta_q(z, \tau) = e^{-\frac{\pi}{2\operatorname{Im}\tau}(z^2 - |z|^2)}\phi_k(x)$ are entire functions (i.e., they solve $\bar{\partial}\theta_q = 0$) and satisfy the periodicity conditions

$$\theta_q(z + 1, \tau) = e^{-2\pi i a}\theta_q(z, \tau), \tag{2.66}$$

$$\theta_q(z + \tau, \tau) = e^{-2\pi i b}e^{-i\pi\tau - 2\pi i z}\theta_q(z, \tau), \tag{2.67}$$

where a, b are real numbers defined by $q = -a\tau + b$. This shows that $\theta_q(z, \tau)$ are the theta functions with characteristics q, and this characteristics is determined by k, which in physics literature is called (Bloch) quasimomentum. Moreover, it has the following series expansion

$$\theta_q(z,\tau) := e^{\pi i(a^2\tau - 2ab - 2az)} \sum_{m=-\infty}^{\infty} e^{2\pi i q m} e^{\pi i m^2 \tau} e^{2\pi i m z}. \qquad (2.68)$$

2.5 MULTI-VORTEX DYNAMICS

Configurations containing several vortices are not, in general, static solutions. Heuristically, this is due to an effective inter-vortex interaction, which causes the vortex centers to move. It is natural, then, to seek an effective description of certain solutions of time-dependent Ginzburg–Landau equations in terms of the vortex locations and their dynamics—a kind of finite-dimensional reduction. In recent years, a number of works have addressed this problem, from different angles, and in different settings. We will first describe results along these lines from Ref. [43] for magnetic vortices in $\mathbb{R}^2$, and then mention some other approaches and results.

Multi-vortex configurations.   Consider test functions describing several vortices, with the centers at points $z_1, z_2, \ldots, z_m$, and with degrees $n_1, n_2, \ldots, n_m$, "glued together." The simplest example is $v_{\underline{z},\chi} = (\Psi_{\underline{z},\chi}, A_{\underline{z},\chi})$, with

$$\Psi_{\underline{z},\chi}(x) = e^{i\chi(x)} \prod_{j=1}^{m} \Psi^{(n_j)}(x - z_j),$$

$$A_{\underline{z},\chi}(x) = \sum_{j=1}^{m} A^{(n_j)}(x - z_j) + \nabla\chi(x),$$

where $\underline{z} = (z_1, z_2, \ldots, z_m) \in \mathbb{R}^{2m}$ and χ is an arbitrary real-valued function yielding the gauge transformation. Since vortices are exponentially localized, for large inter-vortex separations such test functions are approximate—but not exact—solutions of the stationary Ginzburg–Landau equations. We measure the inter-vortex distance by

$$R(\underline{z}) := \min_{j \neq k} |z_j - z_k|,$$

and introduce the associated small parameter $\epsilon = \epsilon(\underline{z}) := R(\underline{z})^{-1/2} e^{-R(\underline{z})}$.

Dynamical problem.   Now consider a time-dependent Ginzburg–Landau equation with an initial condition close to the function $v_{\underline{z}_0,\chi_0} = (\psi_{\underline{z}_0,\chi_0}, A_{\underline{z}_0,\chi_0})$ describing several vortices glued together (if $\kappa > 1/2$, we take $n_j = \pm 1$ since the $|n| \geq 2-$ vortices are then unstable by Theorem 2.1), and ask the following questions:

- Does the solution at a later time t describe well-localized vortices at some locations $\underline{z} = \underline{z}(t)$ (and with a gauge transformation $\chi = \chi(t)$)?
- If so, what is the dynamical law of the vortex centers $\underline{z}(t)$ (and of $\chi(t)$)?

Vortex dynamics results. This section gives a brief description of the vortex dynamics results in Ref. [43]. For related rigorous as well as non-rigorous results, see Refs. [23, 25–27, 48–50, 54, 55, 62–67, 70, 79–82, 91] and for the numerical simulations [94].

Gradient flow (we take, for simplicity, $\gamma = 1, \sigma = 1$ in (2.11)). Consider the gradient flow equations (2.11) with initial data $(\psi(0), A(0))$ close (in the energy norm) to some multi-vortex configuration $v_{\underline{z}_0, \chi_0}$. Then

$$(\Psi(t), A(t)) = v_{\underline{z}(t), \chi(t)} + O(\epsilon \log^{1/4}(1/\epsilon)), \tag{2.69}$$

and the vortex dynamics is governed by the system

$$\gamma_{n_j} \dot{z}_j = -\nabla_{z_j} W(\underline{z}) + O(\epsilon^2 \log^{3/4}(1/\epsilon)). \tag{2.70}$$

Here $\gamma_n > 0$ and $W(\underline{z})$ is the effective vortex interaction energy, of order ϵ, given below.

These statements hold for only as long as the path $\underline{z}(t)$ does not violate a condition of large separation, though in the *repulsive case*, when $\lambda > 1/2$ and $n_j = +1$ (or $n_j = -1$) for all j, the above statements hold for all time t.

Maxwell-Higgs equations. For the Maxwell-Higgs equations (2.12) with initial data $(\Psi(0), A(0))$ close (in the energy norm) to some $v_{\underline{z}_0, \chi_0}$ (and with appropriately small initial momenta),

$$\|(\Psi(t), A(t)) - v_{\underline{z}(t), \chi(t)}\|_{H^1} + \|(\partial_t \psi(t), \partial_t A(t)) - \partial_t v_{\underline{z}(t), \chi(t)}\|_{L^2} = o(\sqrt{\epsilon}) \tag{2.71}$$

with

$$\gamma_{n_j} \ddot{z}_j = -\nabla_{z_j} W(\underline{z}(t)) + o(\epsilon) \tag{2.72}$$

for times up to (approximately) order $\frac{1}{\sqrt{\epsilon}} \log\left(\frac{1}{\epsilon}\right)$.

The effective vortex interaction energy. The interaction energy of a multi-vortex configuration is defined as

$$W(\underline{z}) = E(\underline{z}) - \sum_{j=1}^{m} \mathcal{E}(\Psi^{(n_j)}, A^{(n_j)}), \tag{2.73}$$

where $E(\underline{z})$ is the infimum of the Ginzbuerg–Landau energy, $\mathcal{E}(u)$, over states with simple vortices fixed at the positions $z_1, \ldots, z_m$, with $\underline{z} = (z_1, z_2, \ldots, z_m) \in \mathbb{R}^{2m}$.

Operationally, we can take $E(\underline{z}) = \mathcal{E}(v_{\underline{z},\chi})$. For $\kappa > 1/2$, we have for large $R(\underline{z})$,

$$W(\underline{z}) \sim (\text{const}) \sum_{j \neq k} n_j n_k \frac{e^{-|z_j - z_k|}}{\sqrt{|z_j - z_k|}}.$$

Some ideas behind the proofs (Maxwell-Higgs case).

- *Multi-vortex manifold.* Multi-vortex configurations $v_{\underline{z},\chi}$ comprise an (infinite-dimensional, due to gauge transformations) manifold

$$M := \{v_{\underline{z},\chi} \mid \underline{z} \in \mathbb{R}^{2m}, \chi \in H^2\},$$

  (together with appropriate momenta in the Maxwell-Higgs case) made up of approximate static solutions. The interaction energy (2.73) of a multi-vortex configuration gives rise to a reduced Hamiltonian system on M, using the restriction of the natural symplectic form to M—this is the leading-order vortex motion law.
- *Symplectic orthogonality and effective dynamics.* The *effective dynamics* on M is determined by demanding that the deviation of the solution from M be symplectically orthogonal to the tangent space to M. Informally,

$$(\Psi(t), A(t)) - v_{\underline{z}(t),\chi(t)} \perp \mathbb{J} T_{v_{\underline{z}(t),\chi(t)}} M.$$

  The tangent space is composed of infinitesimal (approximate) symmetry transformations, that is, independent motions of the vortex centers, and gauge transformations.
- *Stability and coercivity.* The manifold M inherits a stability property from the stability of its basic building blocks—the n-vortex solutions—as described in Section 2.3.2. The stability property is reflected in the fact that the linearized operator around a multi-vortex

$$L_{\underline{z},\chi} = \mathcal{E}''(v_{\underline{z},\chi})$$

  is coercive in directions symplectically orthogonal to the tangent space of M:

$$\xi \perp \mathbb{J} T_{v_{\underline{z},\chi}} M \implies \langle \xi, L_{\underline{z},\chi} \xi \rangle > 0.$$

- *Approximately conserved Lyapunov functionals.* Thus the quadratic form $\langle \xi, L_{\underline{z},\chi} \xi \rangle$, where $\xi := (\Psi(t), A(t)) - v_{\underline{z}(t),\chi(t)}$, controls the deviation of the solution from the multi-vortex manifold, and furthermore is approximately conserved—this gives long-time control of the deviation. Finally, approximate conservation of the reduced energy $W(v_{\underline{z}(t),\chi(t)})$ is used to control the difference between the effective dynamics and the leading-order vortex motion law.

2.6   CONCLUSIONS

In the context of the superconductivity and Abelian Yang–Mills–Higgs model of particle physics, we described

- the existence and stability of *magnetic vortices* and *vortex lattices*
- the dynamics of a finite system of vortices.

The presented results revealed the following interesting features:

- a new threshold $\kappa_c(\tau)$ in the Ginzburg–Landau parameter appears in the problem of existence of vortex lattices;
- contrary to what was expected in this area, the stability of Abrikosov lattices is controlled not by the Abrikosov function $\beta(\tau)$, which governs the energetics, but by a new *automorphic function* $\gamma(\tau)$;
- the dynamics of vortices is governed by a finite dimensional dynamical system, giving evolution of vortex centers, coupled to an infinite dimensional one for the vortex gauges. The latter acts within the same equivalence class of solutions, while the former is essentially gradient in the superconductivity case and Hamiltonian for the Abelian Yang–Mills–Higgs model. The effective (static) energy of this dynamical system is given by $W(\underline{z})$.

We gave some indications on how to prove the results described above. In particular, we emphasized that while the proof of existence of the Abrikosov lattices leads to standard theta functions, the proof of stability leads to *theta functions with characteristics*.

Interesting extensions:

- unconventional/high T_c superconductors,
- Weinberg–Salam model of electro-weak interactions,
- microscopic/quantum theory.

APPENDIX 2.A   PARAMETERIZATION OF THE EQUIVALENCE CLASSES [$\mathcal{L}$]

In this appendix, we present some standard results about lattices. Every lattice in $\mathbb{R}^2$ can be written as $\mathcal{L} = \mathbb{Z}\nu_1 + \mathbb{Z}\nu_2$, where (ν_1, ν_2) is a basis in $\mathbb{R}^2$. Given a basis (ν_1, ν_2) in $\mathbb{R}^2$ and identifying $\mathbb{R}^2$ with $\mathbb{C}$, via the map $(x_1, x_2) \to x_1 + ix_2$, we define the complex number $\tau = \nu_2/\nu_1$ called the shape parameter. We can choose a basis so that $\operatorname{Im}\tau > 0$, which we assume from now on. Clearly, τ is independent of rotations and dilatations of the lattice.

Any two bases, (ν_1, ν_2) and (ν'_1, ν'_2), span the same lattice $\mathcal{L}$ if they are related as $(\nu'_1, \nu'_2) = (\alpha\nu_1 + \beta, \gamma\nu_2 + \delta)$, where $\alpha, \beta, \gamma, \delta \in \mathbb{Z}$, and $\alpha\delta - \beta\gamma = 1$ (i.e. the matrix $\begin{pmatrix} \alpha & \beta \\ \gamma & \delta \end{pmatrix}$, an element of the modular group $SL(2, \mathbb{Z})$). Under this map, the shape parameter $\tau = \nu_2/\nu_1$ is being mapped into $\tau' = \nu'_2/\nu'_1$ as $\tau \to \tau' = g\tau$, where $g\tau := \frac{\alpha\tau + \beta}{\gamma\tau + \delta}$. Thus, up to rotation and dilatation, the lattices are in one-to-one correspondence with points τ in the fundamental domain, $\Pi^+/SL(2, \mathbb{Z})$, of the modular group $SL(2, \mathbb{Z})$ acting on the Poincaré half-plane Π^+. Explicitly (see Fig. 2.1),

$$\left\{ \tau \in \mathbb{C} : \operatorname{Im}\tau > 0, \ |\tau| \geq 1, \ -\frac{1}{2} < \operatorname{Re}\tau \leq \frac{1}{2} \right\}. \tag{2.A.1}$$

Furthermore, any quantity, f, which depends of the lattice equivalence classes, can be thought of as a function of τ, $\operatorname{Im}\tau > 0$, invariant under the modular group $SL(2, \mathbb{Z})$ and therefore is determined entirely by its values on the fundamental domain (2.A.1).

APPENDIX 2.B   AUTOMORPHY FACTORS

We list some important properties of g_s:

- If (Ψ, A) satisfies (2.30) with $g_s(x)$, then $T_\chi^{\text{gauge}}(\Psi, A)$ satisfies (2.30) with $g_s(x) \to g'_s(x)$, where

$$g'_s(x) = g_s(x) + \chi(x+s) - \chi(x). \tag{2.B.1}$$

- The functions $g_s(x) = \frac{b}{2}s \wedge x + c_s$, where b satisfies $b|\Omega| \in 2\pi\mathbb{Z}$ and c_s are numbers satisfying $c_{s+t} - c_s - c_t - \frac{1}{2}bs \wedge t \in 2\pi\mathbb{Z}$, satisfies (2.31).
- By the cocycle condition (2.31), for any basis $\{\nu_1, \nu_2\}$ in $\mathcal{L}$, the quantity

$$c(g_s) = \frac{1}{2\pi}(g_{\nu_2}(x+\nu_1) - g_{\nu_2}(x) - g_{\nu_1}(x+\nu_2) + g_{\nu_1}(x)) \tag{2.B.2}$$

is independent of x and of the choice of the basis $\{\nu_1, \nu_2\}$ and is an integer.
- Every exponential g_s satisfying the cocycle condition (2.31) is equivalent to the exponent

$$\frac{b}{2}s \wedge x + c_s, \tag{2.B.3}$$

for b and c_s satisfying $b|\Omega| = 2\pi c(g_s)$ and

$$c_{s+t} - c_s - c_t - \frac{1}{2}bs \wedge t \in 2\pi\mathbb{Z}. \tag{2.B.4}$$

- The condition (2.31) implies the magnetic flux quantization (2.24):

$$\frac{1}{2\pi} \int_\Omega \mathrm{curl} A = \deg \Psi = c(g_s).$$ (2.B.5)

- The exponentials g_s satisfying the cocycle condition (2.31) are classified by the irreducible representation of the group of lattice translations.

Indeed, the first and second statements are straightforward. For the third statement, by the relation (2.31), $g_{\nu_2}(x+\nu_1) + g_{\nu_1}(x) - g_{\nu_1+\nu_2}(x) \in 2\pi\mathbb{Z}$ and $g_{\nu_1}(x+\nu_2) + g_{\nu_2}(x) - g_{\nu_1+\nu_2}(x) \in 2\pi\mathbb{Z}$. Subtracting the second relation from the first shows that $c(g)$ is independent of x and is an integer. The quantity $c(\gamma)$ is called the Chern number.

The fourth property, for example, see Refs. [30, 58, 83, 90], is formulated differently. In the present formulation, this property was shown by A. Weil and generalized in Ref. [38].

To prove the fifth statement, we note that by Stokes' theorem, the magnetic flux through a lattice cell Ω is $\int_\Omega \mathrm{curl} A = \int_{\partial\Omega} A$ and is given by

$$\int_0^1 \left[\nu_1 \cdot (A(a\nu_1 + \nu_2) - A(a\nu_1)) - \nu_2 \cdot (A(a\nu_2 + \nu_1) - A(a\nu_2)) \right] da$$

$$= \int_0^1 \left[\nu_1 \cdot \nabla g_{\nu_2}(a\nu_1) - \nu_2 \cdot \nabla g_{\nu_1}(a\nu_2) \right] da,$$

which, by (2.31), gives $\int_\Omega \mathrm{curl} A = g_{\nu_2}(\nu_1) - g_{\nu_2}(0) - g_{\nu_1}(\nu_2) + g_{\nu_1}(0) \in 2\pi\mathbb{Z}$. The sixth property follows from the fact that c_s's satisfying $c_{s+t} - c_s - c_t - (1/2)bs \wedge t \in 2\pi\mathbb{Z}$ are classified by the irreducible representation of the group of lattice translations.

Given a family g_s of functions satisfying (2.31), equivariant functions $u = (\Psi, A)$ for g_s are identified with *sections of the vector bundle*

$$\mathbb{R}^2 \times (\mathbb{C} \times \mathbb{R}^2)/\mathcal{L},$$

with the base manifold $\mathbb{R}^2/\mathcal{L} = \Omega$ and the projection $p : [(x,u)] \to [x]$, where $[(x,u)]$ and $[x]$ are the equivalence classes of (x,u) and x, under the action of the group $\mathcal{L}$ on $\mathbb{R}^2 \times (\mathbb{C} \times \mathbb{R}^2)$ and on $\mathbb{R}^2$, given by

$$s : (x,u) \to (x+s, T_{g_s(x)}^{\mathrm{gauge}} u) \text{ and } s : x \to x+s,$$

respectively. This implies, in particular, that

- For a family g_s of functions satisfying (2.31), there exists a continuous pair (Ψ, A) satisfying (2.30) with this family.

Remark. In algebraic geometry and number theory, the automorphy factors $e^{ig_s'(x)}$ and $e^{ig_s(x)}$ satisfying $g_s'(x) = g_s(x) + \chi(x+s) - \chi(x)$, for some $\chi(x)$, are said to be equivalent. A function Ψ satisfying $T_s^{\mathrm{trans}}\Psi = e^{ig_s}\Psi$ is called e^{ig_s}−theta function.

Remark. The special form (2.B.3) is related to a general construction of line bundles over the complex torus using symplectic form $\omega(z,w)$ to construct automorphy factors, for example, $g_s(z) = b\omega(z,s) + c_s$, where $b|\mathbb{C}/\mathcal{L}| = 2\pi n$ and c_s satisfies $c_{s+t} - c_s - c_t - \frac{b}{2}\omega(s,t) \in 2\pi\mathbb{Z}$. The Chern number, $c(\gamma)$, is expressed in terms of ω as

$$c(\gamma) = b\omega(\nu_1, \nu_2), \qquad (2.B.6)$$

where $\{\nu_1, \nu_2\}$ is a basis of $\mathcal{L}$.

Acknowledgments

The author is grateful to Stephen Gustafson, Yuri Ovchinnikov, and Tim Tzaneteas for many fruitful discussions and collaboration, and to the anonymous referee, for many constructive remarks that led to improvement of the presentation. Author's research is supported in part by NSERC under Grant NA 7901.

REFERENCES

1. A.A. Abrikosov, "On the magnetic properties of superconductors of the second group," *Sov. Phys., JETP* **5**, 1174–1182 (1957).

2. A. Aftalion, X. Blanc, and F. Nier, "Lowest Landau level functional and Bargmann spaces for Bose Einsein condensates," *J. Funct. Anal.* **241**, 661–702 (2006).

3. A. Aftalion and S. Serfaty, "Lowest Landau level approach in superconductivity for the Abrikosov lattice close to H_{c_2}," *Selecta Math.* (N.S.) 13, 183–202 (2006).

4. S. Alama, L. Bronsard and E. Sandier, "On the shape of interlayer vortices in the Lawrence–Doniach model," *Trans. AMS* **360**(1), 1–34 (2008).

5. S. Alama, L. Bronsard, and E. Sandier, "Periodic minimizers of the anisotropic Ginzburg–Landau model," *Calc. Var. Partial Differ. Equ.* **36**, (3), 399–417 (2009).

6. Y. Almog, "On the bifurcation and stability of periodic solutions of the Ginzburg-Landau equations in the plane," *SIAM J. Appl. Math.* **61**, 149–171 (2000).

7. Y. Almog, "Abrikosov lattices in finite domains," *Comm. Math. Phys.* **262**, 677–702 (2006).

8. H. Aydi and E. Sandier, "Vortex analysis of the periodic Ginzburg-Landau model," *Ann. Inst. H. Poincaré Anal. Non Linéaire* **26** no. 4, 1223–1236 (2009).

9. E. Barany, M. Golubitsky, and J. Turksi, "Bifurcations with local gauge symmetries in the Ginzburg-Landau equations," *Phys. D* **67**, 66–87 (1993).

10. P. Baumann, D. Phillips, and Q. Shen, "Singular limits in polymerized liquid crystals," *Proc. Roy. Soc. Edinb.* **A 133** (1), 11–34 (2003).

11. M.S. Berger and Y. Y. Chen, "Symmetric vortices for the nonlinear Ginzburg-Landau equations of superconductivity, and the nonlinear desingularization phenomenon," *J. Funct. Anal.* **82**, 259–295 (1989).

12. F. Bethuel, H. Brezis, and F. Hélein, *Ginzburg-Landau Vortices*, Birkhauser, Boston, 1994.

13. F. Bethuel, G. Orlandi, D. Smets, "Dynamics of multiple degree Ginzburg-Landau vortices," *Comm. Math. Phys.* **272** (1), 229–261 (2007).

14. E.B. Bogomol'nyi, "The stability of classical solutions," *Yad. Fiz.* **24**, 861–870 (1976).

15. A. Boutet de Monvel-Berthier, V. Georgescu, and R. Purice, "A boundary value problem related to the Ginzburg-Landau model," *Comm. Math. Phys.* **142** (1), 1–23 (1991).

16. H. Brezis, Degree theory: old and new. Topological nonlinear analysis, II (Frascati, 1995), 87Đ108, Progr. Nonlinear Differential Equations Appl., 27, Birkhauser Boston, MA, 1997.

17. J. Burzlaff and V. Moncrief, "The global existence of time-dependent vortex solutions," *J. Math. Phys.* **26**, 1368–1372 (1985).

18. S.J. Chapman, "Nucleation of superconductivity in decreasing fields," *Eur. J. Appl. Math.* **5** 449–468 (1994).

19. S.J. Chapman "Asymptotic analysis of the Ginzburg-Landau model of superconductivity: reduction to a free boundary model," *Quart. Appl. Math.* **53**, 601–627 (1995).

20. S.J. Chapman, "Hierarchy of models for type-II superconductors," *SIAM Rev.*, **42**, (4), 555–598 (2000).

21. S.J. Chapman, S.D. Howison, and J.R. Ockendon, "Macroscopic models for superconductivity," *SIAM Rev.* **34** (4), 529–560 (1992).

22. R.M. Chen and D. Spirn, "Symmetric Chern-Simons vortices," *Comm. Math. Phys.* **285**, 1005–1031 (2009).

23. J. Colliander and R. Jerrard, "Vortex dynamics for the Ginzburg-Landau Schrödinger equation," *Int. Math. Res. Not.* **7**, 333–358 (1998).

24. M. Comte and M. Sauvageot, "On the hessian of the energy form in the Ginzburg-Landau model of superconductivity." *Rev. Math. Phys.* **16** (4), 421–450 (2004).

25. S. Demoulini and D. Stuart, "Gradient flow of the superconducting Ginzburg-Landau functional on the plane," *Comm. Anal. Geom.* **5**, 121–198 (1997).

26. S. Demoulini and D. Stuart, "Adiabatic limit and the slow motion of vortices in a Chern-Simons-Schroedinger system," *Comm. Math. Phys.* **290** (2), 597–632 (2009).

27. Q. Du and F.-H. Lin, "Ginzburg-Landau vortices: dynamics, pinning, and hysteresis," *SIAM J. Math. Anal.* **28**, (6), 1265–1293 (1997).

28. M. Dutour, "Phase diagram for Abrikosov lattice," *J. Math. Phys.* **42**, 4915–4926 (2001).

29. M. Dutour, *Bifurcation vers l'tat d'Abrikosov et diagramme des phases*, Thesis Orsay, http://www.arxiv.org/abs/math-ph/9912011. (accessed October 31, 2014).

30. G. Eilenberger, "Zu Abrikosovs Theorie der periodischen Lösungen der GL-Gleichungen für Supraleiter 2," *Z. Phys.* **180**, 32–42 (1964).

31. S. Fournais and B. Helffer, "Spectral methods in surface superconductivity," *Prog. Nonlin. Diff. Equ.* **77**, 324p (2010).

32. S. Fournais and A. Kachmar, "Nucleation of bulk superconductivity close to critical magnetic field," *Adv. Math.* **226**, (2), 1213–1258 (2011).

33. S. Fournais and A. Kachmar, "The ground state energy of the three dimensional Ginzburg-Landau functional. Part I: Bulk regime," *Comm. Partial Diff. Equ.* **38**, (2), 339–383 (2013).

34. R.L. Frank, C. Hainzl, R. Seiringer, and J.P. Solovej, "Microscopic derivation of Ginzburg-Landau theory," *J. AMS* **25** (3), 667–713 (2012).

35. V.L. Ginzburg and L.D. Landau, "On the theory of superconductivity," *Zh. Eksp. Teor. Fiz.* **20**, 1064–1082 (1950).

36. L.P. Gorkov, "Microscopic derivation of the Ginzburg-Landau equations in the theory of superconductivity," *Sov. Phys. JETP* **36**, 635 (1959).

37. L.P. Gork'ov and G.M. Eliashberg, "Generalization of the Ginzburg-Landau equations for non-stationary problems in the case of alloys with paramagnetic impurities," *Sov. Phys. JETP* **27** (2), 328–334 (1968).

38. R.C. Gunning, "The structure of factors of automorphy," *Am. J. Math.*, **78**, (2), 357–382 (1956).

39. R.C. Gunning, *Riemann Surfaces and Generalized Theta Functions*, Springer, New York, 1976.

40. S.J. Gustafson, "Dynamical stability of magnetic vortices," *Nonlinearity* **15**, 1717–1728 (2002).

41. S.J. Gustafson and I.M. Sigal, "The stability of magnetic vortices," *Comm. Math. Phys.* **212**, 257–275 (2000).

42. S.J. Gustafson and I.M. Sigal. *Mathematical Concepts of Quantum Mechanics*. Springer-Verlag, Berlin, 2nd Edition, 2011.

43. S.J. Gustafson and I.M. Sigal, "Effective dynamics of magnetic vortices," *Adv. Math.* **199**, 448–498 (2006).

44. S.J. Gustafson and F. Ting. "Dynamic stability and instability of pinned fundamental vortices," *J. Nonlin. Sci.* **19**, 341–374 (2009).

45. S.J. Gustafson, I.M. Sigal, and T. Tzaneteas, "Statics and dynamics of magnetic vortices and of Nielsen-Olesen (Nambu) strings," *J. Math. Phys.* **51**, 015217 (2010).

46. L. Jacobs and C. Rebbi, "Interaction of superconducting vortices," *Phys. Rev.* **B19**, 4486–4494 (1979).

47. A. Jaffe and C. Taubes, *Vortices and Monopolos: Structure of Static Gauge Theories*. Progress in Physics 2, Birkhäuser, Boston, 1980.

48. R. Jerrard, "Vortex dynamics for the Ginzburg-Landau wave equation," *Calc. Var. Partial Diff. Equ.* **9** (8), 683–688 (1999).

49. R. Jerrard and M. Soner, "Dynamics of Ginzburg-Landau vortices," *Arch. Rational Mech. Anal.* **142** (2), 99–125 (1998).

50. R. Jerrard and D. Spirn, "Refined Jacobian estimates and Gross-Piaevsky vortex dynamics," *Arch. Rat. Mech. Anal.* **190**, 425–475 (2008).

51. W.H. Kleiner, L. M. Roth, and S.H. Autler, "Bulk solution of Ginzburg-Landau equations for type II superconductors: upper critical field region," *Phys. Rev.* **133**, A1226–A1227 (1964).

52. S. Komineas and N. Papanicolaou, "Vortex dynamics in two-dimensional antiferromagnets," *Nonlinearity* **11**, 265–290 (1998).

53. G. Lasher, "Series solution of the Ginzburg-Landau equations for the Abrikosov mixed state," *Phys. Rev.* **140**, A523–A528 (1965).

54. F.-H. Lin and J. Xin, "On the incompressible fluid limit and the vortex motion law of the nonlinear Schrödinger equation," *Comm. Math. Phys.* **200**, 249–274 (1999).

55. N. Manton, "A remark on the scattering of BPS monopoles," *Phys. Lett.* **B 110** (1), 54–56 (1982).

56. N. Manton and P. Sutcliffe, *Topological Solitons*, Cambridge Monographs on Mathematical Physics, Cambridge University Press, Cambridge, 2004.

57. S. Nonnenmacher and A. Voros, "Chaotic eigenfunctions in phase space," *J. Stat. Phys.* **92**, 431–518 (1998).

58. F. Odeh, "Existence and bifurcation theorems for the Ginzburg-Landau equations," *J. Math. Phys.* **8**, 2351–2356 (1967).

59. Yu. N. Ovchinnikov, "Structure of the superconducting state of superconductors near the critical field H_{c2} for values of the Ginzburg-Landau parameter κ close to unity," *JETP* **85**(4), 818–823 (1997).

60. Yu. N. Ovchinnikov, "Generalized Ginzburg-Landau equation and properties of superconductors for values of Ginzburg-Landau parameter κ close to 1," *JETP* **88**(2), 398–405 (1999).

61. Yu. Ovchinnikov and I.M. Sigal, "Symmetry breaking solutions to the Ginzburg-Landau equations," *JETP* **99** (5), 1090–1107 (2004).

62. N. Papanicolaou and T.N. Tomaras, "On dynamics of vortices in a nonrelativistic Ginzburg-Landau model," *Phys. Lett.* **179**, 33–37 (1993).

63. N. Papanicolaou and T.N. Tomaras, "Dynamics of magnetic vortices," *Nucl Phys.* **B360**, 425–462 (1991); "Dynamics of interacting magnetic vortices in a model Landau-Lifshitz equation," *Phys. D* **80**, 225–245 (1995).

64. L. Peres and J. Rubinstein, "Vortex dynamics in $U(1)$ Ginzburg-Landau models," *Phys. D* **64**, 299–309 (1993).

65. L.M. Pismen and D. Rodriguez, "Mobility of singularities in dissipative Ginzburg-Landau equations," *Phys. Rev. A* **42**, 2471 (1990).

66. L.M. Pismen and J. Rubinstein, "Motion of vortex lines in the Ginzburg-Landau model," *Phys. D* **47**, 353–360 (1991).

67. L.M. Pismen, D. Rodriguez and L. Sirovich, "Motion of interacting defects in the Ginzburg-Landau model," *Phys. Rev. A* **44**, 798 (1991).

68. B. Plohr, *The existence, regularity, and behavior of isotropic solutions of classical Gauge field theories*, Princeton thesis (1980).

69. J. Rubinstein, "Six Lectures on Superconductivity," *Boundaries, Interfaces, and Transitions. CRM Proc. Lec. Notes* **13**, 163–184 (1998).

70. E. Sandier and S. Serfaty, "Gamma-convergence of gradient flows with applications to Ginzburg-Landau," *Comm. Pure Appl. Math.* **57**, (12), 1627–1672 (2004).

71. E. Sandier and S. Serfaty, "Vortices in the Magnetic Ginzburg-Landau Model," *Progress in Nonlinear Differential Equations and their Applications,* **70**, Birkhäuser, Berlin, 2007.

72. E. Sandier and S. Serfaty, "From the Ginzburg-Landau model to vortex lattice problems," *Comm. Math. Phys.* **313**, (3), 635–743 (2012).

73. M. Sauvageot, "Classification of symmetric vortices for the Ginzburg-Landau equation," *Diff. Int. Equ.* **19**, (7), 721–760 (2006).

74. A. Schmidt, "A time dependent Ginzburg-Landau equation and its application to the problem of resistivity in the mixed state," *Phys. Kondens. Materie* **5**, 302–317 (1966).

75. I.M. Sigal and F. Ting, "Pinning of magnetic vortices," *Algebra Anal.* **16**, 239–268 (2004).

76. I.M. Sigal and T. Tzaneteas, "Abrikosov vortex lattices at weak magnetic fields," *J. Funct. Anal.*, **263**, 675–702 (2012).

77. I.M. Sigal and T. Tzaneteas, "Stability of Abrikosov lattices under gauge-periodic perturbations," *Nonlinearity* **25**, 1–24 (2012).

78. I.M. Sigal and T. Tzaneteas. "On stability of Abrikosov lattices," in preparation.

79. D. Spirn, "Vortex dynamics of the full time-dependent Ginzburg-Landau equations," *Comm. Pure Appl. Math.* **55**, (5), 537–581 (2002).

80. D. Spirn, "Vortex dynamics of the Ginzburg-Landau-Schrödinger equations," *SIAM. J. Math. Anal.* **34**, 1435–1476 (2003).

81. G.N. Stratopoulos, and T.M. Tomaras, "Vortex pairs in charged fluids," *Phys. Rev. B* **N17**, 12493–12504 (1996).

82. D. Stuart, "Dynamics of Abelian Higgs vortices in the near Bogomolny regime," *Adv. Math. Phys.* **159**, 51–91 (1994).

83. P. Takáč, "Bifurcations and vortex formation in the Ginzburg-Landau equations," *Z. Angew. Math. Mech.* **81**, 523–539 (2001).

84. C. Taubes, "Arbitrary *n*-vortex solutions to the first order Ginzburg-Landau equations," *Comm. Math. Phys.* **72**, 277 (1980).

85. C. Taubes, "On the equivalence of the first and second order equations for gauge theories," *Comm. Math. Phys.* **75**, 207 (1980).

86. D.R. Tilley and J. Tilley, *Superfluidity and Superconductivity*. 3rd edition. Institute of Physics Publishing, Bristol and Philadelphia, 1990.

87. F. Ting and J.-C. Wei, "Finite-energy degree-changing non-radial magnetic vortex solutions to the Ginzburg-Landau equations," *Commun. Math. Phys.* **317**(1), 69–97 (2013).

88. M. Tinkham, *Introduction to Superconductivity*, McGraw-Hill, New York, 1996.

89. T. Tzaneteas and I. M. Sigal, "Abrikosov lattice solutions of the Ginzburg-Landau equations," In *Spectral Theory and Geometric Analysis. Contemp. Math.*, **535**, 195–213, (2011).

90. T. Tzaneteas and I.M. Sigal, "On Abrikosov lattice solutions of the Ginzburg-Landau equations," *Math. Model. Nat. Phenom.*, **8**(05), 190–305 (2013), arXiv1112.1897, 2011.

91. E. Weinan, "Dynamics of vortices in Ginzburg-Landau theories with applications to superconductivity," *Phys. D* **77**, 383–404 (1994).

92. E. Weinberg, "Multivortex solutions of the Ginzburg-Landau equations," *Phys. Rev. D* **19**, 3008–3012 (1979).

93. E. Witten, "From superconductors and four-manifolds to weak interactions," *Bull. AMS* **44** (3), 361–391 (2007).

94. Y. Zhang, W. Bao, and Q. Du, "The dynamics and interaction of quantized vortices in Ginzburg-Landau-Schrödinger equation," *SIAM J. Appl. Math.* **67**, (6), 1740–1775 (2007).

3

NUMERICAL CHALLENGES IN A CHOLESKY-DECOMPOSED LOCAL CORRELATION QUANTUM CHEMISTRY FRAMEWORK

DAVID B. KRISILOFF[1], JOHANNES M. DIETERICH[2],
FLORIAN LIBISCH[2], AND EMILY A. CARTER[2,3]

[1]*Department of Chemistry, Princeton University, Princeton, NJ, USA*
[2]*Department of Mechanical and Aerospace Engineering, Princeton University, Princeton, NJ, USA*
[3]*Program in Applied and Computational Mathematics, and Andlinger Center for Energy and the Environment, Princeton University, Princeton, NJ, USA*

3.1 INTRODUCTION

For more than 60 years, chemists have used variants of molecular orbital (MO) theory as a framework to model and predict chemical behavior. Variational MO theory, otherwise known as Hartree-Fock (HF) theory, solves Schrödinger's equation using a wavefunction where each electron's spatial distribution is described by a single, one-electron wavefunction called an MO. The electron is said to *occupy* the MO. In HF theory, the motion of the electrons is uncorrelated—the spatial probability distribution of each electron is independent of all other electrons. This lack of correlation is unphysical since electrons interact via a distance-dependent Coulomb potential: each electron's probability distribution should be a function of the coordinates of all the other electrons. So-called correlated wavefunction methods improve

Mathematical and Computational Modeling: With Applications in Natural and Social Sciences, Engineering, and the Arts, First Edition. Roderick Melnik.

upon HF theory by including electron correlation. They typically express electron correlation in the many-electron wavefunction via consideration of electronic excitations from occupied to unoccupied MOs. The resulting electron configurations are allowed to mix into the ground state configuration to obtain a better description of the many-body wavefunction. It can be shown that such configuration interaction reduces electron–electron repulsion, thereby lowering the energy toward the exact solution of the Schrödinger equation. Correlated wavefunction methods deliver highly accurate molecular properties at a high computational cost. For instance, the "gold standard" of quantum chemistry for molecules at their equilibrium geometries is Coupled-Cluster theory [1] at the single and double excitation levels with perturbative triple excitations, CCSD(T). However, CCSD(T) scales painfully as $\mathcal{O}(\mathcal{N}^7)$ with the size of the basis set. In addition to the scaling of the floating point operations, correlated wavefunctions typically also require large intermediary data structures stored in memory or on disk, greatly adding to their computational cost. Hence, recent research efforts have focused on approximations to correlated wavefunction approaches that reduce their computational cost. These techniques hope to exploit sophisticated strategies to reduce the computational demands of solving Schrödinger's equation, allowing the treatment of larger chemical systems than currently feasible.

Local (electron) correlation methods are one approach to finding less computationally expensive methods. Correlation of electrons in molecules decays rapidly, $1/r^6$, as a function of distance. Pulay first suggested exploiting this fact: by removing numerically unimportant long-range correlations from correlated wavefunction techniques, the required computations are reduced without sacrificing accuracy [2–6]. This approach, called the local electron correlation approximation, has been applied to most electron correlation methods including Møller-Plesset Perturbation theory [7,8], Coupled-Cluster theory [9, 10], and Multireference Configuration Interaction theory (MRCI) [11–20].

The various correlated wavefunction techniques are appropriate for different situations. Coupled-Cluster techniques provide the most accurate solution to Schrödinger's equation for closed shell spin singlet or doublet molecules at their equilibrium geometry. However, for molecules where near-degeneracies are present and where multiple HF solutions contribute significantly such as during the breaking and forming of bonds in chemical reactions or in molecules containing multiple unpaired electrons, the more accurate method to employ is MRCI. The latter allows easy inclusion of all important HF solutions as references from which excitations are then considered and is thus well suited to describe multireference problems (*vide infra*). Note that practical applications of MRCI need to employ a size extensivity error correction [21].

In the following sections, we discuss our local electron correlation approximation to MRCI and our numerical implementation of this method in our code called TigerCI. In Section 3.2, we describe the basic approach our algorithm takes to MRCI and the local electron correlation approximation. In Section 3.3, we give an overview of the numerical importance of the major components in our method. In Section 3.4, we review the Cholesky decomposition of the two-electron integrals.

Finally, we describe rebuilding the two-electron integrals after decomposition (Section 3.6), buffering and storage of the associated data structures (Section 3.7) and the algorithm to evaluate matrix-vector products (Section 3.8).

3.2   LOCAL MRSDCI

3.2.1   MRSDCI

We use the most common variant of MRCI called Multireference Singles and Doubles Configuration Interaction (MRSDCI). MRSDCI starts with HF theory. The HF wavefunction is composed of antisymmetrized products of MOs, one-electron wavefunctions. The antisymmetrization—required by the Pauli Exclusion Principle—is typically achieved by using a determinantal form for the many-electron wavefunction, referred to as a Slater determinant [22]. Each MO is expressed as a variationally optimized linear combination of basis functions, each of which is typically a sum of products of Gaussians and spherical harmonics centered at the atomic nuclei. The multiple reference configurations in MRSDCI consist of different ways of occupying the MO basis with electrons, including both spin and spatial degrees of freedom. The MOs are classified into three categories. (i) MOs that are always occupied in each reference are called inactive orbitals. (ii) MOs whose occupation varies in different references are called active orbitals. (iii) MOs that are not occupied in any reference are called external orbitals. The inactive and active [(i) + (ii)] orbitals are often referred together as the internal orbitals.

The MRSDCI wavefunction consists of a linear combination of the references and all electron configurations generated by excitations (of at most two electrons) from any reference configuration into previously unoccupied MOs in either the active or external orbital sets. The MRSDCI wavefunction includes every single and double electron excitation from each reference. Thus, a single or double excitation is a new configuration where respectively one or two electrons are moved into a different MO.

$$\Psi_{\mathrm{MRSDCI}} = \sum_{\mathrm{ref}} c_{\mathrm{ref}} \psi_{\mathrm{ref}} + \sum_{i,a} c_i^a \psi_i^a + \sum_{i<j,a<b} c_{ij}^{ab} \psi_{ij}^{ab} \qquad (3.1)$$

Single excitations are denoted as ψ_i^a where an electron in orbital i has been excited to orbital a and double excitations are denoted as ψ_{ij}^{ab} where two electrons from orbitals i and j have been excited to orbitals a and b.

Using the MRSDCI wavefunction, solving Schrödinger's equation becomes an eigenvalue problem.

$$H\psi = E\psi \qquad (3.2)$$

In typical molecular cases, the MRSDCI wavefunction involves millions to billions of configurations, making the Hamiltonian extremely expensive to store and diagonalize. Fortunately, molecular applications do not require the entire eigenvalue spectrum, only the few lowest energy eigenvalues. An iterative eigenvalue solver, typically

Davidson's method [23], can solve for the lowest few states without storing the entire Hamiltonian. Iterative eigenvalues solvers require evaluation of a matrix-vector product

$$\sigma = H\psi, \tag{3.3}$$

for a given ψ. Since the Hamiltonian is never stored, the matrix elements are computed on-the-fly during the product. The matrix-vector product is the most time-consuming step during an MRSDCI calculation. Each Hamiltonian matrix element between two configurations (formally configuration state functions, CSFs) p and q is written as

$$H_{pq} = \sum_{kl} A_{pq}^{kl}(k|l) + \frac{1}{2} \sum_{ijkl} B_{pq}^{ijkl}(ij|kl) \tag{3.4}$$

where $(i|j)$ and $(ij|kl)$ are the one- and two-electron integrals and A_{pq}^{kl} and B_{pq}^{ijkl} are the coupling coefficients.

$$(i|j) = \langle i| h_1 |j\rangle \tag{3.5}$$

$$(ij|kl) = \int \int \frac{\phi_i(r_1)\phi_j(r_1)\phi_k(r_2)\phi_l(r_2)}{|r_1 - r_2|} dr_1 dr_2 \tag{3.6}$$

$$A_{pq}^{kl} = \langle p| E_{kl} |q\rangle \tag{3.7}$$

$$B_{pq}^{ijkl} = \langle p| E_{ij}E_{kl} - \delta_{jk}E_{il} |q\rangle \tag{3.8}$$

h_1 is the one-electron portion of the Hamiltonian. E_{ij} is a replacement operator that replaces orbital j with orbital i. The electron integrals evaluate the electron kinetic energy, electron-nuclear attraction, and electron–electron repulsion for the given spatial distributions of the electrons represented by the MOs. The coupling coefficients are abstract quantities relating the permutations of orbital occupations between p and q and are reducible to simple algebraic formulas [24].

The two inputs to the Hamiltonian matrix elements, the two-electron integrals, and the coupling coefficients are handled very differently. The two-electron integrals are computed once before the calculation begins. They are often first calculated in the atomic orbital (AO) basis, the basis of the one-electron MOs, and then transformed into the MO basis. The coupling coefficients are easier to calculate. They are computed on-the-fly during the matrix-vector product .

3.2.2 Symmetric group graphical approach

Performing the matrix-vector product (Equation 3.3) on a modern computer is not a trivial task. Efficient performance occurs when the problem is handled globally where operations are performed on groups of similar configurations instead of handling each configuration individually. The Symmetric Group Graphical Approach (SGGA) [24–26] is one way of organizing the matrix-vector product efficiently. In the

SGGA, each configuration is built as an antisymmetrized product of a spatial (orbital occupations) and spin function. Using such configurations, the coupling coefficients in Equation 3.4 can be expressed in terms of the representation matrices of the symmetric group S_N. Furthermore, the SGGA finds similar pairs of configurations by representing both the spin and spatial spaces as separate graphs. Here we only discuss the organization of the spatial space. For a complete description of the spin space and for all the nuances of the SGGA method, see Ref. [24].

SGGA graphs allow for efficient book-keeping of all configurations: each configuration in the wavefunction is represented as an array

$$T = [N_0, N_1, \ldots, N_n]. \tag{3.9}$$

The ith element of this array represents the sum of number of electrons in the configuration up to and including the ith orbital. N_0 is always zero and N_n, where n is the number of orbitals, is always the total number of electrons in the molecule. Such an array construct allows for easy placement of each configuration on a graph, with the number of electrons on the horizontal axis and the orbitals on the vertical axis (see, for example, Fig. 3.1). This is the SGGA occupation graph, with each array being a directed path on the graph from the head at zero electrons, zero orbitals to the tail at number of electrons, number of orbitals. Each path on the graph can be assigned a unique identifier referred to as a weight.

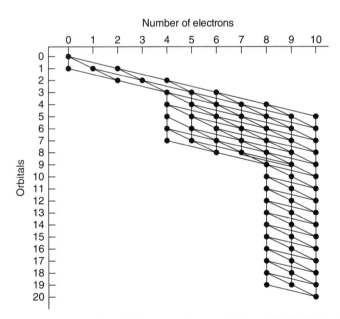

FIGURE 3.1 An example of an SGGA occupation graph for an MRSDCI calculation. The calculation includes 10 electrons in 3 inactive orbitals, 5 active orbitals and 12 external orbitals. Each path from $(0,0)$ to $(10,20)$ represents a configuration in the wavefunction.

We can characterize the interaction H_{pq} between two configurations p and q with the graph. If a single configuration is a directed path on the graph, then two configurations represent a loop on the graph. Different types of loops are associated with different types of coupling coefficients. The occupation graph allows us to search for all loops with the same type of coupling coefficients and avoid recalculating those coupling coefficients for each pair of configurations. Loops are classified by the number of orbitals where the electron occupation differs between the two configurations making the loop. If a loop has configurations that differ in occupations of two orbitals it is called a two-segment loop. Three- and four-segment loops are also possible. When discussing the scaling of our algorithm, we will refer to the individual contributions of handling the two-, three-, and four-segment loops.

Finally, the SGGA also takes advantage of the differences in the internal and external orbitals. Since we include up to double excitations, the external orbitals contain at most two electrons. This is permutationally much simpler than the internal space, which can be very complicated. Hence the contributions of the internal and external space are handled separately. Our implementation of the SGGA only constructs the graph for the internal space and handles the external space implicitly. Individual SGGA routines (for which we present benchmark results below) are named by the type of loop and number of internal or external orbitals in the loop. For instance, the one-external two-segment routine handles all two-segment loops with one external orbital.

3.2.3   Local electron correlation approximation

Within configuration interaction (CI), different excited configurations in the wavefunction represent different electron correlations. To remove negligible long-range correlations, we need to identify the corresponding excitations. These excited configurations involve exciting electrons to or from spatially distant orbitals. Identifying long-range excitations requires a localized MO basis set because in a delocalized basis no two orbitals are ever far apart. There are multiple approaches available to localize the internal orbitals [27–29]. In our TigerCI code, we use the Pipek-Mezey method [27] to localize the internal orbitals. Localizing external orbitals is more difficult: we use the Local Orthogonal Virtual Orbitals [30] approach for this purpose.

Because double excitations are the most numerous (and thus the most numerically demanding), we remove only double excitations when considering long-range correlation. For each ψ_{ij}^{ab}, we consider the internal space where the excitation originates (the holes i and j) and the external space where the electrons are excited to (the orbitals a and b). We truncate double excitations using two related approximations: Weak Pairs (WP) and Truncation of Virtuals (TOV). The WP approximation [6] truncates excitations if the two holes i and j are spatially far apart. This approximation significantly speeds up a CI calculation [15]. However, quantum chemistry basis sets typically have many more external orbitals than internal orbitals. Therefore, the external space also needs to be truncated to increase computational efficiency. We define the set of virtual orbitals close to a single internal orbital i as the domain of i. In the TOV approximation, a double excitation ψ_{ij}^{ab} is truncated if either a or b is not in the union of the domains of i and j. This truncation of the external space greatly reduces the computational cost of local MRSDCI compared to canonical MRSDCI [13].

Both the WP and TOV approximations require defining a quantitative distance between two orbitals. In TigerCI we generate spheres around the *center* of each orbital to estimate their spatial distribution and use the spheres to calculate distances [13]. Briefly, for each orbital we calculate the norm of the orbital over each atomic center. We sort the absolute value of each atom's contribution into descending order and then add up the largest contributions until a threshold value is reached. The threshold is typically 0.6–0.8 as each orbital is normalized to 1. The largest contributors are then used to define the sphere's location and size. The averaged location of the important centers, weighted by their contribution to the norm, is the center of the sphere. The sphere's radius is set to the largest distance between any two important centers, ensuring that the sphere encompasses all the important atoms. Very well-localized orbitals may involve only one important atom. In those cases, a default radius size is used instead. Active orbitals, that is, orbitals with varying occupation in the references, can use a cylinder with hemispherical caps instead of a sphere. This avoids discontinuities in potential energy surfaces that arise in local correlation methods as the molecular geometry shifts and individual configurations are discretely added and removed from the wavefunction.

Our implementation of both the WP and TOV approximations works with the SGGA graph. Indeed, one of the advantages of a graphical approach is its ability to easily remove configurations from the wavefunction. Each full path in the internal space, with internal holes at orbitals i and j, is connected to an entire block $\{\psi_{ij}^{xy}\}$ of configurations in the total wavefunction: they feature the same internal space but different external orbitals x, y to which the electrons excite. In our implementation, we assign each full path in the internal space a unique identifier or weight. This weight then applies to the entire block $\{\psi_{ij}^{xy}\}$.

If i and j do not interact under the WP approximation, then the entire block of configurations is truncated. We accomplish this by altering the mapping: the weight for the removed internal configuration maps to a specific value for removed configurations. The TOV approximation removes individual configurations from $\{\psi_{ij}^{xy}\}$ by placing restrictions on the external orbitals x and y. We maintain a list of allowed external orbitals for specific i and j hole orbitals which determines the size of $\{\psi_{ij}^{xy}\}$.

Our implementation of the WP and TOV approximations runs into ambiguity in some multireference calculations. It is possible in multireference calculations to have an excited configuration that could originate from a double excitation out of more than one reference configuration, which would make it difficult to apply the WP and TOV approximations to determine whether or not this configuration should be kept in the wavefunction (see, for example, Table 3.1). The SGGA only includes these excited configurations once in the wavefunction, as there is only one full path on the graph corresponding to this excited configuration no matter how many references could have contributed. However, the holes and electrons for such an excitation are different depending on which reference configuration is considered. This issue has

TABLE 3.1 An example of a doubly excited configuration that could arise from more than one reference configuration

	Internal orbital occupations
Excited configuration	2020202
Reference 1	2220102
Reference 2	2121102
Reference 3	2021112

These configurations are more complicated to handle under the local electron correlation approximation.

been mentioned previously [13] without a thorough resolution. Here we present a general solution. For the WP approximation, we only truncate a configuration if all the possible excitations that could have created the configuration should be removed under the WP approximation. The TOV approximation is more troublesome. Each internal configuration has a list of allowed external orbitals that can be excited to and a list of hole pairs that could give rise to that internal configuration (see Table 3.1 for a simple example). Any external orbital close enough to any hole pair is added to the list of allowed orbitals. This creates some nonlocal character. For instance, if external orbital a is local to one hole pair and external orbital b is local to another hole pair, a double excitation to a and b would be included. This is true even if a and b are not close to each other. Ideally such cross terms (double excitations to two orbitals local to two different, spatially distant hole pairs) would not be included in the wavefunction. Removing these cross terms remains a goal for the future. Note that in cases where a configuration could not only arise from a double excitation, but also from a single excitation from a different reference, the configuration is never removed—we do not truncate single excitations.

3.2.4 Algorithm summary

Altogether our implementation of local MRSDCI can be summarized in the following steps:

1. Construct a localized MO basis set.
2. Handle the two-electron integrals (*vide infra*).

   (a) Calculate the two-electron integrals in the AO basis set.
   (b) Decompose the two-electron integrals (*vide infra*).
   (c) Transform the decomposed two-electron integrals into the MO basis set.
   (d) Calculate the two-electron integrals in the MO basis set in the order required by the SGGA.

3. Construct the SGGA spatial and spin graphs.

4. Remove configurations under the local approximation.

5. Use Davidson's method to solve Schrödinger's equation, using the SGGA to handle the matrix-vector product.

3.3 NUMERICAL IMPORTANCE OF INDIVIDUAL STEPS

In the following sections, we will benchmark TigerCI and decompose the timings into their individual components. We are interested in the scaling behavior with respect to system size and the number of references. This requires a controllable environment that allows us to scale these quantities without additional complications. Elongated alkane chains (linear hydrocarbon molecules) allow easy control over the scaling with the system size. Alkane chains represent optimal systems for our local approximations as they minimize the number of interactions due to their topology. Therefore, these calculations represent the best case scenario. Concerning the scaling with the number of references, we introduce the full set of reference configurations as dictated by the active space setting. All calculations we present here for testing are based on MOs derived from the HF reference wave function in order to minimize differences between different calculations. Consequently, artifacts due to, for example, nonlocalizable internal orbitals do occur.

In general, all calculations have been run three times to obtain averaged times for wall time, CPU time, and standard deviations. All error bars given in the graphs represent one sigma from the mean.

Our algorithm can be seen as multiple steps that each pose their own challenges. In principle, we can separate the numerical load into the following parts: Cholesky decomposition of the two-electron integrals, transform of the Cholesky vectors from the AO into the MO basis, integral assembly and iterative solution of the Schrödinger equation using the SGGA engine. Other steps such as the localization of the MOs do not represent a significant numerical challenge and therefore will not be accounted for.

Each of these steps has its own profile with respect to load on the central processing unit (CPU), the memory, and the storage subsystem. We will address each of these defining properties in the following paragraphs. Additionally, the scaling behavior of the different algorithmic components with respect to system size and number of references is discussed. Table 3.2 provides an overview of the different components and their scaling characteristics.

Starting with the single reference case of pentadecane $C_{15}H_{32}$, note that the *Pareto principle*, which states that 20% of the algorithm takes 80% of the resources, does not hold true for this typical calculation setup. Both the integral assembly step and the SGGA engine take around 40% of the wall time, followed by the Cholesky decomposition with 16.7%.

Table 3.2 provides clear evidence that the SGGA portion of the algorithm quickly becomes dominant with increasing numbers of references. This behavior is expected since all other components are independent of the number of references. In the case with 19 references (a complete active space of four electrons in four orbitals,

TABLE 3.2 Overview of the importance of different algorithmic components of our local multireference correlation framework with respect to system size and number of reference configurations

System	$C_{15}H_{32}$	$C_{15}H_{32}$	$C_{15}H_{32}$	$C_{35}H_{72}$
No. of references	1	3	19	1
Cholesky decomposition	16.7	7.0	0.4	40.0
Cholesky transform	4.7	2.0	0.1	9.0
integral assembly	40.1	15.9	0.9	26.9
SGGA	38.5	75.1	98.5	24.2

All values are percent of the total wall time.

denoted as CAS [4e,4o]), the SGGA subsystem takes in excess of 98% of the total wall time, now turning it into a clear hot-spot. The scaling with system size shows a different picture. Due to its unfavorable $\mathcal{O}(\mathcal{N}^3)$ scaling and high associated prefactor, the Cholesky decomposition will likely dominate the wall time for typical large calculations.

We discuss the resource requirements with respect to memory and I/O in more detail in the sections below, where warranted.

3.4 CHOLESKY DECOMPOSITION

The two-electron integrals represent one of the largest computational hurdles in a correlated wavefunction calculation. As shown in Equation 3.6, the number of integrals scales $\mathcal{O}(\mathcal{N}^4)$ with molecular size. Worse, the integrals are usually calculated in the AO basis set and then transformed into the MO basis set which scales as $\mathcal{O}(\mathcal{N}^5)$. Therefore, all reduced scaling methods need to find an approximate formulation of the two-electron integrals.

Though the number of two-electron integrals grows as $\mathcal{O}(\mathcal{N}^4)$, not all of them are significant. The overlap between two Gaussian-type orbitals, $\phi_i(r_1)\phi_j(r_1)$ decays rapidly with distance of the centers of the two functions i and j. There are numerous techniques to avoid calculating numerically insignificant integrals. The most common technique uses the Cauchy-Schwarz (CS) inequality to calculate an upper bound on blocks of two-electron integrals [31]. Those blocks whose upper bounds fall below a screening threshold are set to zero and never calculated. Further advances have focused on producing a tighter upper bound [7,9]. A relatively recent approach eschews the upper bound for an estimate of the integral's magnitude [32–34].

The cost of the two-electron integrals can be further reduced by exploiting their linear dependencies. The two-electron integrals are typically stored in a matrix $V_{ij,kl} = (ij|kl)$. The product basis $\phi_i\phi_j$ contains numerous linear dependencies. We can reduce the size of the two-electron integral matrix by decomposing it into a linearly independent basis. There are two approaches used: density fitting and the Cholesky decomposition. In this work, we use the Cholesky decomposition; however, there is no *a priori* reason the density fitting approach would not work as well.

The use of a Cholesky decomposition of the two-electron integral matrix was first proposed in 1977 [35] but was not practically formulated until 2003 [36]. Two factors make applying the decomposition difficult: the two-electron integral matrix is positive semi-definite (instead of positive definite) and calculating two-electron integrals is most efficient in blocks. The Cholesky decomposition is only stable for positive semi-definite matrices when pivoting is used. In such a formulation, the matrix is decomposed column by column and a symmetric pivot occurs at the beginning of each iteration. The decomposition stops when the pivoting falls below a predetermined threshold. Therefore, the Cholesky decomposition requires a single column of the two-electron integral matrix at a time. Unfortunately, calculating two-electron integrals is most efficient when done in blocks where integrals involving atomic orbitals with the same contracted Gaussians are calculated at the same time. Calculating each column of the two-electron integral matrix individually is therefore significantly less efficient than calculating blocks of columns. The algorithm devised by Koch et al. [36] provides a compromise by only partially pivoting. Each time the decomposition pivots, a block of columns in the matrix (containing the pivot) is calculated and some of those columns are decomposed.

The partial pivoting strategy seeks a balance between the needs of the Cholesky decomposition and the calculation of the two-electron integral matrix elements. However, this compromises has some drawbacks. First, the partial pivoting strategy increases the rank of the Cholesky decomposition (the number of Cholesky vectors). This increases the cost of subsequent calculations. Second, the partial pivoting strategy is unusual, and to the best of our knowledge, unavailable in any standard linear algebra program library. In fact, it was only recently that the full pivoted Cholesky decomposition was added to LAPACK [37]. Finally, it should be noted that while an error analysis of the fully pivoted Cholesky decomposition has been performed [38], no similar analysis has been done for the partially pivoted algorithm.

We have investigated using the LAPACK-pivoted Cholesky decomposition to decompose the two-electron integral matrix. This approach provides a more compact decomposition (by exploiting full pivoting). It is also beneficial from a software engineering standpoint: the complexity of the Cholesky decomposition is now entirely handled by a reliable and efficient external library, LAPACK. The drawback of this approach is that the LAPACK algorithm requires the entire matrix to be stored in memory. This means all the integrals need to be calculated before the decomposition. Fewer integral calculations are required in the scheme of Koch et al. because the integrals are calculated as needed (in blocks of columns) and the decomposition ends before all columns are decomposed.

After outlining the details of our LAPACK-based Cholesky decomposition of the two-electron integral matrix, we will compare the results of the LAPACK-based scheme with our previous implementation of the Koch et al. algorithm. We have chosen DPSTF2 (BLAS level 2) as the Cholesky decomposition routine instead of DPSTRF (BLAS level 3). In our testing, the BLAS level 2 routine is faster. The two routines perform the decomposition via different algorithms and different memory access patterns. We suspect the performance difference is heavily dependent on the matrix size and underlying hardware.

1. Calculate the diagonal elements, $V_{ik,ik} = (ik|ik)$.
2. Prescreen the two-electron integrals using the diagonal elements and the same prescreening criteria as Koch et al. [36].
3. Calculate the unique two-electron integrals employing CS screening for each block of integrals.
4. Use LAPACK's DPSTF2 to perform the Cholesky decomposition.

We expect this approach should produce a more compact decomposition while requiring a slightly larger amount of computer time. Indeed, the new approach generates a more compact decomposition (see Fig. 3.2). In general, using full pivoting always produces a lower rank decomposition than a partial pivoting approach. Counter to our expectations, the new algorithm appears faster than the previous partial pivoting approach (Fig. 3.3) even though the new algorithm requires more integral calculations. Since the new approach spends more time calculating the matrix elements than the old approach (see Fig. 3.3), the unexpected wall times must be due to the time for the decomposition. Indeed, our implementation of Koch et al.'s approach was not heavily optimized, unlike the LAPACK DPSTF2 routine. Surprisingly, the dramatic speed up of the decomposition outweighs the increased time spent constructing matrix elements. We do not expect this to be the case when using a heavily optimized version of Koch et al.'s approach.

All the algorithms discussed so far have been in-core, that is, they have assumed that there was sufficient memory to store all the necessary data structures. It is a fact that the two-electron integrals for large chemical systems may not always fit in memory. For instance, after prescreening with a Cholesky decomposition threshold of $1 \cdot 10^{-7}$, the two-electron integral matrix for pentacosane ($C_{50}H_{102}$) requires more than 100 GB of space. Efficient out-of-core Cholesky decomposition algorithms

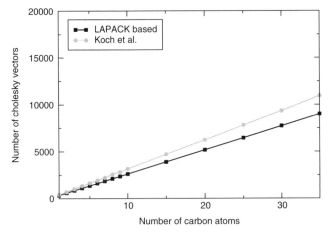

FIGURE 3.2 Using full pivoting during the Cholesky decomposition (LAPACK routine) provides a more compact decomposition than partial pivoting (the Koch et al. scheme).

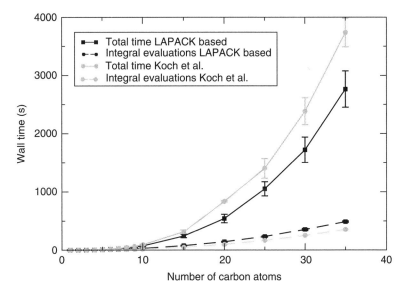

FIGURE 3.3 Using the Koch et al. scheme and the new LAPACK-based scheme produces similar overall run times. The new LAPACK-based scheme requires less time for the Cholesky decomposition but more integral evaluations. Plotted values are the averaged wall times over five calculations; error bars correspond to one standard deviation.

are a point of active research [39]. Unfortunately, the recursive data structures that characterize the most efficient out-of-core Cholesky decompositions would appear to prevent symmetric pivoting. To the best of our knowledge, only one out-of-core pivoted Cholesky decomposition algorithm, by Aquilante et al., has been published [40]. However, the problem of a pivoted out-of-core Cholesky decomposition is similar to that of a partially pivoted, out-of-core LU decomposition [41].

We have developed an out-of-core pivoted Cholesky decomposition based on our in-core decomposition scheme. Like our in-core decomposition, the entire matrix is calculated before decomposition, thereby allowing the use of full pivoting. The additional I/O incurred by storing the matrix on disk is quite expensive as shown in Figure 3.4. An *efficient*, fully-pivoted, out-of-core Cholesky decomposition very much remains an open problem.

3.5   TRANSFORMATION OF THE CHOLESKY VECTORS

The Cholesky decomposed two-electron integrals (Cholesky vectors) are originally computed in the AO basis set. For use in the SGGA, we require the two-electron integrals in the MO basis set. Therefore, we transform the Cholesky vectors from the AO to MO basis. Formally the transformation can be written as:

$$T_{i\nu}^m = \sum_{\mu} C_{i\mu} L_{\mu\nu}^m \tag{3.10}$$

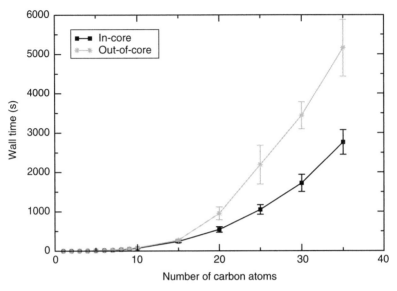

FIGURE 3.4   The out-of-core pivoted Cholesky decomposition is significantly less efficient than its in-core counterpart due to the increased I/O. Plotted values are the averaged wall times over five calculations; error bars correspond to one standard deviation.

$$T_{ij}^m = \sum_{\nu} C_{i\nu} T_{j\nu}^m \qquad (3.11)$$

The indices i and j represent the MO basis and the μ and ν indices represent the AO basis. The $L_{\mu\nu}^m$ is the original AO basis Cholesky vector and T_{ji}^m is the final MO basis Cholesky vector. The transformation matrix element $C_{i\mu}$ contains the coefficient of AO μ in MO i.

Figure 3.5 shows that the Cholesky transform scales as $\mathcal{O}(\mathcal{N}^{3.48895})$. This scaling is impressive compared to the $\mathcal{O}(\mathcal{N}^5)$ scaling of transforming the two-electron integrals without performing the Cholesky decomposition. The high scaling is directly related to the number of Cholesky vectors, indexed by m in Equation 3.11. Depending on the system studied, the number of Cholesky vectors scales somewhere between linearly and quadratically with system size [18, 42]. The Cholesky transform is one of the highest scaling pieces of TigerCI, and of those pieces (*vide infra*), it is the most expensive.

3.6   TWO-ELECTRON INTEGRAL REASSEMBLY

The Cholesky decomposition provides a compact representation of the two-electron integrals and dramatically speeds up the basis set transformation of the (decomposed) integrals. However, the SGGA equations are written in terms of the canonical, that is, not decomposed, two-electron integrals. After the Cholesky vectors are transformed

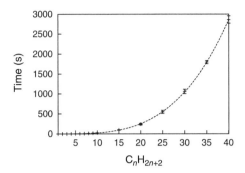

FIGURE 3.5 Scaling of the wall time of the Cholesky vector basis set transform with the size of the system. Full line is a fit of the wall time against $f(x) = a \cdot x^b$ with $a = 0.0073$ and $b = 3.48895$.

from the AO into the MO basis, we reassemble the four-index two-electron integrals:

$$(ab|cd) = \sum_{i}^{N} L_{ab}^{i} L_{cd}^{i} \tag{3.12}$$

with the integral $(ab|cd)$ for the molecular orbitals a, b, c, d and the sum over N transformed Cholesky vectors L_{ab}^{i} for the orbitals a, b, and L_{cd}^{i} for orbitals c, d, respectively.

All possible combinations of the molecular orbitals form a tesseract of high symmetry [22]. The following integrals are equivalent:

$$(ab|cd) = (ab|dc) = (ba|cd) = (ba|dc) = (cd|ab) = (cd|ba) = (dc|ab) = (dc|ba) \tag{3.13}$$

Clearly, only calculating the unique integrals, as opposed to all integrals, provides a significant reduction in the prefactor of the integral assembly. The address of any one integral is easily determined for the evenly shaped tesseract. Nevertheless, such an algorithm would scale $\mathcal{O}(N^5)$ and would result in a rather sparse tesseract since MOs that are far apart would ideally result in a zero integral.

As previously discussed, our local approximations exploit the spatial separation of localized MOs and can be used here to filter out a significant number of integrals. This reduces the prefactor and, more importantly, the scaling of the integral assembly.

A naive implementation would simply feature four nested loops, filtering out calculations based on pairwise local approximations. Although certainly beneficial in terms of simplicity, two difficulties arise with such an approach. First, certain parts of our CI algorithm later require only certain parts of the integral tesseract and need them in a different order. For example, a subroutine may need only the integrals for all a, b, c, d in the external space and in the order $(ac|bd)$ instead of $(ab|cd)$. Second, the fact that the local approximation filters out integrals compresses the tesseract into a nonevenly shaped entity. While this entity still exhibits the same symmetry, the unevenness makes it significantly more difficult to actually exploit the integral symmetry for reasons we will highlight later.

The first issue can be solved trivially; our current implementation [18] breaks the total integral tesseract into classes based on their later usage within the SGGA. They are calculated individually and stored in the right order with respect to later usage. To reduce I/O, we fill a buffer with Cholesky vectors which we subsequently use in a BLAS level 2 operation to obtain a series of integrals in one sweep. We only store the non-zero integrals, yet we calculate and store all redundant integrals on disk. This is justified as we must avoid computationally expensive mapping of a set of a, b, c, d to an integral identifier in later parts of the code.

The scaling of this code is greatly reduced in comparison to the canonical ($\mathcal{O}(\mathcal{N}^5)$) limit (Fig. 3.6), being close to quadratic for both CPU and wall time. Although both prefactors seem small (<2), we should keep in mind that not all possible integral symmetries are exploited—redundant calculations are performed.

As can be seen from Figure 3.7, for simple single reference calculations, the integral assembly takes a significant part of the total wall and CPU time. As the wall times are very small for the first few data points, we will restrict our analysis to the more important later data points. The integral assembly varies from 20% to 40% of the overall wall time, decaying to 30% for the largest systems. Eventually, the Cholesky decomposition with its unfavorable $\mathcal{O}(\mathcal{N}^3)$ scaling will dominate the overall single-reference runtime, yet we consider the presented size regime to be the one currently relevant for actual applications. What these results clearly demonstrate is that for single reference calculations, a reduction of the integral assembly cost by means of symmetry exploitation would be beneficial.

The importance of the integral assembly decreases with more references; see Figures 3.8 and 3.9. For the CAS[2e,2o] case, our maximal observed fraction is 20% in wall and CPU time. Integral reassembly takes less than 10% of the wall time for

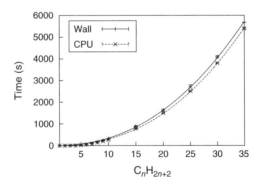

FIGURE 3.6 Scaling of the CPU and wall time of the integral assembly with the size of the system. Full line is a fit of the wall time against $f(x) = a \cdot x^b$ with $a = 1.97026$ and $b = 2.2425$, and dashed line is a fit against the CPU time with $a = 1.39125$ and $b = 2.32556$.

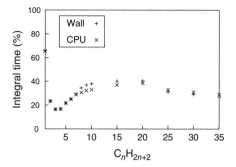

FIGURE 3.7 Fraction of the total wall and CPU time of the TigerCI code that can be attributed to the integral assembly for single reference calculations.

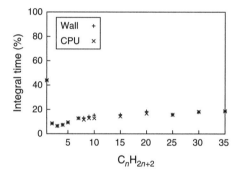

FIGURE 3.8 Fraction of the total wall and CPU time of the TigerCI code attributed to the integral assembly for CAS[2e,2o] calculations.

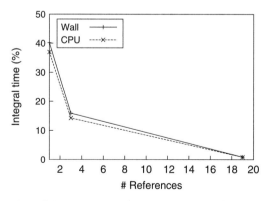

FIGURE 3.9 Fraction of the total wall and CPU time of the TigerCI code attributed to the integral assembly for $C_{15}H_{32}$ from single reference to CAS[4e,4o].

the CAS[4e,4o] data points. The integral assembly does not depend on the number of references and scales solely with the number of orbitals.

Nevertheless, we are able to clearly define the remaining challenges within this part of the algorithm. First, we ideally would like to reduce the scaling of this code portion to linear if possible, which will require a different scheme for the integral evaluation. Second, we would like to reduce the prefactor by properly exploiting all the existing symmetries within the original tesseract.

The latter is, as previously indicated, hindered by the needs of subsequent parts of the algorithm and the local approximations. Since later parts of the algorithm require integrals in non-canonical order, we keep redundant parts on disk in order to have them sequentially accessible. At the same time, the reduction of the tesseract to an uneven entity makes the calculation of only nonredundant integrals and their correct redundant storage very hard in one sweep. An obvious approach to circumvent this may be to first only calculate the nonredundant integrals and subsequently sort them into their redundant on-disk representation. With such a scheme though one does need to be aware of both the I/O overhead and extra mapping imposed by the sorting.

3.7 INTEGRAL AND EXECUTION BUFFER

The performance of any CI code depends on the efficient storage of the two-electron integrals [43]. The simplest solution, storing the integrals individually in random access files, produces far too much I/O. Luckily, the SGGA does not read the integrals randomly, but instead jumps between different contiguous blocks of integrals. We can exploit this access pattern to reduce I/O, providing the design principle of our integral storage. Instead of storing individual integrals, we store blocks of integrals. The integral block size can be aligned with the block size of the underlying file system to reduce and optimize the number of I/O operations. Obviously, this does not entirely remove the stress that our code exercises on the underlying storage systems; we note that it may be beneficial under some circumstances to recalculate the integrals on-the-fly where needed, a so-called *integral-direct mode*.

Additionally, we want to maximize the usage of available fast memory, thereby reducing the I/O to slower storage options like hard drives. In order to facilitate this, we developed a generic C++ library, including a Fortran interface for codes like ours. It is entirely generic, meaning that arbitrary data can be stored following the same basic ideas. Everything is based on blocks of data, where individual entries can be requested using integer file and entry identifiers. The library employs a tiered data model, meaning that it tries to keep as many blocks of data simultaneously in memory as allowed following a *last in first out* (LIFO) policy. In order to bound the used storage on a more fine-grained level, different files are grouped into different pools with individual memory limits. Within TigerCI, we use one pool for all integral files and additional pools for auxiliary data. The auxiliary data forms an execution buffer that stores the order of different SGGA operations along with some intermediate data.

3.8   SYMMETRIC GROUP GRAPHICAL APPROACH

As previously discussed, the core part of our local correlation code is the SGGA engine. It is not a monolithic entity but decomposes into multiple parts of varying difficulty and numerical character. We will discuss the scaling behavior and importance of each individual part in what follows. The current implementation of our SGGA engine uses BLAS statements wherever beneficial and tries to reduce floating-point operations as well as to keep memory allocations as local and tight as possible. Of course, there is always room for optimization in any computer code.

We start our discussion with the two-segment loops: for numerical treatment, we separate them into contributions entirely from the external space, contributions entirely from the internal space, and mixed contributions with one orbital in the internal and one in the external space (see Ref. [24] for a discussion of the parts of the SGGA). The purely external contributions stem from integrals of type $(ii|ab)$ and $(ia|ib)$, where a, b are in the external space and i is in the internal space. The purely internal contributions stem from integrals $(ik|jk)$ and the mixed contributions stem from $(ik|ak)$ type integrals, where i, j, k are in the internal and a is in the external space.

As we can see from Figures 3.10 to 3.16, all of the two-segment loops exhibit rather similar scaling characteristics. All of them scale approximately cubically with system size but with such small prefactors that they do not contribute significantly to the overall wall or CPU time. No significant difference between the CPU and wall time can be observed, proving that these code parts do not feature significant I/O but are memory and CPU bound. Their scaling with respect to the number of references is significantly more favorable than their scaling with system size and is below quadratic. The combined two-segment loops never contribute more than 7% to the overall runtime within our tested size and reference regime. Therefore, they will only play a significant role within our local correlation framework in the limit of large

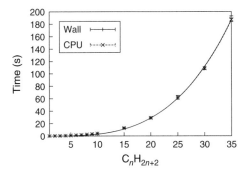

FIGURE 3.10   Scaling of the CPU and wall time of the purely external SGGA part with the size of the system. Full line is a fit of the wall time against $f(x) = a \cdot x^b$ with $a = 0.00130$ and $b = 3.34105$, dashed line is a fit against the CPU time with $a = 0.00134$ and $b = 3.33139$.

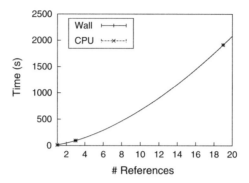

FIGURE 3.11   Scaling of the total wall and CPU time of the purely external two-segment loops for $C_{15}H_{32}$ from single reference to CAS [4e,4o]. Full line is a fit of the wall time against $f(x) = a \cdot x^b$ with $a = 15.6015$ and $b = 1.63368$, dashed line is a fit against the CPU time with $a = 15.5645$ and $b = 1.63372$.

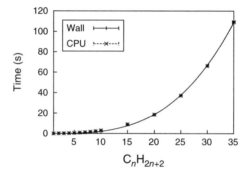

FIGURE 3.12   Scaling of the CPU and wall time of the purely internal two-segment loops with the size of the system. Full line is a fit of the wall time against $f(x) = a \cdot x^b$ with $a = 0.00150726$ and $b = 3.14717$, dashed line is a fit against the CPU time with $a = 0.00150384$ and $b = 3.14716$.

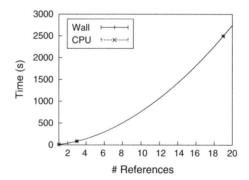

FIGURE 3.13   Scaling of the total wall and CPU time of the purely internal two-segment loops for $C_{15}H_{32}$ from single reference to CAS [4e,4o]. Full line is a fit of the wall time against $f(x) = a \cdot x^b$ with $a = 10.9037$ and $b = 1.84591$, dashed line is a fit against the CPU time with $a = 10.8738$ and $b = 1.84601$.

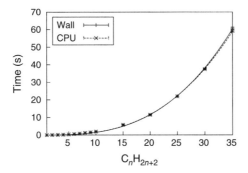

FIGURE 3.14 Scaling of the CPU and wall time of the one-external two-segment loops with the size of the system. Full line is a fit of the wall time against $f(x) = a \cdot x^b$ with $a = 0.00163536$ and $b = 2.95769$, dashed line is a fit against the CPU time with $a = 0.00200265$ and $b = 2.89419$.

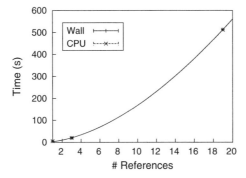

FIGURE 3.15 Scaling of the total wall and CPU time of the one-external two-segment loops for $C_{15}H_{32}$ from single reference to CAS [4e,4o]. Full line is a fit of the wall time against $f(x) = a \cdot x^b$ with $a = 3.10999$ and $b = 1.73426$, dashed line is a fit against the CPU time with $a = 3.09931$ and $b = 1.73476$.

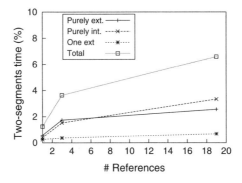

FIGURE 3.16 Fraction of the total wall time of the TigerCI code that can be attributed to the two-segment loops for $C_{15}H_{32}$ from single reference to CAS [4e,4o].

systems with a large set of references. Then they might also compete with cubically scaling, but reference-agnostic, Cholesky decomposition.

Obviously, we can expect the remaining parts of the SGGA dealing with three- and four-segment loops to play a more important role for the overall runtime, that is, they have larger scaling prefactors, as they cover larger integral and excitation spaces. At the same time, a larger spread between different subparts of the SGGA can be expected, for example, between four-internal and four-external parts. We therefore will discuss the parts in ascending order of external orbitals, starting with the part based on only internal integrals and ending with the entirely external parts. All wall time fractions are decomposed in terms of three- and four-segment loops and summarized in Figures 3.17 and 3.18, respectively. Again, we follow the naming conventions for both integral classes and different segment loops as defined in the original publications [15, 24].

As can be seen from Figures 3.19 to 3.20, the four-internal loops scale vary favorably, sub-quadratic, both with respect to the system size and the number of references.

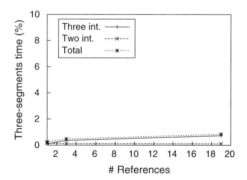

FIGURE 3.17 Fraction of the total wall time of the TigerCI code that can be attributed to the three-segment loops for $C_{15}H_{32}$ from single reference to CAS [4e,4o].

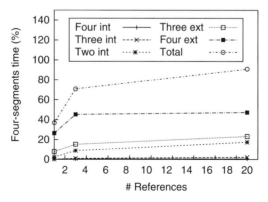

FIGURE 3.18 Fraction of the total wall time of the TigerCI code that can be attributed to the four-segment loops for $C_{15}H_{32}$ from single reference to CAS [4e,4o].

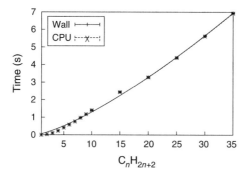

FIGURE 3.19 Scaling of the CPU and wall time of the four-internal four-segment loops with the size of the system. Full line is a fit of the wall time against $f(x) = a \cdot x^b$ with $a = 0.0594834$ and $b = 1.33929$, dashed line is a fit against the CPU time with $a = 0.0596011$ and $b = 1.33792$.

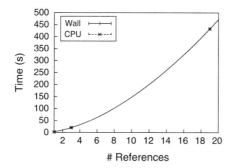

FIGURE 3.20 Scaling of the total wall and CPU time of the four-internal four-segment loops for $C_{15}H_{32}$ from single reference to CAS [4e,4o]. Full line is a fit of the wall time against $f(x) = a \cdot x^b$ with $a = 3.15824$ and $b = 1.6717$, dashed line is a fit against the CPU time with $a = 3.14937$ and $b = 1.67195$.

Additionally, the prefactors are small enough so that this part of the SGGA algorithm does not significantly contribute to the overall runtime (always below 1%).

As can be seen from Figures 3.21 to 3.24, all four-internal routines scale sub-quadratically with respect to both the system size and the number of references, as do all but one of the three-internal routines (the exception being the three-internal three-segment loops, Fig. 3.21). Again, the small prefactor keeps the contributions to the total runtime at approximately 1% and 2%, respectively, for the tested size regime.

For the two-internal loops, we have both three-segment and four-segment loops. Just as for the three-internal loops, we can see from Figures 3.25 to 3.28 that the two-internal three-segment loops have a significantly smaller prefactor than the three-internal four-segment loops, by multiple orders of magnitude. At the same time, the two-internal three-segment loops scale almost $\mathcal{O}(\mathcal{N}^4)$ with respect to the system size as opposed to sub-quadratic for the two-internal four-segment loops. Therefore, in the limit of large systems, the two-internal three-segment loops will contribute significantly to the total

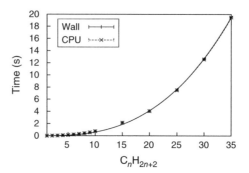

FIGURE 3.21 Scaling of the CPU and wall time of the three-internal three-segment loops with the size of the system. Full line is a fit of the wall time against $f(x) = a \cdot x^b$ with $a = 0.00101718$ and $b = 2.77356$, dashed line is a fit against the CPU time with $a = 0.00103956$ and $b = 2.76609$.

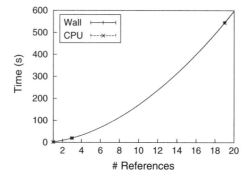

FIGURE 3.22 Scaling of the total wall and CPU time of the three-internal three-segment loops for $C_{15}H_{32}$ from single reference to CAS [4e,4o]. Full line is a fit of the wall time against $f(x) = a \cdot x^b$ with $a = 2.65693$ and $b = 1.8083$, dashed line is a fit against the CPU time with $a = 2.65453$ and $b = 1.80774$.

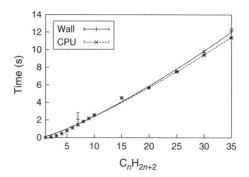

FIGURE 3.23 Scaling of the CPU and wall time of the three-internal four-segment loops with the size of the system. Full line is a fit of the wall time against $f(x) = a \cdot x^b$ with $a = 0.127988$ and $b = 1.27759$, dashed line is a fit against the CPU time with $a = 0.136562$ and $b = 1.24573$.

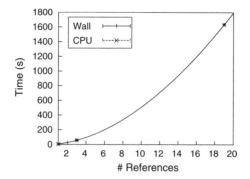

FIGURE 3.24 Scaling of the total wall and CPU time of the three-internal four-segment loops for $C_{15}H_{32}$ from single reference to CAS [4e,4o]. Full line is a fit of the wall time against $f(x) = a \cdot x^b$ with $a = 7.65376$ and $b = 1.82175$, dashed line is a fit against the CPU time with $a = 7.63234$ and $b = 1.82196$.

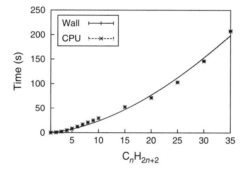

FIGURE 3.25 Scaling of the CPU and wall time of the two-internal four-segment loops with the size of the system. Full line is a fit of the wall time against $f(x) = a \cdot x^b$ with $a = 0.481148$ and $b = 1.69438$, dashed line is a fit against the CPU time with $a = 0.480805$ and $b = 1.69358$.

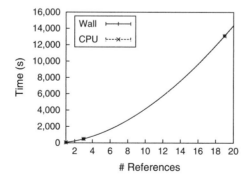

FIGURE 3.26 Scaling of the total wall and CPU time of the two-internal four-segment loops for $C_{15}H_{32}$ from single reference to CAS [4e,4o]. Full line is a fit of the wall time against $f(x) = a \cdot x^b$ with $a = 67.1057$ and $b = 1.79154$, dashed line is a fit against the CPU time with $a = 67.1055$ and $b = 1.79075$.

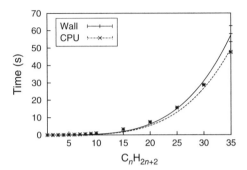

FIGURE 3.27 Scaling of the CPU and wall time of the two-internal three-segment loops with the size of the system. Full line is a fit of the wall time against $f(x) = a \cdot x^b$ with $a = 4.28565\text{e-}05$ and $b = 3.96736$, dashed line is a fit against the CPU time with $a = 3.70181\text{e-}05$ and $b = 3.96736$.

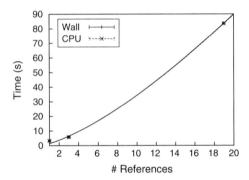

FIGURE 3.28 Scaling of the total wall and CPU time of the two-internal three-segment loops for $C_{15}H_{32}$ from single reference to CAS [4e,4o]. Full line is a fit of the wall time against $f(x) = a \cdot x^b$ with $a = 1.4224$ and $b = 1.3841$, dashed line is a fit against the CPU time with $a = 1.41405$ and $b = 1.38508$.

runtime, albeit irrelevant within our tested size regime. Again, both contributions scale sub-quadratically with respect to the number of references. Unlike the previous loops, the two-integral four-segment loops have a significant prefactor of almost 70, causing this part to contribute significantly to the total runtime: close to 20%. Two challenges can therefore be readily identified: First, we would like to reduce the scaling with respect to system size of the two-internal three-segment loops in order to reduce the asymptotic scaling of our framework. Second, the large prefactor of the two-internal four-segment loops warrants further optimization of this algorithmic part.

Unsurprisingly, the trend of increasing contribution to the total runtime with the number of external orbitals in the four-segment loops holds true for the three-external four-segment loops as depicted in Figures 3.29 to 3.30. Although this part scales favorably with respect to the system size [approximately $\mathcal{O}(\mathcal{N}^{2.5})$] and sub-quadratic

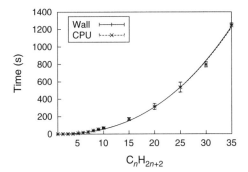

FIGURE 3.29 Scaling of the CPU and wall time of the three-external four-segment loops with the size of the system. Full line is a fit of the wall time against $f(x) = a \cdot x^b$ with $a = 0.186781$ and $b = 2.47377$, dashed line is a fit against the CPU time with $a = 0.198114$ and $b = 2.45425$.

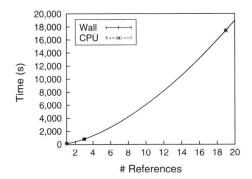

FIGURE 3.30 Scaling of the total wall and CPU time of the three-external four-segment loops for $C_{15}H_{32}$ from single reference to CAS [4e,4o]. Full line is a fit of the wall time against $f(x) = a \cdot x^b$ with $a = 138.695$ and $b = 1.64186$, dashed line is a fit against the CPU time with $a = 138.695$ and $b = 1.64109$.

with the number of references, the even larger prefactor causes this part to contribute significantly (up to 25% within the tested regime) to the overall runtime.

None of the parts discussed so far play an important role for the total runtime in the limit of a single reference and our medium-sized pentadecane test system. Also, none of the parts discussed so far showed a significant difference between the CPU and the wall time, showing that our buffering library effectively took care of any I/O bottlenecks within the tested regimes using buffer sizes of 500 MB for the integrals and 50 MB for the execution buffer.

Next we consider the part of the SGGA with the highest number of external orbitals, the four-external four-segment loops. These, for the first time, contribute significantly for single-reference calculations (Fig. 3.31) with more than 20% of the wall time. Again the scaling with system size is almost quadratic and the scaling

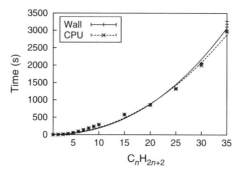

FIGURE 3.31 Scaling of the CPU and wall time of the four-external four-segment loops with the size of the system. Full line is a fit of the wall time against $f(x) = a \cdot x^b$ with $a = 1.09712$ and $b = 2.23242$, dashed line is a fit against the CPU time with $a = 1.65515$ and $b = 2.09877$.

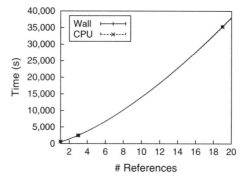

FIGURE 3.32 Scaling of the total wall and CPU time of the four-external four-segment loops for $C_{15}H_{32}$ from single reference to CAS [4e,4o]. Full line is a fit of the wall time against $f(x) = a \cdot x^b$ with $a = 511.649$ and $b = 1.43882$, dashed line is a fit against the CPU time with $a = 511.649$ and $b = 1.43772$.

with the number of references is sub-quadratic (Fig. 3.32). Nevertheless, the large prefactor causes this part to be a hotspot of our overall local correlation framework.

Summarizing the SGGA analysis, we can draw a couple of key conclusions. Since the SGGA can be decomposed into multiple parts of vastly varying difficulty, the prefactors vary by multiple orders of magnitude between the two-internal three-segment loops and the four-external four-segment loops. Likewise, the scaling with respect to system size varies from sub-quadratic to almost quartic for the different parts. Since the latter is only the case for the routines with the smallest prefactors, this only plays a role in the limit of very large systems that are not computationally feasible to study at the moment. Also, this analysis and detection of challenges within our SGGA engine will allow for targeted improvements in the scaling behavior, reducing or removing said bottlenecks in the future. On a positive note, our SGGA implementation already now scales sub-quadratically for all parts with respect to the number of references, making complicated local multireference calculations feasible at the present time.

3.9 SUMMARY AND OUTLOOK

We have reported the current status of our open source, local correlation framework, TigerCI. In particular, we identified and analyzed the numerical challenges arising within such an approach. We see potential for improvements in the integral decomposition scheme currently handled by a Cholesky decomposition, especially for the out-of-core case. Although the Cholesky decomposition exhibits the expected cubic scaling with respect to system size, the timing of this part of the algorithm is constant with respect to the number of references. Therefore, it loses importance with the inclusion of more references, playing only an important role in large single-reference applications.

Another challenge is the integral assembly and the efficient exploitation of the high symmetry exhibited by these integrals.

The core part of our algorithm, the SGGA engine, exhibits very different behavior in terms of scaling and prefactor depending upon which subpart is studied in detail. Although we found a formal $\mathcal{O}(\mathcal{N}^4)$ asymptotic scaling for the SGGA with system size, this holds true only for a small subset of the routines involving only a minor fraction of the total runtime. The SGGA parts that dominate in our relevant size regime scale cubically or better with system size. Also, the SGGA exhibits a very favorable sub-quadratic scaling with the number of references in all parts. In general, we are confident that our local correlation code will allow for routine simulations of large molecules with a large number of references.

Further algorithmic developments therefore will concentrate on solving the numerical challenges identified above, parallelizing the code in order to more effectively exploit modern systems and optimizing our local approximations to reduce the scaling of the code. Independent of these efforts, we envision to make TigerCI accessible for other researchers in the field, introducing it to the toolbox of computational chemistry.

Acknowledgments

This work has been funded in part by the National Science Foundation and the Office of Naval Research. JMD acknowledges a German academic exchange service (DAAD) fellowship. All calculations were carried out using Princeton's TIGRESS High Performance Computing resources.

REFERENCES

1. I. Shavitt and R. J. Bartlett, *Many-Body methods in Chemistry and Physics: MBPT and Coupled-Cluster Theory*. Cambridge: Cambridge University Press, 2009. [Online]. Available: http://books.google.com/books?id=gU1eHAAACAAJ

2. P. Pulay, "Localizability of dynamic electron correlation," *Chemical Physics Letters*, vol. 100, no. 2, pp. 151–154, 1983. [Online]. Available: http://dx.doi.org/10.1016/0009-2614(83)80703-9

3. S. Saebø and P. Pulay, "Local configuration interaction: An efficient approach for larger molecules," *Chemical Physics Letters*, vol. 113, no. 1, pp. 13–18, 1985. [Online]. Available: http://dx.doi.org/10.1016/0009-2614(85)85003-X

4. S. Saebø and P. Pulay, "Fourth-order Moller Plessett perturbation theory in the local correlation treatment. I. Method," *The Journal of Chemical Physics*, vol. 86, no. 2, p. 914–922, 1987. [Online]. Available: http://dx.doi.org/10.1063/1.452293

5. S. Saebø and P. Pulay, "The local correlation treatment. II. Implementation and tests," *The Journal of Chemical Physics*, vol. 88, no. 3, p. 1884–1890, 1988. [Online]. Available: http://dx.doi.org/10.1063/1.454111

6. S. Saebo and P. Pulay, "Local treatment of electron correlation," *Annual Review of Physical Chemistry*, vol. 44, no. 1, pp. 213–236, 1993. [Online]. Available: http://dx.doi.org/10.1146/annurev.pc.44.100193.001241

7. M. Schütz, G. Hetzer, and H.-J. Werner, "Low-order scaling local electron correlation methods. I. Linear scaling local MP2," *The Journal of Chemical Physics*, vol. 111, no. 13, p. 5691–5705, 1999. [Online]. Available: http://dx.doi.org/10.1063/1.479957

8. R. B. Murphy, M. D. Beachy, R. A. Friesner, and M. N. Ringnalda, "Pseudospectral localized Moller Plessett methods: Theory and calculation of conformational energies," *The Journal of Chemical Physics*, vol. 103, no. 4, p. 1481–1490, 1995. [Online]. Available: http://dx.doi.org/10.1063/1.469769

9. M. Schütz and H.-J. Werner, "Low-order scaling local electron correlation methods. IV. Linear scaling local coupled-cluster (LCCSD)," *The Journal of Chemical Physics*, vol. 114, no. 2, p. 661–681, 2001. [Online]. Available: http://dx.doi.org/10.1063/1.1330207

10. W. Li, P. Piecuch, and J. Gour, "Local correlation calculations using standard and renormalized coupled-cluster methods," in *AIP Conference Proceedings*, vol. 1102, p. 68. September 23–27, 2008, Shanghai, China. College park, MD: AIP Publishing, 2009. [Online]. Available: http://dx.doi.org/10.1063/1.3108393

11. S. Hoyau, D. Maynau, and J.-P. Malrieu, "A regionally contracted multireference configuration interaction method: General theory and results of an incremental version" *The Journal of Chemical Physics*, vol. 134, no. 5, p. 054125, 2011. [Online]. Available: http://dx.doi.org/10.1063/1.3533777

12. N. Ben Amor, F. Bessac, S. Hoyau, and D. Maynau, "Direct selected multireference configuration interaction calculations for large systems using localized orbitals" *The Journal of Chemical Physics*, vol. 135, no. 1, p. 014101, 2011. [Online]. Available: http://dx.doi.org/10.1063/1.3600351

13. D. Walter, A. Venkatnathan, and E. A. Carter, "Local correlation in the virtual space in multireference singles and doubles configuration interaction," *The Journal of Chemical Physics*, vol. 118, no. 18, p. 8127–8139, 2003. [Online]. Available: http://dx.doi.org/10.1063/1.1565314

14. D. Walter, A. B. Szilva, K. Niedfeldt, and E. A. Carter, "Local weak-pairs pseudospectral multireference configuration interaction," *The Journal of Chemical Physics*, vol. 117, no. 5, p. 1982–1993, 2002. [Online]. Available: http://dx.doi.org/10.1063/1.1487816

15. D. Walter and E. A. Carter, "Multi-reference weak pairs local configuration interaction: Efficient calculations of bond breaking," *Chemical Physics Letters*, vol. 346, no. 1–2, pp. 177–185, 2001. [Online]. Available: http://dx.doi.org/10.1016/S0009-2614(01)00966-6

16. A. Venkatnathan, A. B. Szilva, D. Walter, R. J. Gdanitz, and E. A. Carter, "Size extensive modification of local multireference configuration interaction" *The Journal of Chemical Physics*, vol. 120, no. 4, pp. 1693–704, 2004. [Online]. Available: http://dx.doi.org/10.1063/1.1635796

17. T. S. Chwee, A. B. Szilva, R. Lindh, and E. A. Carter, "Linear scaling multireference singles and doubles configuration interaction" *The Journal of Chemical Physics*, vol. 128, no. 22, p. 224106, 2008. [Online]. Available: http://dx.doi.org/10.1063/1.2937443

18. T. S. Chwee and E. A. Carter, "Cholesky decomposition within local multireference singles and doubles configuration interaction" *The Journal of Chemical Physics*, vol. 132, no. 7, p. 074104, 2010. [Online]. Available: http://dx.doi.org/10.1063/1.3315419

19. T. S. Chwee and E. A. Carter, "Density fitting of two-electron integrals in local multireference single and double excitation configuration interaction calculations," *Molecular Physics*, vol. 108, no. 19–20, pp. 2519–2526, 2010. [Online]. Available: http://dx.doi.org/10.1080/00268976.2010.508052

20. D. B. Krisiloff and E. A. Carter, "Approximately size extensive local multireference singles and doubles configuration interaction" *Physical Chemistry Chemical Physics*, vol. 14, no. 21, pp. 7710–7717, 2012. [Online]. Available: http://dx.doi.org/10.1039/C2CP23757A

21. P. Szalay, "Configuration interaction: Corrections for size-consistency," in *Encyclopedia of Computational Chemistry* (P. von Rague Schleyer, ed.). Chichester: Wiley Inter Science, 2005. [Online]. Available: http://dx.doi.org/10.1002/0470845015.cn0066

22. A. Szabó and N. S. Ostlund, *Modern Quantum Chemistry: Introduction to Advanced Electronic Structure Theory*. New York: McGraw-Hill, 1982. [Online]. Available: http://books.google.com/books?id=6mV9gYzEkgIC

23. E. Davidson, "The iterative calculation of a few of the lowest eigenvalues and corresponding eigenvectors of large real-symmetric matrices," *Journal of Computational Physics*, vol. 17, p. 87, 1975. [Online]. Available: http://dx.doi.org/10.1016/0021-9991(75)90065-0

24. W. Duch and J. Karwowski, "Symmetric group approach to configuration interaction methods," *Computer Physics Reports*, vol. 2, no. 3, pp. 93–170, 1985. [Online]. Available: http://dx.doi.org/10.1016/0167-7977(85)90001-2

25. W. Duch and J. Karwowski, "Symmetric group graphical approach to the direct configuration interaction method," *International Journal of Quantum Chemistry*, vol. 22, no. 4, pp. 783–824, 1982. [Online]. Available: http://dx.doi.org/10.1002/qua.560220411

26. W. Duch and J. Karwowski, "A multireference direct CI program based on the symmetric group graphical approach," *Theoretica Chimica Acta*, vol. 71, no. 2–3, pp. 187–199, 1987. [Online]. Available: http://dx.doi.org/10.1007/BF00526416

27. J. Pipek and P. G. Mezey, "A fast intrinsic localization procedure applicable for ab initio and semiempirical linear combination of atomic orbital wave functions," *The Journal of Chemical Physics*, vol. 90, no. 9, p. 4916, 1989. [Online]. Available: http://dx.doi.org/10.1063/1.456588

28. S. Boys, "Construction of some molecular orbitals to be approximately invariant for changes from one molecule to another," *Reviews of Modern Physics*, vol. 32, no. 2, p. 296, 1960. [Online]. Available: http://dx.doi.org/10.1103/RevModPhys.32.296

29. C. Edmiston and K. Ruedenberg, "Localized atomic and molecular orbitals," *Reviews of Modern Physics*, vol. 35, no. 3, p. 457, 1963. [Online]. Available: http://dx.doi.org/10.1103/RevModPhys.35.457

30. J. E. Subotnik, A. D. Dutoi, and M. Head-Gordon, "Fast localized orthonormal virtual orbitals which depend smoothly on nuclear coordinates," *The Journal of Chemical Physics*, vol. 123, no. 11, p. 114108, 2005. [Online]. Available: http://dx.doi.org/10.1063/1.2033687

31. M. Haser and R. Ahlrichs, "Improvements on the direct SCF method," *Journal of Computational Chemistry*, vol. 10, no. 1, pp. 104–111, 1989. [Online]. Available: http://dx.doi.org/10.1002/jcc.540100111

32. D. S. Lambrecht and C. Ochsenfeld, "Multipole-based integral estimates for the rigorous description of distance dependence in two-electron integrals," *The Journal of Chemical Physics*, vol. 123, no. 18, p. 184101, 2005. [Online]. Available: http://dx.doi.org/10.1063/1.2079967

33. S. A. Maurer, D. S. Lambrecht, D. Flaig, and C. Ochsenfeld, "Distance-dependent Schwarz-based integral estimates for two-electron integrals: Reliable tightness vs. rigorous upper bounds," *The Journal of Chemical Physics*, vol. 136, no. 14, p. 144107, 2012. [Online]. Available: http://dx.doi.org/10.1063/1.3693908

34. D. S. Lambrecht, B. Doser, and C. Ochsenfeld, "Rigorous integral screening for electron correlation methods" *The Journal of Chemical Physics*, vol. 123, no. 18, p. 184102, 2005. [Online]. Available: http://dx.doi.org/10.1063/1.2079987

35. N. H. F. Beebe and J. Linderberg, "Simplifications in the generation and transformation of two-electron integrals in molecular calculations," *International Journal of Quantum Chemistry*, vol. 12, no. 4, pp. 683–705, 1977. [Online]. Available: http://dx.doi.org/10.1002/qua.560120408

36. H. Koch, A. Sanchez de Meras, and T. B. Pedersen, "Reduced scaling in electronic structure calculations using Cholesky decompositions," *The Journal of Chemical Physics*, vol. 118, no. 21, p. 9481, 2003. [Online]. Available: http://dx.doi.org/10.1063/1.1578621

37. S. Hammarling, N. Higham, and C. Lucas, "LAPACK-style codes for pivoted Cholesky and QR updating," in *Applied Parallel Computing. State of the Art in Scientific Computing*, ser. Lecture Notes in Computer Science, (B. Kågström, E. Elmroth, J. Dongarra, and J. Wasniewski, eds.) Berlin: Springer, 2007, vol. 4699, pp. 137–146. [Online]. Available: http://dx.doi.org/10.1007/978-3-540-75755-9

38. N. J. Higham, "Analysis of the Cholesky decomposition of a semi-definite matrix," in *Reliable Numerical Computation*, (M. G. Cox and S. J. Hammarling, eds.) New York: Oxford University Press, 1990, pp. 161–185. [Online]. Available: http://eprints.ma.man.ac.uk/1193/

39. N. Béreux, "Out-of-core implementations of Cholesky factorization: Loop-based versus recursive algorithms," *SIAM Journal on Matrix Analysis and Applications*, vol. 30, no. 4, pp. 1302–1319, 2009. [Online]. Available: http://dx.doi.org/10.1137/06067256X

40. F. Aquilante, L. Boman, J. Boström, H. Koch, R. Lindh, A. S. de Merás, and T. B. Pedersen, *Linear-Scaling Techniques in Computational Chemistry and Physics*. Dordrecht: Springer, 2011. [Online]. Available: http://dx.doi.org/10.1007/978-90-481-2853-2

41. J. Dongarra, S. Hammarling, and D. Walker, "Key concepts for parallel out-of-core LU factorization," *Computers & Mathematics with Applications*, vol. 35, no. 7, pp. 13–31, 1998. [Online]. Available: http://dx.doi.org/10.1016/S0898-1221(98)00029-7

42. H. Harbrecht, M. Peters, and R. Schneider, "On the low-rank approximation by the pivoted Cholesky decomposition," *Applied Numerical Mathematics*, vol. 62, no. 4, pp. 428–440, 2012. [Online]. Available: http://dx.doi.org/10.1016/j.apnum.2011.10.001

43. H. Dachsel, H. Lischka, R. Shepard, J. Nieplocha, and R. J. Harrison, "A massively parallel multireference configuration interaction program: The parallel COLUMBUS program," *Journal of Computational Chemistry*, vol. 18, no. 3, pp. 430–448, 1997. [Online]. Available: http://dx.doi.org/10.1002/(SICI)1096-987X(199702)18:3<430::AID-JCC12> 3.3.CO;2-J

4

GENERALIZED VARIATIONAL THEOREM IN QUANTUM MECHANICS

MEL LEVY[1,2,3] AND ANTONIOS GONIS[4]

[1]*Department of Chemistry, Duke University, Durham, NC, USA*
[2]*Department of Physics, North Carolina A&T State University, Greensboro, NC, USA*
[3]*Department of Chemistry and Quantum Theory Group, School of Sciences and Engineering, Tulane University, New Orleans, LO, USA*
[4]*Physics and Life Sciences, Lawrence Livermore National Laboratory, Livermore, CA, USA*

4.1 INTRODUCTION

Solutions of the Schrödinger equation in quantum mechanics provide the necessary information for ascertaining the electronic structures of atoms, molecules, and solids. With this in mind, the well-known Rayleigh-Ritz variational principle allows the determination of the ground-state energy of the Hamiltonian describing a system of N interacting electrons through the minimization of the energy functional,

$$E_0 = \underbrace{\text{Min}}_{\Psi} E[\Psi] \tag{4.1}$$

where,

$$E[\Psi] = \frac{\langle \Psi | H | \Psi \rangle}{\langle \Psi | \Psi \rangle}, \tag{4.2}$$

Mathematical and Computational Modeling: With Applications in Natural and Social Sciences, Engineering, and the Arts, First Edition. Roderick Melnik.

with H denoting the Hamiltonian of the system, and where $\langle \Psi | H | \Psi \rangle$ denotes the expectation value given as an integral over the coordinates and summation over the spins of the N electrons in the system. The trial wave functions, $| \Psi \rangle$, belong to the antisymmetric part of the Hilbert space defined by the system of N particles.

In the conventional applications of the principle, the wave functions, $\Psi(x_1, x_2, \ldots, x_N)$, where x_i is a compound label for the spatial and spin coordinates of the ith electron, have as many coordinates as those defining the Hamiltonian. We clearly have

$$E[\Psi] \geq E_0, \tag{4.3}$$

where E_0 is the ground-state energy of the N-electron system.

Now, consider an electronic system with Hamiltonian, $H(\mathbf{r}_1, \mathbf{r}_2, \ldots, \mathbf{r}_N)$, expressed in terms of single-particle coordinates, $\mathbf{r}_j$, and an arbitrary, trial antisymmetric wave function of $N + M$ electrons, $\Psi(x_1, x_2, \ldots, x_{N+M})$. Note that the number of electrons in the wave function *exceeds* the number of electrons in the Hamiltonian. Nevertheless, we prove that the familiar variational inequality, (4.3), still applies. That is,

$$\int \cdots \int \Psi^*(x_1, x_2, \ldots, x_{N+M}) H(\mathbf{r}_1, \mathbf{r}_2, \ldots, \mathbf{r}_N)$$
$$\times \Psi(x_1, x_2, \ldots, x_{N+M}) dx_1 dx_2 \cdots dx_{N+M}$$
$$\geq E_0 \tag{4.4}$$

Observe that the integrations denote integrations over spatial coordinates and summations over the spins of all $N + M$ electrons, even though H contains only N electrons.

In the following sections, we provide two alternative proofs of the inequality in (4.4).

4.2 FIRST PROOF

We begin by expanding Ψ in the form of its *natural expansion* [1,3],

$$\Psi(x_1, x_2, \ldots, x_{N+M})$$
$$= \sum_{k=0}^{\infty} \Phi_k(x_1, x_2, \ldots, x_N) Q_k(x_{N+1}, \ldots, x_{N+M}), \tag{4.5}$$

where Φ_k and Q_k are complementary natural functions of N and M electrons, respectively. Given the eigenstates of the Hamiltonian,

$$H(\mathbf{r}_1, \mathbf{r}_2, \ldots, \mathbf{r}_N) \Psi_i(x_1, \ldots, x_N) = E_i \Psi_i(x_1, \ldots, x_N), \tag{4.6}$$

with

$$E_0 \leq E_1 \leq E_2 \ldots, \tag{4.7}$$

expand the $\Phi_k(x_1, x_2, \ldots, x_N)$ in Equation 4.5 in terms of $\Psi_i(x_1, \ldots, x_N)$,

$$\Phi_k = \sum_{i=0}^{\infty} c_{ik} \Psi_i, \tag{4.8}$$

where the c_{ik} are expansion coefficients.

The result is that Ψ can be expressed in the form

$$\begin{aligned}
\Psi(x_1, x_2, \ldots, x_{N+M}) \\
= \sum_{i=0}^{\infty} \Psi_i(x_1, \ldots, x_N) G_i(x_{N+1}, \ldots, x_{N+M}),
\end{aligned} \tag{4.9}$$

where

$$G_i = \sum_{k=0}^{\infty} c_{ik} Q_k. \tag{4.10}$$

The G_i are generalized M-electron Dyson functions. (The familiar Dyson orbitals are functions involving only one particle.) Now, from Equation 4.9, we have

$$\begin{aligned}
\langle \Psi(x_1, \ldots, x_{N+M}) | H(\mathbf{r}_1, \mathbf{r}_2, \ldots, \mathbf{r}_N) | \Psi(x_1, \ldots, x_{N+M}) \rangle \\
= \sum_{i=0}^{\infty} \Psi_i(x_1, \ldots, x_N) | H(\mathbf{r}_1, \mathbf{r}_2, \ldots, \mathbf{r}_N) | \Psi_i(x_1, \ldots, x_N) \rangle \\
\times \langle G_i(x_{N+1}, \ldots, x_{N+M}) G_i(x_{N+1}, \ldots, x_{N+M}) \rangle \\
= \sum_{i=0}^{\infty} E_i \langle G_i | G_i \rangle.
\end{aligned} \tag{4.11}$$

Because $\langle G_i | G_i \rangle \geq 0$ for all i, we have,

$$\langle \Psi | H | \Psi \rangle \geq E_0 \sum_{i=0}^{\infty} \langle G_i | G_i \rangle. \tag{4.12}$$

Moreover, from Equation 4.9 and by using the normalization of the wave function, $\langle \Psi | \Psi \rangle = 1$, and the orthonormality of the Ψ_i, we have

$$\sum_{i=0}^{\infty} \langle G_i | G_i \rangle = 1. \tag{4.13}$$

The combination of Equations 4.12 and 4.13 proves the desired result (Eq. 4.4).

4.3 SECOND PROOF

We express Ψ as

$$\Psi(x_1,\ldots,x_{N+M}) = D^{1/2}(x_{N+1},\ldots,x_{N+M})B(x_1,\ldots,x_N), \qquad (4.14)$$

where

$$D(x_{N+1},\ldots,x_{N+M}) = \int \cdots \int \Psi^*\Psi dx_1 \ldots dx_N. \qquad (4.15)$$

Since B is defined by the expression

$$B(x_1,\ldots,x_{N+M}) = \frac{\Psi(x_1,\ldots,x_{N+M})}{[\int \cdots \int \Psi^*\Psi dx_1 \ldots dx_N]^{1/2}}, \qquad (4.16)$$

and observing that the following integration extends only over $dx_1 \ldots dx_N$, it is important to note that

$$\int \cdots \int B(x_1,\ldots,x_{N+M})$$
$$\times B(x_1,\ldots,x_{N+M})dx_1 \ldots dx_N = 1. \qquad (4.17)$$

Now,

$$\int \cdots \int \Psi^*H\Psi dx_1 \ldots dx_N$$
$$= D \int \cdots \int BHBdx_1 \ldots dx_N. \qquad (4.18)$$

Note that D multiplies the integration because it involves only electrons $N+1$ through $N+M$.

Next, utilize the fact that

$$\int \cdots \int BHBdx_1 \ldots dx_N \geq E_0, \qquad (4.19)$$

because Equation 4.17 states that B is normalized to unity over the range of integration of electrons 1 through N.

The combination of Equations 4.18 and 4.19 gives

$$\int \cdots \int \Psi^*H\Psi dx_1 \ldots dx_N \geq DE_0. \qquad (4.20)$$

Equation 4.20 results, in part, because $D \geq 0$ for all values of its coordinates, $x_{N+1},\ldots,x_{N+M}$.

Next, integrate Equation 4.20 over the remaining coordinates, $x_{N+1},\ldots,x_{N+M}$, giving

$$\int\cdots\int \Psi^* H\Psi\, dx_1\ldots dx_{N+M}$$
$$\geq E_0 \int\cdots\int D\, dx_{N+1}\ldots dx_{N+M}. \tag{4.21}$$

Further, because $\langle\Psi|\Psi\rangle = 1$,

$$\int\cdots\int D\, dx_{N+1}\ldots dx_{N+M} = 1. \tag{4.22}$$

It, therefore, follows that

$$\int\cdots\int \Psi^* H\Psi\, dx_1\ldots dx_{N+M} \geq E_0, \tag{4.23}$$

which is the desired result.

4.4 CONCLUSIONS

This chapter reports a proof of a fundamental element of quantum mechanics: the generalization of the variational theorem for Hamiltonians of N-electron systems to wave functions of dimensions higher than N. There are a number of applications. For instance, it can be shown that the combination of Equation 4.4 with the fact that the trial wave functions are antisymmetric and H is symmetric enables one to derive bounds involving the kinetic energy and potential energy components of the total ground-state energy. These bounds are useful for helping one to obtain accurate approximations to these quantities. Finally, we note that the theorem derived here generalizes an earlier one that was restricted to the case $M = 1$, for which only a very brief outline of its proof was given [2].

Acknowledgments

Mel Levy thanks Professors Paul Ayers, Viktor Staroverov, Ian Hamilton, and Roderick Melnik, for their wonderful hospitality during his visit to Waterloo. Work at LLNL was performed under the auspices of the U.S. Department of Energy under Contract DE-AC52-07NA27344 with LLNS, LLC.

REFERENCES

1. B. C. Carlson and Joseph M. Keller. Eigenvalues of density matrices. *Phys. Rev.*, 121:659–661, 1961.
2. Mel Levy and Andreas Görling. Bounds for the exchange and correlation potentials. *Phys. Rev. A*, 51:2851–2856, 1995.
3. Per-Olov Löwdin. Quantum theory of many-particle systems. I. Physical interpretations by means of density matrices, natural spin-orbitals, and convergence problems in the method of configurational interaction. *Phys. Rev.*, 97:1474–1489, 1955.

SECTION 3

MATHEMATICAL AND STATISTICAL MODELS IN LIFE AND CLIMATE SCIENCE APPLICATIONS

5

A MODEL FOR THE SPREAD OF TUBERCULOSIS WITH DRUG-SENSITIVE AND EMERGING MULTIDRUG-RESISTANT AND EXTENSIVELY DRUG-RESISTANT STRAINS

Julien Arino[1] and Iman A. Soliman[2]

[1]*Department of Mathematics, University of Manitoba, Winnipeg, Canada*
[2]*Department of Mathematics, Cairo University, Giza, Egypt*

5.1 INTRODUCTION

Tuberculosis (TB) is a global issue, being the second highest cause of infectious disease-induced mortality after HIV/AIDS [21]. It is a disease of poverty that strikes mostly vulnerable populations [24]. If treatment is available and treatment regimens are followed seriously, most individuals recover. The same is not true of individuals with active TB who are not treated; in this case, tuberculosis is fatal in up to 50% of cases [16]. This further accentuates inequalities when facing the disease.

Because of the immense impact it has had on society for hundreds of years, TB has been the object of a considerable volume of work. The complexity of TB transmission and the diversity of patient life histories it involves, in particular the potential lifelong incubation period, mean that TB has been the object of many mathematical modeling

Mathematical and Computational Modeling: With Applications in Natural and Social Sciences, Engineering, and the Arts, First Edition. Roderick Melnik.

studies. It is beyond the scope of the present work to review these studies; see, for example, Refs. [1, 11, 25] and the references therein.

One aspect of TB epidemiology that has recently become very important is concernsdrug resistance. As often with the use of drugs, selective pressure on *Mycobacterium tuberculosis* (the causative agent of TB) due to the use of antituberculosis drugs has led to the emergence of antituberculosis drugs-resistant strains [26]. The situation further evolved in recent years, with the detection of mycobacteria in the 1990s, resistant to more than one of the drugs typically used to combat the infection. A *M. tuberculosis* strain is called multidrug resistant (MDR) if it is resistant at least to isoniazid and rifampicin [9]. The incidence of MDR-TB is not homogeneous. In a 2004 study [17], it was noted that the incidence of MDR-TB was generally low save for a few hot spots in China, Estonia, Latvia, and Russia. In 2010, the situation remained similar, with more countries reporting imported cases [31]. Further evolutions in drug resistance were noted in the 2000s, with mycobacteria resistant to second-line antituberculosis drugs [27]. The definition of extensively drug resistant (XDR) strains was then specified to consist of strains resistant to rifampicin and isoniazid, any fluoroquinolone, and one of the three injectable drugs, capreomycin, kanamycin, and amikacin [9]; see in particular the reviews in Refs. [18, 22]. Drug-resistant TB (M/XDR-TB for short) makes it a considerable challenge to control TB, since treatment is less efficacious for a patient infected with MDR-TB [18] and can even be unsuccessful for patients with XDR-TB.

Although several mathematical models have considered multiple strains of TB (see a review in Ref. [14]), few consider explicitly MDR- and XDR-TB and their emergence in a population as a consequence of treatment. The model in Ref. [4] uses a variation on previous models for TB and considers both nosocomial and community propagation. The models in Refs. [6, 13] are in the spirit of the model presented here: they consider multiple strains of TB and the evolution between these strains. They were, however, the object of little analytical work. The model in Ref. [7] considers a simple mechanism for the emergence of resistance: individuals with active drug-sensitive TB are treated at the rate ϕ; of those, a proportion r develops drug-resistant TB because of treatment failure (the remaining $1 - r$ are removed from the system).

In the present work, we formulate a model for the spread of drug-resistant (MDR and XDR) tuberculosis in a population. We assume that drug resistance can emerge as a consequence of treatment. The model is derived from earlier models given in Ref. [10]. Conditions are investigated which lead to the existence of a so-called backward bifurcation, where subthreshold endemic equilibria exist. The global stability of the disease-free equilibrium (DEF) is established when parameter values preclude the existence of a backward bifurcation.

5.1.1 Model formulation

The population of interest is divided into eight compartments depending on the epidemiological status of individuals. The number in each compartment at time t is given by the following variables.

1. $S(t)$ is the susceptible population, individuals who have never encountered TB.
2. $L_s(t)$ are the individuals infected with the drug-sensitive TB strain but who are in a latent stage, that is, who are neither showing symptoms nor infecting others.
3. $L_m(t)$ individuals are latently infected with MDR-TB.
4. $L_x(t)$ individuals are those who are latently infected with XDR-TB.
     Individuals in all three latent stages L_s, L_m, and L_x make up the so-called *latent tuberculosis infections* (LTBI). It is assumed that LTBI with a drug-sensitive strain are treated, while latent infections with multidrug-resistant or extensively resistant strains are not treated.
5. $I_s(t)$ are individuals infected with the drug-sensitive TB strain who are infectious to others (and most likely, showing symptoms as well).
6. $I_m(t)$ are those individuals who are infectious with the MDR-TB strain.
7. $I_x(t)$ individuals are infectious with the XDR-TB strain.
     Individuals in all three infectious stages I_s, I_m, and I_x make up the so-called *active TB* cases. All active TB cases are offered treatment.
8. $R(t)$ are those individuals for whom treatment was successful.

If this does not lead to ambiguities, we omit the time dependence of state variables. The total population N is given by

$$N = S + L_s + L_m + L_x + I_s + I_m + I_x + R.$$

We assume that flows between classes take the form indicated in the flow diagram in Figure 5.1. We formulate the model by reasoning as follows. Rather than stating all hypotheses at a time, we state them when they are needed. To simplify notation, we use the generic index $r \in \{s, m, x\}$ in state variables and parameters to refer to strains.

Susceptible population.   The evolution of the number of susceptible individuals in the population is governed by the following equation:

$$S' = b - dS - \beta_s \frac{SI_s}{N} - \beta_m \frac{SI_m}{N} - \beta_x \frac{SI_x}{N}, \tag{5.1a}$$

where b is the rate at which new individuals join the susceptible population (recruitment) and β_r are coefficients indicating the rates at which new infections arise given contacts between susceptible and infectious individuals in the different infectious classes. Note that incidence is here assumed to be proportional.

When a susceptible individual becomes infected with TB, that person leaves the S compartment and transitions to the LTBI or active TB compartments corresponding to the strain harbored by the individual that infected them, that is, to compartment L_r or I_r, respectively. A proportion λ_r of new infections by strain r transitions to LTBI compartment L_r, the remaining $1 - \lambda_r$ moves directly to infectious compartment I_r through so-called *fast infections*. This first infection with TB is called a *primary infection*.

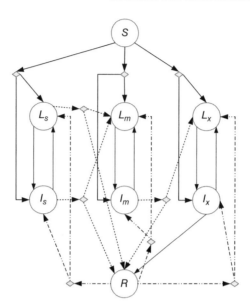

FIGURE 5.1 Simplified flow diagram of the model with drug-sensitive (L_s, I_s), MDR (L_m, I_m), and XDR (L_x, I_x) TB strains. Because of notational burden, birth and death are not shown and rates are not indicated. Diamonds indicate that the given flow is further divided between the indicated outcomes. Plain thick arrows indicate primary infections. Dash-dotted thick arrows indicate infections of previously successfully treated patients. Dotted arrows indicate treatment pathways that potentially lead to an increase in the resistance of the myobacteria in a given patient. Finally, thin plain arrows indicate other flows within the system, including exogenous reinfection.

Individuals latently infected with the drug-sensitive strain. The number $L_s(t)$ of drug-sensitive LTBI is increased by primary infections with the drug-sensitive strain I_s at the rate $\lambda_s \beta_s S I_s / N$ and by natural recovery of individuals in the drug-sensitive infectious compartment I_s at the per capita rate γ_s. Treatment, natural death, and natural progression to the infectious stage (due to a weakening immune system) also decrease the population in the L_s compartment.

Our model also incorporates *exogeneous reinfection*. Exogeneous reinfection is an important process in TB transmission; it happens when an individual already bearing the mycobacterium gets infected again [12, 30]. Here, we assume that two types of individuals can be subject to reinfection: latently infected and treated patients. Exogeneous reinfection of L_s individuals occurs because of contacts with individuals carrying the same strain I_s or a different strain I_m or I_x, following which the reinfected individual transitions from L_s to the corresponding infectious compartment, leading to a decrease of L_s at the rate $\alpha_{sr} \beta_r$. Similarly, a treated individual that comes in contact with an infectious individual can become reinfected. We assume that treatment reduces the probability of such a reinfection; we take the efficiency of treatment to be $1 - \sigma_s \in (0, 1)$. Thus, exogeneous reinfection of treated individuals increases L_s at the rate $\sigma_s \lambda_s \beta_s R I_s / N$.

$$L'_s = \lambda_s \beta_s (S + \sigma_s R)\frac{I_s}{N} + \gamma_s I_s - \{d + \varepsilon_s + t_{\ell s}\} L_s$$
$$- (\alpha_{ss}\beta_s I_s + \alpha_{sm}\beta_m I_m + \alpha_{sx}\beta_x I_x)\frac{L_s}{N}. \tag{5.1b}$$

Individuals latently infected with the M/XDR-TB strains. As with the drug-sensitive case, the number L_m of individuals latently infected with MDR-TB increases when individuals in the S compartment are infected with MDR-TB. Reinfection of L_s individuals by an MDR-TB strain occurs at the rate $\alpha_{sm}\beta_m L_s I_m/N$, decreasing the population in L_s at that rate and increasing that in L_m at the rate $\lambda_m \alpha_{sm}\beta_m L_s I_m/N$, with the remaining $(1 - \lambda_m)\alpha_{sm}\beta_m L_s I_m/N$ making a fast transition to I_m.

Individuals also become latently infected with the MDR-TB strain when they develop resistance to drugs. This occurs to individuals infected with the drug-sensitive strain at rates $(1-p_1)t_{\ell s}$ and $(1-p_2)t_{is}$ for latently infected and infectious individuals, respectively. The number of individuals in L_m decreases because of reinfection with I_x and exogenous reinfection at the rates $\alpha_{mx}\beta_x$ and $\alpha_{mm}\beta_m$, respectively. Note that we assume that L_m individuals, being already infected by an MDR-TB strain, cannot be exogenously reinfected by an individual carrying a drug sensitive strain, as this would "downgrade" the strain they are carrying.

We assume that while treatment is offered to drug-sensitive strain carriers both in the LTBI and active TB stages, it is only offered to MDR-TB and MDR-TB-infected individuals with active TB, not to those with LTBI. Contrary to other treatment rates, the rate t_{ix} of treatment of active XDR TB infections is the rate of successful treatment, not just the rate of treatment.

The rate of change of L_m is then given by

$$L'_m = \lambda_m \beta_m (S + \sigma_m R)\frac{I_m}{N} + \lambda_m \alpha_{sm}\beta_m \frac{L_s I_m}{N}$$
$$+ \gamma_m I_m + (1 - p_1)t_{\ell s}L_s + (1 - p_2)t_{is}I_s \tag{5.1c}$$
$$- (\alpha_{mm}\beta_m I_m + \alpha_{mx}\beta_x I_x)\frac{L_m}{N} - \{d + \varepsilon_m\} L_m$$

and the rate of change of L_x is

$$L'_x = \lambda_x \beta_x (S + \sigma_x R)\frac{I_x}{N} + \lambda_x \beta_x (\alpha_{sx}L_s + \alpha_{mx}L_m)\frac{I_x}{N}$$
$$+ \gamma_x I_x + (1 - p_3)t_{im}I_m - \alpha_{xx}\beta_x \frac{L_x I_x}{N} - \{d + \varepsilon_x\} L_x. \tag{5.1d}$$

Individuals infectious with drug-sensitive TB. To describe the rate of change of the number of individuals in infectious compartment I_s, we note that natural recovery, natural death, death due to TB, and failure of treatment that causes resistance to drugs in I_s are the only reasons to leave I_s at rates γ_s, d_s, δ_s, and t_s, respectively. All other flows, that is, exogenous reinfection in L_s, fast infection in S or R, and individuals who become infectious in L_s, feed into I_s at rate $\alpha_{ss}\beta_s$, $(1-\lambda_s)\beta_s$ and ε_s respectively.

$$I'_s = (1-\lambda_s)\beta_s(S+\sigma_s R)\frac{I_s}{N} + \alpha_{ss}\beta_s\frac{L_s I_s}{N} + \varepsilon_s L_s$$
$$- \{d+\delta_s+t_{is}+\gamma_s\}I_s. \tag{5.1e}$$

Individuals infectious with M/XDR-TB. Exogenous reinfection in L_m, fast infection or reinfection in S, R or L_s, and individuals who become infectious in L_m feed into I_m at the corresponding rates; see Table 5.1. Natural recovery, natural death, death

TABLE 5.1 Model parameters

Parameter	Interpretation
	Demography
b	Birth/recruitment rate
d	*per capita* natural death rate
	Disease dynamics
β_r	Transmission coefficient for strain r
λ_r	Proportion of newly infected individuals developing LTBI with strain r
$1-\lambda_r$	Proportion of newly infected individuals progressing to active TB with strain r due to fast infection
ε_r	*per capita* rate of endogenous reactivation of L_r
$\alpha_{r_1 r_2}$	Proportion of exogenous reinfection of L_{r_1} due to contact with I_{r_2}
γ_r	*per capita* rate of natural recovery to the latent stage L_r
δ_r	*per capita* rate of death due to TB of strain r
	Treatment related
$t_{\ell s}$	*per capita* rate of treatment for L_s
t_{ir}	*per capita* rate of treatment for I_r. Note that t_{ix} is the rate of successful treatment of I_x
$1-\sigma_r$	Efficiency of treatment in preventing infection with strain r
p_1	Probability of treatment success for L_s
$1-p_1$	Proportion of treated L_s moved to L_m due to incomplete treatment or lack of strict compliance in the use of drugs
p_2	Probability of treatment success for I_s
$1-p_2$	Proportion of treated I_s moved to L_m due to incomplete treatment or lack of strict compliance in the use of drugs
p_3	Probability of treatment success for I_m
$1-p_3$	Proportion of treated I_m moved to L_x due to incomplete treatment or lack of strict compliance in the use of drugs

The notation $r, r_1, r_2 \in \{s,m,x\}$ is used.

due to TB, and failure in treatment that causes resistance to drugs in I_m lead to the decrease in I_m.

$$I'_m = (1 - \lambda_m)\beta_m(S + \sigma_m R)\frac{I_m}{N} + \alpha_{mm}\beta_m\frac{L_m I_m}{N} + (1 - \lambda_m)\beta_m\alpha_{sm}\frac{L_s I_m}{N}$$
$$+ \varepsilon_m L_m - \{d + \delta_m + t_{im} + \gamma_m\}I_m. \tag{5.1f}$$

Similarly, the rate of change of I_x is given by

$$I'_x = \alpha_{xx}\beta_x\frac{L_x I_x}{N} + (1 - \lambda_x)\beta_x\left(\frac{SI_x}{N} + \sigma_x\frac{RI_x}{N} + \alpha_{sx}\frac{L_s I_x}{N} + \alpha_{mx}\frac{L_m I_x}{N}\right)$$
$$+ \varepsilon_x L_x - \{d + \delta_x + t_{ix} + \gamma_x\}I_x. \tag{5.1g}$$

Treated individuals. Finally the rate of change of R depends positively on the proportion of individuals in L_s, I_s, I_m, and I_x who successfully got treated and negatively on reinfection with the sensitive, MDR and XDR strains, and natural death.

$$R' = p_1 t_{\ell s} L_s + p_2 t_{is} I_s + p_3 t_{im} I_m + t_{ix} I_x$$
$$- (\sigma_s \beta_s I_s + \sigma_m \beta_m I_m + \sigma_x \beta_x I_x)\frac{R}{N} - dR. \tag{5.1h}$$

Table 5.1 lists all parameters and their interpretation. Model (5.1) is considered together with nonnegative initial conditions.

5.1.2 Mathematical Analysis

To simplify notation, define $\mathcal{X} := (S, L_s, L_m, L_x, I_s, I_m, I_x, R)^T$. Where needed, we write x_i, $i = 1, \ldots, 8$, the components of $\mathcal{X}$ (with the order the same as that in $\mathcal{X}$). We denote

$$\mathcal{I} := (L_s, L_m, L_x, I_s, I_m, I_x)^T$$

the infected variables.

5.1.2.1 Basic properties of solutions

Proposition 5.1 *Given nonnegative initial conditions, solutions to (5.1) exist and are unique for all $t \geq 0$. Furthermore, the positive orthant $\mathbb{R}^8_+$ is positively invariant under the flow of (5.1).*

Proof: Since the vector field in (5.1) consists of sums of constants and rational polynomial functions of the state variables and that we show later that the total population N is positive, it is differentiable. Hence solutions to (5.1) exist and are unique.

To prove the nonnegativity of solutions, first consider S; setting $S = 0$ in (5.1a), we get

$$S' = b > 0.$$

This implies that for nonnegative initial conditions, $S(0) \geq 0$, $S(t)$ is positive for all $t > 0$. For all other state variables, the vector field is nonnegative on the boundary of the orthant. It follows that solutions remain nonnegative for nonnegative initial conditions. As $S(t) > 0$ for all $t > 0$, we have $N(t) > 0$ for all $t > 0$.

From now on, we assume that $S(0) > 0$. Note that it is also easy to show that if initial conditions are positive, solutions remain positive for all t.

Proposition 5.2 *Given nonnegative initial conditions, solutions to (5.1) are bounded for all $t \geq 0$. Furthermore, the closed set*

$$\Omega := \left\{ \mathcal{X} \in \mathbb{R}^8_+ : S + L_s + L_m + L_x + I_s + I_m + I_x + R \leq \frac{b}{d} \right\} \tag{5.2}$$

attracts the flow of (5.1) for any initial condition in $\mathbb{R}^8_+$.

Proof: To establish boundedness, remark that the rate of change of the total population is given by

$$N' = b - dN - \delta_s I_s - \delta_m I_m - \delta_x I_x \leq b - dN. \tag{5.3}$$

This implies that $N(t)$ is bounded above by solutions of the differential equation $\Psi' = b - d\Psi$, that is, $N(t) \leq \max(\Psi(0), b/d)$, with, for all sufficiently large t, $N(t) \leq b/d$. Whence, since $\mathcal{X}$ is nonnegative, $\mathcal{X}$ is also bounded. Now consider Ω given by (5.2). We have that Ω is positively invariant. Moreover, for any solution outside Ω, that is, $N \geq b/d$, by (5.3), $N' < 0$. Thus Ω attracts all solutions of (5.1) with initial condition in $\mathbb{R}^8_+$.

5.1.2.2 Nature of the disease-free equilibrium   The system is at an equilibrium if the time derivatives in (5.1) are zero. An equilibrium is a DFE if $\mathcal{I} = 0$. From (5.1h), this implies that $R = 0$. Thus, at a DFE, (5.1) is such that $S = N = b/d$; the DFE is unique and given by

$$\mathcal{E}_{\mathrm{DFE}} = \left(\frac{b}{d}, 0, 0, 0, 0, 0, 0, 0 \right). \tag{5.4}$$

5.1.2.3 Local asymptotic stability of the DFE   The local asymptotic stability of the DFE is investigated using the next-generation method [15, 29]. The aim of the method is to produce a number, the *basic reproduction number*, usually denoted $\mathcal{R}_0$, that governs the local asymptotic stability of the DFE. To derive a formula for $\mathcal{R}_0$ using the next-generation method, we follow Ref. [29] and write the dynamics of the infected classes $\mathcal{I}$ as $\mathcal{I}' = \mathcal{F} - \mathcal{V}$, where $\mathcal{F}$ has the new infections into the infected classes and here takes the form

$$
\mathcal{F} := \begin{pmatrix}
\lambda_s \beta_s (S + \sigma_s R)\dfrac{I_s}{N} \\
\lambda_m \beta_m (S + \sigma_m R)\dfrac{I_m}{N} \\
\lambda_x \beta_x (S + \sigma_x R)\dfrac{I_x}{N} \\
(1-\lambda_s)\beta_s (S + \sigma_s R)\dfrac{I_s}{N} \\
(1-\lambda_m)\beta_m (S + \sigma_m R)\dfrac{I_m}{N} \\
(1-\lambda_x)\beta_x (S + \sigma_x R)\dfrac{I_x}{N}
\end{pmatrix}.
$$

The vector $-\mathcal{V}$ (not shown here) has all other flows within and out of the infected classes $\mathcal{I}$. The matrix of new infections F and the matrix of transfers between compartments V are the Jacobian matrices obtained by taking the Fréchet derivatives of $\mathcal{F}$ and $\mathcal{V}$ with respect to the infected variables $\mathcal{I}$ and evaluating them at the DFE. They take the form

$$
F = \begin{pmatrix} 0 & F_{12} \\ 0 & F_{22} \end{pmatrix}, \qquad V = \begin{pmatrix} V_{11} & V_{12} \\ V_{21} & V_{22} \end{pmatrix}, \tag{5.5}
$$

where

$$
F_{12} = \begin{pmatrix} \lambda_s \beta_s & 0 & 0 \\ 0 & \lambda_m \beta_m & 0 \\ 0 & 0 & \lambda_x \beta_x \end{pmatrix},
$$

$$
F_{22} = \begin{pmatrix} (1-\lambda_s)\beta_s & 0 & 0 \\ 0 & (1-\lambda_m)\beta_m & 0 \\ 0 & 0 & (1-\lambda_x)\beta_x \end{pmatrix},
$$

$$
V_{11} = \begin{pmatrix} d+\varepsilon_s+t_{\ell s} & 0 & 0 \\ -(1-p_1)t_{\ell s} & d+\varepsilon_m & 0 \\ 0 & 0 & d+\varepsilon_x \end{pmatrix},
$$

$$
V_{12} = \begin{pmatrix} -\gamma_s & 0 & 0 \\ -(1-p_2)t_{is} & -\gamma_m & 0 \\ 0 & -(1-p_3)t_{im} & -\gamma_x \end{pmatrix},
$$

$$V_{21} = \begin{pmatrix} -\varepsilon_s & 0 & 0 \\ 0 & -\varepsilon_m & 0 \\ 0 & 0 & -\varepsilon_x \end{pmatrix},$$

$$V_{22} = \begin{pmatrix} d+\delta_s+t_{is}+\gamma_s & 0 & 0 \\ 0 & d+\delta_m+t_{im}+\gamma_m & 0 \\ 0 & 0 & d+\delta_x+t_{ix}+\gamma_x \end{pmatrix}.$$

Then the basic reproduction number $\mathcal{R}_0$ for system (5.1) is the spectral radius of the next-generation matrix FV^{-1} and is given by

$$\mathcal{R}_0 = \rho(FV^{-1}) = \max(\mathcal{R}_{0s}, \mathcal{R}_{0m}, \mathcal{R}_{0x}), \tag{5.6}$$

where

$$\mathcal{R}_{0s} = \frac{\beta_s(\varepsilon_s + (1-\lambda_s)(d+t_{\ell s}))}{(\varepsilon_s + d + t_{\ell s})(t_{is}+\delta_s+d)+\gamma_s(t_{\ell s}+d)}$$

$$\mathcal{R}_{0m} = \frac{\beta_m(\varepsilon_m + (1-\lambda_m)d)}{(\varepsilon_m + d)(t_{im}+\delta_m+d)+d\gamma_m}$$

and

$$\mathcal{R}_{0x} = \frac{\beta_x(\varepsilon_x + (1-\lambda_x)d)}{(\varepsilon_x + d)(t_{ix}+\delta_x+d)+d\gamma_x}$$

are the basic reproduction numbers for the drug-sensitive, MDR and XDR strains, respectively.

The method in Ref. [29] thus transforms the problem of local asymptotic stability of the DFE of (5.1) into that of the local asymptotic stability of $\mathcal{I} = 0$ in the reduced model $\mathcal{I}' = \mathcal{F} - \mathcal{V}$. The linearization of the latter problem at $\mathcal{I} = 0$, with the noninfected variables (S and R here) taking their values at the DFE, then leads to the linear system $\mathcal{I}' = (F - V)\mathcal{I}$. It is proved in Ref. [29] that for matrices F and V obtained with this method, there holds that

$$\max\{\Re(\lambda), \lambda \in \mathrm{Sp}(F - V)\} < 0 \Leftrightarrow \max\{|\lambda|, \lambda \in \mathrm{Sp}(FV^{-1})\} < 1,$$

where $\mathrm{Sp}(M)$ is the spectrum of matrix M. Thus the local asymptotic stability of the DFE is governed by the location inside the complex unit ball of the eigenvalues of FV^{-1}. This is summarized in the next result.

Lemma 5.1   *The DFE (5.4) of (5.1) is locally asymptotically stable if $\mathcal{R}_0 < 1$ and unstable if $\mathcal{R}_0 > 1$, where $\mathcal{R}_0$ is defined by (5.6).*

5.1.2.4   Existence of subthreshold endemic equilibria   Lemma 5.1 establishes conditions under which the DFE is locally asymptotically stable and unstable. This does not provide a full picture of the behavior near $\mathcal{R}_0 = 1$, though. Indeed, the direction and stability of the branch of equilibria that bifurcates at $\mathcal{R}_0 = 1$ are unknown without further analysis.

The classic situation is depicted in Figure 5.2, with a negative (and therefore biologically irrelevant) equilibrium entering the positive orthant at $\mathcal{R}_0 = 1$ and exchanging stability with the DFE through a transcritical bifurcation. The situation depicted in Figure 5.3 has come to be known as a "backward bifurcation" after the seminal work of Hadeler and van and Driessche [19]. In this case, as $\mathcal{R}_0$ increases from small values, the system first undergoes a saddle-node bifurcation when $\mathcal{R}_0 = \mathcal{R}_c$, where $\mathcal{R}_c < 1$ is some critical value. Then as $\mathcal{R}_0$ continues to increase, the lower unstable branch of equilibria

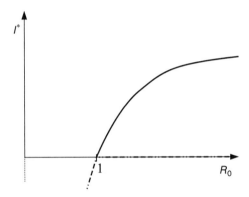

FIGURE 5.2   Forward bifurcation. The two most common bifurcation scenario at $\mathcal{R}_0 = 1$, where the value of $I^* = \|\mathcal{I}^*\|$ is plotted as a function of $\mathcal{R}_0$. Thick continuous lines indicate that the equilibrium is locally asymptotically stable and dashed lines indicate an unstable equilibrium.

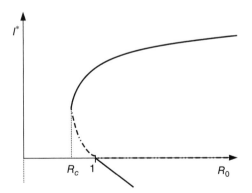

FIGURE 5.3   Backward bifurcation. The two most common bifurcation scenario at $\mathcal{R}_0 = 1$, where the value of $I^* = \|\mathcal{I}^*\|$ is plotted as a function of $\mathcal{R}_0$. Thick continuous lines indicate that the equilibrium is locally asymptotically stable and dashed lines indicate an unstable equilibrium.

undergoes a transcritical bifurcation and leaves the positive orthant. Castillo-Chavez and Song [11] and van den Driessche and Watmough [29] have provided methods to characterize the direction of the bifurcation that takes place at $\mathcal{R}_0 = 1$.

The model from which the current system is derived is studied in Ref. [10]. In that case, it is shown that under certain conditions, the system undergoes a backward bifurcation. As the present model is a refinement of this model, such behavior can be expected here.

In the absence of exogeneous reinfection by the XDR-TB strain ($\alpha_{xx} = 0$) or when the proportion $1 - \lambda_x$ of infected individuals making a fast transition to I_x is large, there is no backward bifurcation. Otherwise, conditions can be obtained under which such a bifurcation occurs. This is shown in the following theorem.

Theorem 5.1   *If $\alpha_{xx} = 0$ or*

$$\alpha_{xx} \leq 1 - \lambda_x, \tag{5.7}$$

then (5.1) undergoes a forward bifurcation at $\mathcal{R}_0 = 1$. Otherwise, the model has a backward bifurcation at $\mathcal{R}_0 = 1$ if

$$(\alpha_{xx} + \lambda_x - 1)(\lambda_x \beta_x + \gamma_x)d \left(1 - \frac{\varepsilon_x}{(\varepsilon_x + d)^2} + \frac{\varepsilon_x}{(\varepsilon_x + d)} \right)$$
$$> \lambda_x(1 - \sigma_x) \left(\frac{\varepsilon_x}{(\varepsilon_x + d)} + t_{ix} \right) + (1 - \lambda_x)d \tag{5.8}$$

Proof: The proof uses the center manifold techniques of Refs. [11, 29]. Consider the model when $\mathcal{R}_0 = 1$, and using β_x as the bifurcation parameter, then

$$\beta_x = \frac{d^2 + (t_{ix} + \delta_x + \varepsilon_x + \gamma_x)d + \varepsilon_x(t_{ix} + \delta_x)}{(\varepsilon_x + (1 - \lambda_x)d)}. \tag{5.9}$$

Checking the eigenvalues of the Jacobian of model (5.1) evaluated at the DFE, $\mathcal{E}^*$, and β_x shows that 0 is a simple eigenvalue and all other eigenvalues have a negative real parts. Hence we can use Ref. [11, theorem 4.1]. The Jacobian of model (5.1) has a right eigenvector $\mathbf{w}$ (corresponding to the zero eigenvalue) given by

$$\mathbf{w} = \left[-\frac{k\beta_x}{d}, 0, 0, \frac{k(\lambda_x \beta_x + \gamma_x)}{d + \varepsilon_x}, 0, 0, k > 0, \frac{kt_{ix}}{d} \right]^T, \tag{5.10}$$

and a left eigenvector $\mathbf{v}$ given by

$$\mathbf{v} = \left[-\frac{k\beta_x}{d}, 0, 0, \frac{k(\lambda_x \beta_x + \gamma_x)}{d + \varepsilon_x}, 0, 0, k > 0, \frac{kt_{ix}}{d} \right]^T, \tag{5.11}$$

where k is a positive parameter.

To use Ref. [11, theorem 4.1], it is convenient to write the vector field in (5.1) as

$$\mathcal{X}' = (f_1, f_2, f_3, f_4, f_5, f_6, f_7, f_8)^T.$$

Using this notation, the parameter a used in Ref. [11, theorem 4.1] takes the form

$$
\begin{aligned}
a := {} & \sum_{i,j,k=1}^{8} v_k \varepsilon_i \varepsilon_j \frac{\partial^2 f_k}{\partial x_i \partial x_j}(\mathcal{E}^*, \beta_x), \\
= {} & \frac{2v_7 k^2 \beta_x d}{b}\left[-(\lambda_x \beta_x + \gamma_x)\frac{\lambda_x + \alpha_{xx}}{(d + \varepsilon_x)^2} - \frac{\lambda_x}{(d + \varepsilon_x)} - (\lambda_x \beta_x + \gamma_x)\frac{1 - \lambda_x - \alpha_{xx}}{(d + \varepsilon_x)} \right. \\
& \left. - t_{ix}\lambda_x \beta_x \frac{1 - \sigma_x}{d(d + \varepsilon_x)} - (1 - \lambda_x) - t_{ix}(1 - \lambda_x)\frac{1 - \sigma_x}{d} \right],
\end{aligned}
$$

which is strictly negative if $\alpha_{xx} = 0$ or inequality (5.7) holds. The parameter b in Ref. [11, theorem 4.1] takes the form

$$
b = \sum_{i,k=1}^{8} v_k \varepsilon_i \frac{\partial^2 f_k}{\partial x_i \partial \beta_x}(\mathcal{E}^*, \beta_x) = v_7 k \frac{d(1 - \lambda_x) + \varepsilon_x}{d + \varepsilon_x} > 0.
$$

Therefore, by Ref. [11, theorem 4.1], (5.1) has a forward bifurcation at $\mathcal{R}_0 = 1$. If condition (5.7) is broken, then (5.1) undergoes a backward bifurcation if condition (5.8) holds.

Theorem 5.1 shows that (5.1) undergoes a backward bifurcation only if the XDR strain itself undergoes a backward bifurcation, regardless of the type of bifurcations of the other two strains. Indeed, as can be noticed in the model, there is directed movement of individuals between the strains, starting with drug-sensitive TB and terminating with the XDR strain. Because of that movement, whether or not the first two strains are in backward bifurcation, the whole model develops a backward bifurcation only if the terminal strain undergoes a backward bifurcation.

5.1.2.5 *Global stability of the DFE when the bifurcation is "forward"* We now investigate the global asymptotic stability of the DFE under conditions that preclude a backward bifurcation.

Theorem 5.2 *Assume that*

$$0 \leq \alpha_{xx} \leq 1 - \lambda_x, \tag{A_1}$$

$$0 \leq \alpha_{mm} \leq 1 - \lambda_m, \tag{A_2}$$

$$0 \leq \alpha_{ss} \leq 1 - \lambda_s. \tag{A_3}$$

Then the DFE (5.4) of (5.1) is globally asymptotically stable when $\mathcal{R}_0 < 1$.

Proof: We prove the global stability of the DFE by showing that if $\mathcal{R}_0 < 1$, then $\lim_{t \to \infty} X(t) = \mathcal{E}_{\mathrm{DFE}}$. The attractivity of $\mathcal{E}_{\mathrm{DFE}}$, together with the local asymptotic stability of the DFE implied by the fact that $\mathcal{R}_0 < 1$, gives the result.

Since (5.1) is not of type K, a standard comparison theorem is not applicable. Let $\tau_n \to \infty$ be a sequence such that $L_s(\tau_n) \to L_s^\infty := \limsup_{t \to \infty} L_s(t)$. Thus, $L_s(\tau_n)' \to 0$ using lemma 2.1 in Ref. [28]. Then Equation 5.1b gives

$$0 = \lambda_s \beta_s \frac{S(\tau_n) + \sigma_s R(\tau_n)}{N(\tau_n)} I_s(\tau_n) - \alpha_{ss} \beta_s \frac{L_s(\tau_n)}{N(\tau_n)} I_s(\tau_n) - \alpha_{sm} \beta_m \frac{I_m(\tau_n)}{N(\tau_n)} L_s(\tau_n)$$

$$- \alpha_{sx} \beta_x \frac{I_x(\tau_n)}{N(\tau_n)} L_s(\tau_n) - \{d + \varepsilon_s + t_{\ell s}\} L_s^\infty + \gamma_s I_s(\tau_n)$$

$$\leq \lambda_s \beta_s \frac{S(\tau_n) + \sigma_s R(\tau_n)}{N(\tau_n)} I_s(\tau_n) - \{d + \varepsilon_s + t_{\ell s}\} L_s^\infty + \gamma_s I_s(\tau_n).$$

Since $\frac{S(t) + \sigma_s R(t)}{N(t)} < 1$ and $I_s(t) \leq I_s^\infty$ at any t, it follows that

$$L_s^\infty \leq \frac{\lambda_s \beta_s + \gamma_s}{d + \varepsilon_s + t_{\ell s}} I_s^\infty. \tag{5.12}$$

Now let $s_n \to \infty$ be a sequence such that $I_s(s_n) \to I_s^\infty$; this again implies that $I_s(s_n)' \to 0$ [28, lemma 2.1]. Then Equation 5.1e gives

$$0 = \alpha_{ss} \beta_s \frac{L_s(\tau_n)}{N(\tau_n)} I_s^\infty + (1 - \lambda_s) \beta_s \frac{S(\tau_n) + \sigma_s R(\tau_n)}{N(\tau_n)} I_s^\infty + \varepsilon_s L_s(s_n)$$

$$- \{d + \delta_s + t_{is} + \gamma_s\} I_s^\infty.$$

Using Assumption (A$_3$),

$$0 \leq (1 - \lambda_s) \beta_s \frac{S(\tau_n) + \sigma_s R(\tau_n) + L_s(\tau_n)}{N(\tau_n)} I_s^\infty + \varepsilon_s L_s(s_n)$$

$$- \{d + \delta_s + t_{is} + \gamma_s\} I_s^\infty.$$

For simplicity, define $a_1 := (d + \delta_s + t_{is} + \gamma_s)$ and $a_2 := (d + \varepsilon_s + t_{\ell s})$. The fact that for all t, $\frac{S(t) + \sigma_s R(t) + L_s(t)}{N(t)} < 1$ and $L_s(t) \leq L_s^\infty$, together with Equation 5.12, implies that

$$0 \leq \left[(1 - \lambda_s) \beta_s - a_1 + \frac{\lambda_s \beta_s \varepsilon_s + \gamma_s \varepsilon_s}{a_2} \right] I_s^\infty$$

$$\leq [a_2(1 - \lambda_s)\beta_s - a_1 a_2 + \lambda_s \beta_s \varepsilon_s + \gamma_s \varepsilon_s] \frac{1}{a_2} I_s^\infty$$

$$\leq \frac{\mathcal{R}_{0s} - 1}{a_2(a_1 a_2 - \varepsilon_s \gamma_s)} I_s^\infty. \tag{5.13}$$

Since $\mathcal{R}_0 = \max\{\mathcal{R}_{0s}, \mathcal{R}_{0m}, \mathcal{R}_{0x}\}$, $\mathcal{R}_0 < 1$ implies that $\mathcal{R}_{0s} < 1$. Therefore, (5.13) implies that $I_s^\infty = 0$. Hence, $\lim_{t \to \infty} I_s(t) = 0$. Similarly, using Assumptions (A$_1$) and (A$_2$), we can prove the following inequalities involving I_m and I_x:

$$0 \leq \frac{\mathcal{R}_{0m} - 1}{a_3(a_3a_4 - \varepsilon_m\gamma_m)} I_m^\infty$$

$$0 \leq \frac{\mathcal{R}_{0x} - 1}{a_5(a_5a_6 - \varepsilon_x\gamma_x)} I_x^\infty , \tag{5.14}$$

where

$$a_3 := d + \varepsilon_m, \qquad a_4 := d + t_{im} + \delta_m + \gamma_m, \tag{5.15a}$$
$$a_5 := d + \varepsilon_x, \qquad a_6 := d + t_{ix} + \delta_x + \gamma_x. \tag{5.15b}$$

The inequalities (5.14) imply that $I_m^\infty = I_x^\infty = 0$ when $\mathcal{R}_0 < 1$; therefore, $\lim_{t\to\infty} I_m(t) = \lim_{t\to\infty} I_x(t) = 0$. As a consequence, using (5.3), the total population $N(t)$ converges to b/d. To finish the proof, we study system (5.1) after N, I_s, I_m, and I_x have converged, thereby reducing (5.1) to the following system

$$S' = b - dS \tag{5.16a}$$
$$L_s' = -\{d + \varepsilon_s + t_{\ell s}\} L_s \tag{5.16b}$$
$$L_m' = -\{d + \varepsilon_m\} L_m + (1 - p_1)t_{\ell s}L_s \tag{5.16c}$$
$$L_x' = -\{d + \varepsilon_x\} L_x \tag{5.16d}$$
$$R' = p_1 t_{\ell s} L_s - dR. \tag{5.16e}$$

System (5.16) is linear and clearly limits to $(b/d, 0, 0, 0, 0)$. Finally, when $\mathcal{R}_0 < 1$, the DFE is locally asymptotically stable. As a consequence, the DFE is globally asymptotically stable when $\mathcal{R}_0 < 1$.

In the absence of exogenous reinfection, it was established in Ref. [10] that under certain conditions, the DFE of a drug-sensitive TB model and of a two-strains TB model was globally asymptotically stable. With exogenous reinfection, in Ref. [10] a drug-sensitive TB strain only model was considered; a backward bifurcation phenomenon because of exogenous reinfection was shown to exist. Theorem 5.2 shows that even with exogenous reinfection occurring, in the case of (5.1), there is a range of values of the exogenous reinfection parameter for which the model undergoes a forward bifurcation with all solutions going to the DFE. Outside that range, the system can undergo a backward bifurcation if (5.8) holds, as established in Theorem 5.1.

5.1.2.6 Strain-specific global stability in "forward" bifurcation cases Conditions in Theorem 5.2 mean that while the existence of a backward bifurcation in system (5.1) depends on the existence of a backward bifurcation in the terminal XDR strain (Theorem 5.1), the global asymptotic stability of the DFE of model (5.1) requires the DFE to be globally asymptotically stable for each strain. In view of this, we now investigate the global asymptotic stability of the DFE in specific strains. We show only the case of the XDR-TB strain, the drug-resistant and MDR-TB strains proceed similarly. The submodel for XDR-TB in the absence of the other strains is easily shown to live on the invariant hyperplane $\{\mathcal{X} \in \mathbb{R}_+^8 ; L_s = L_m = I_s = I_m = 0\}$

and takes the following form

$$S' = b - dS - \beta_x \frac{SI_x}{N}, \tag{5.17a}$$

$$L'_x = \lambda_x \beta_x (S + \sigma_x R) \frac{I_x}{N} - \alpha_{xx} \beta_x \frac{L_x I_x}{N} - \{d + \varepsilon_x\} L_x + \gamma_x I_x, \tag{5.17b}$$

$$I'_x = (1 - \lambda_x) \beta_x (S + \sigma_x R) \frac{I_x}{N} + \alpha_{xx} \beta_x \frac{L_x I_x}{N} + \varepsilon_x L_x$$
$$\quad - \{d + \delta_x + t_{ix} + \gamma_x\} I_x, \tag{5.17c}$$

$$R' = t_{ix} I_x - \sigma_x \beta_x \frac{RI_x}{N} - dR. \tag{5.17d}$$

Theorem 5.3   *Under assumption* (A$_1$), *the DFE* $(b/d, 0, 0, 0)$ *of the XDR-TB sub-model* (5.17) *is globally asymptotically stable when* $\mathcal{R}_{0x} < 1$.

Proof: Similarly to the proof of Theorem 5.2, we prove the global stability of the DFE by showing that, if $\mathcal{R}_0 < 1$, then

$$\lim_{t \to \infty} S(t) = \frac{b}{d}, \lim_{t \to \infty} I_x(t) = \lim_{t \to \infty} L_x(t) = \lim_{t \to \infty} R(t) = 0.$$

Here again, (5.17) is not of type K and a standard comparison theorem cannot be used. Let $\tau_n \to \infty$ be the sequence such that $L_x(\tau_n) \to L_x^\infty$. Then $L_x(\tau_n)' \to 0$ using lemma 2.1 in Ref. [28]. Then equation 5.17b gives

$$0 \leq \lambda_x \beta_x I_x(\tau_n) - \{d + \varepsilon_x\} L_x^\infty + \gamma_x I_x(\tau_n),$$

that is,

$$L_x^\infty \leq \frac{\lambda_x \beta_x + \gamma_x}{d + \varepsilon_x} I_x^\infty. \tag{5.18}$$

Now let $s_n \to \infty$ be a sequence such that $I_x(s_n) \to I_x^\infty$, implying that $I_x(s_n)' \to 0$ [28, lemma 2.1]. Then Equation 5.17c gives

$$0 < \alpha_{xx} \beta_x \frac{L_x(s_n)}{N(s_n)} I_x^\infty + (1 - \lambda_x) \beta_x \frac{S(s_n) + \sigma_x R(s_n)}{N(s_n)} I_x^\infty + \varepsilon_x L_x(s_n)$$
$$\quad - \{d + \delta_x + t_{ix} + \gamma_x\} I_x^\infty.$$

Using Assumption (A$_3$), it follows from $\left(\frac{S(t) + \sigma_s R(t) + L_s(t)}{N(t)} \right) < 1$, $L_s(t) \leq L_s^\infty$ and Equation 5.18 that

$$0 \leq \frac{\mathcal{R}_{0x} - 1}{a_5(a_5 a_6 - \varepsilon_s \gamma_s)} I_s^\infty, \tag{5.19}$$

where a_5 and a_6 are given by (5.15b). If $\mathcal{R}_{0x} < 1$, then $I_x^\infty = 0$. Hence $\lim_{t \to \infty} I_x(t) = 0$. Moreover, the total population N converges to b/d. We then study (5.17) after convergence of N and I_x, reducing it to the following model

$$
\begin{aligned}
S' &= b - dS \\
L_x' &= -\{d + \varepsilon_x\} L_x \\
R' &= -dR.
\end{aligned}
\tag{5.20}
$$

The proof is finished by remarking that (5.20) is linear and limits to $(b/d, 0, 0)$.

The same method of proof can be used to show that the disease-free equilibria for the drug sensitive and MDR-TB subsystems are globally asymptotically stable under assumptions (A_3) and (A_2) when $\mathcal{R}_{0s} < 1$ and $\mathcal{R}_{0m} < 1$, respectively.

5.2 DISCUSSION

In this chapter, a model to study the emergence and propagation of drug-resistant TB, both MDR and XDR, is developed and analyzed. The most important results proved are as follows:

1. System (5.1) has a globally asymptotically stable DFE when $\mathcal{R}_0 < 1$ under suitable conditions (Theorem 5.2).
2. If condition (A_1) in Theorem 5.2 does not hold, system (5.1) can undergo a backward bifurcation. The existence of subthreshold endemic equilibria is governed by the bifurcation structure of the "top level" model, namely, that for XDR-TB.

Note that the model presented here has not been validated in the usual sense. Indeed, in epidemiology, validating models is a hard task. There are many factors that contribute to this difficulty. The data available to carry out the task is of varying quality. For instance, TB data of relatively good quality is mostly available for countries with strong healthcare systems, which see very few "homegrown" M/XDR-TB cases (most are imported), so estimating the rates of treatment-induced progression from drug-sensitive TB to MDR-TB and onward to XDR-TB is hard. Validation is further complicated because estimating parameters of the contact rate function is, at best, guesswork. Indeed, even for a disease such as TB that has been studied intensely for over a century, while the general mechanisms of transmission are well known, the specifics are to a large extent unknown. For instance, it is known that repeated and prolonged contacts favor the person-to-person transmission of TB. But a quantification of what constitutes this type of contacts is unknown. As a consequence, this model, like most that came before it and many that will follow, remains largely an intellectual exercise. However, this does not make it irrelevant. Mathematical epidemiology has contributed a lot to the understanding of disease transmission processes and their control; see, for instance, the discussions in Refs. [3, 20]. The contribution of the present model is in laying out a potential scenario for the occurrence and spread of M/XDR-TB in a population and the elucidation of some of the basic properties of this model.

Our mathematical analysis revealed the potential presence of subthreshold endemic equilibria. Despite having been investigated since before the work of Hadeler and van der Driessche [19], little progress has been made in the understanding of the reasons for the presence of such subthreshold endemic equilibria in epidemic models (other than the mathematical reasons, which are straightforward and have to do with the degree of the multivariable polynomial, one must solve to find endemic equilibria). We know of no work that would satisfactorily address the pressing question of the determination of the direction of a bifurcation in real data; given the quality of epidemic data and the uncertainty on parameters, it is in general impossible to decide whether one is observing an endemic situation with $\mathcal{R}_0 < 1$ or $\mathcal{R}_0 > 1$. The situation is not as clear cut as it was which led to the publication [23], since backward bifurcations were identified in more realistic models in more feasible parameter regions, but, in our view, backward bifurcations are mostly anecdotal. Their presence should however be established, they forbid most type of global stability analysis; to the best of our knowledge, the work by Arino et al. [2] is the only work in which it was proved that when $\mathcal{R}_c < \mathcal{R}_0 < 1$, all solutions not starting on the stable manifold of the unstable subthreshold equilibrium are attracted to one of the locally stable equilibria.

This is a preliminary analysis. The model is quite complicated with a large number of nonlinearities, and considerations beyond the simple case of the disease-free equilibrium for all three strains are quite involved and will be considered in further work. Further work will also involve parametrizing the model and, if possible, comparing it with data.

Note. It has come to the authors' attention during revisions of this manuscript that the model studied here is very similar to a model of Bhunu [5], with some subtle differences. The model in Ref. [5] adds the possibility for individuals with active XDR-TB to be quarantined but does not consider treatment of individuals latently infected with the drug-sensitive strain nor natural recovery of individuals with active TB. Also, we assume that any type of reinfection, not only of treated individuals, can lead to slow or fast progression to infectiousness. These differences imply that while the classical analysis ($\mathcal{R}_0$ and nature of the bifurcation at $\mathcal{R}_0 = 1$) proceeds quite similarly, there are little differences in the results. Also, Bhunu [5] considered boundary equilibria, which are barely discussed here in Theorem 5.3. On the other hand, our results in the case of $\mathcal{R}_0 < 1$ are global when a backward bifurcation is ruled out. Thus, the analyses here and in Ref. [5] complement each other. The fact that two models starting with the same basic building blocks (the SLIR TB model [8], the SLIT model [10] and their progeny) and description of the epidemiology of M/XDR-TB are so similar is also encouraging and should probably be interpreted as a first step in the validation of the model of Bhunu (and ours).

REFERENCES

1. J.P. Aparicio and C. Castillo-Chávez. Mathematical modelling of tuberculosis epidemics. *Math Biosci Eng*, 6(2):209–237, 2009.

2. J. Arino, C.C. McCluskey, and P. van den Driessche. Global results for an epidemic model with vaccination that exhibits backward bifurcation. *SIAM J Appl Math*, 64(1):260–276, 2003.

3. J. Arino, C. Bauch, F. Brauer, S.M. Driedger, A.L. Greer, S.M. Moghadas, N.J. Pizzi, B. Sander, A. Tuite, P. van den Driessche, J. Watmough, J. Wu, and P. Yan. Pandemic influenza: modelling and public health perspectives. *Math Biosci Eng*, 8(1):1–20, 2011.

4. S. Basu and A.P. Galvani. The transmission and control of XDR TB in South Africa: an operations research and mathematical modelling approach. *Epidemiol Infect*, 136(12):1585–1598, 2008.

5. C.P. Bhunu. Mathematical analysis of a three-strain tuberculosis transmission model. *App Math Model*, 35:4647–4660, 2011.

6. S.M. Blower and T. Chou. Modeling the emergence of the 'hot zones': tuberculosis and the amplification dynamics of drug resistance. *Nat Med*, 10(10):1111–1116, 2004.

7. S.M. Blower and J.L. Gerberding. Understanding, predicting and controlling the emergence of drug-resistant tuberculosis: a theoretical framework. *J Mol Med (Berl)*, 76(9):624–636, 1998.

8. S.M. Blower, A.R. McLean, T.C. Porco, P.M. Small, P.C. Hopewell, M.A. Sanchez, and A.R. Moss. The intrinsic transmission dynamics of tuberculosis epidemics. *Nat Med*, 1(8):815–821, 1995.

9. G.L. Calligaro and K. Dheda. Drug-resistant tuberculosis. *Cont Med Edu*, 31(9), 2013.

10. C. Castillo-Chavez and Z. Feng. To treat or not to treat: the case of tuberculosis. *J Math Biol*, 35(6):629–656, 1997.

11. C. Castillo-Chavez and B. Song. Dynamical models of tuberculosis and their applications. *Math Biosci Eng*, 1(2):361–404, 2004.

12. C.-Y. Chiang and L.W. Riley. Exogenous reinfection in tuberculosis. *Lancet Infect Dis*, 5(10):629–636, 2005.

13. T. Cohen and M. Murray. Modeling epidemics of multidrug-resistant *M. tuberculosis* of heterogeneous fitness. *Nat Med*, 10(10):1117–1121, 2004.

14. T. Cohen, C. Dye, C. Colijn, B. Williams, and M. Murray. Mathematical models of the epidemiology and control of drug-resistant TB. *Expert Rev Respir Med*, 3(1):67–79, 2009.

15. O. Diekmann and J.A.P. Heesterbeek. *Mathematical epidemiology of infectious diseases: model building, analysis and interpretation*. New York: John Wiley & Jons, Inc., 2000.

16. C. Dye, S. Scheele, P. Dolin, V. Pathania, and M.C. Raviglione. Consensus statement. global burden of tuberculosis: estimated incidence, prevalence, and mortality by country. who global surveillance and monitoring project. *JAMA*, 282(7):677–686, 1999.

17. M.A. Espinal. The global situation of MDR-TB. *Tuberculosis*, 83(1–3):44–51, 2003.

18. N.R. Gandhi, P. Nunn, K. Dheda, H.S. Schaaf, M. Zignol, D. van Soolingen, P. Jensen, and J. Bayona. Multidrug-resistant and extensively drug-resistant tuberculosis: a threat to global control of tuberculosis. *Lancet*, 375(9728):1830–1843, 2010.

19. K.P. Hadeler and P. van den Driessche. Backward bifurcation in epidemic control. *Math Biosci*, 146(1):15–35, 1997.

20. A. Huppert and G. Katriel. Mathematical modelling and prediction in infectious disease epidemiology. *Clin Microbiol Infect*, 19(11):999–1005, 2013.

21. Institute for Health Metrics and Evaluation. *The global burden of disease: generating evidence, guiding policy.* Seattle, WA: University of Washington, 2013.

22. M. Jassal and W.R. Bishai. Extensively drug-resistant tuberculosis. *Lancet Infect Dis*, 9(1):19–30, 2009.

23. M. Lipsitch and M.B. Murray. Multiple equilibria: Tuberculosis transmission require unrealistic assumptions. *Theor Popul Biol*, 63:169–170, 2003.

24. J.-P. Millet, A. Moreno, L. Fina, L. del Baño, A. Orcau, P.G. de Olalla, and J.A. Caylà. Factors that influence current tuberculosis epidemiology. *Eur Spine J*, 22 (Suppl 4):539–548, 2013.

25. C. Ozcaglar, A. Shabbeer, S.L. Vandenberg, B. Yener, and K.P. Bennett. Epidemiological models of mycobacterium tuberculosis complex infections. *Math Biosci*, 236(2):77–96, 2012.

26. A. Pablos-Méndez, M.C. Raviglione, A. Laszlo, N. Binkin, H.L. Rieder, F. Bustreo, D.L. Cohn, C.S. Lambregts-van Weezenbeek, S.J. Kim, P. Chaulet, and P. Nunn. Global surveillance for antituberculosis-drug resistance, 1994-1997. *N Engl J Med*, 338(23):1641–1649, 1998.

27. N.S. Shah, A. Wright, G.-H. Bai, L. Barrera, F. Boulahbal, N. Martín-Casabona, F. Drobniewski, C. Gilpin, M. Havelková, R. Lepe, R. Lumb, B. Metchock, F. Portaels, M.F. Rodrigues, S. Rüsch-Gerdes, A. Van Deun, V. Vincent, K. Laserson, C. Wells, and J.P. Cegielski. Worldwide emergence of extensively drug-resistant tuberculosis. *Emerg Infect Dis*, 13(3):380–387, 2007.

28. H.R. Thieme. Persistence under relaxed point-dissipativity (with applications to an endemic model). *SIAM J Math Anal*, 24:407–435, 1993.

29. P. van den Driessche and J. Watmough. Reproduction numbers and sub-threshold endemic equilibria for compartmental models of disease transmission. *Math Biosci*, 180:29–48, 2002.

30. A. van Rie, R. Warren, M. Richardson, T.C. Victor, R.P. Gie, D.A. Enarson, N. Beyers, and P.D. van Helden. Exogenous reinfection as a cause of recurrent tuberculosis after curative treatment. *N Engl J Med*, 341(16):1174–1179, 1999.

31. World Health Organization. *Multidrug and extensively drug-resistant TB (M/XDR-TB): 2010 global report on surveillance and response.* Geneva World Health Organization, 2010.

6

THE NEED FOR MORE INTEGRATED EPIDEMIC MODELING WITH EMPHASIS ON ANTIBIOTIC RESISTANCE

Eili Y. Klein[1,2], Julia Chelen[1], Michael D. Makowsky[1], and Paul E. Smaldino[1]

[1]Center for Advanced Modeling in the Social, Behavioral, and Health Sciences, Department of Emergency Medicine, Johns Hopkins University, Baltimore, MD, USA
[2]Center for Disease Dynamics, Economics & Policy, Washington, DC, USA

6.1 INTRODUCTION

Antibiotic resistance has become one of the greatest threats to public and patient health. Pathogens resistant to antibiotics can significantly decrease a physician's ability to treat infection and increase the probability of mortality in patients [3]. Estimates are that, per annum, a minimum of two million Americans contract antibiotic resistant infections, resulting in 23,000 deaths [18]. Decreases in the efficacy of antibiotics threaten to reverse a variety of major medical gains [60, 85]. For example, the ability to perform transplants and other surgical procedures are dependent on antibiotic effectiveness [28] and would be severely hampered in a post-antibiotic world. Overall, the annual economic cost to the US health care system of antibiotic-resistant infections is estimated to be $21–$34 billion [34, 64, 70]. Given the importance of this problem, from standpoints both of human health and economics, there is much to be gained from better understanding how resistant bacterial pathogens evolve and

Mathematical and Computational Modeling: With Applications in Natural and Social Sciences, Engineering, and the Arts, First Edition. Roderick Melnik.
© 2015 John Wiley & Sons, Inc. Published 2015 by John Wiley & Sons, Inc.

persist in human populations. Integrated computational models encompassing aspects of human behavior can improve our understanding of the evolution and the spread of resistance, offering a clearer picture of how ecology and epidemiology interact to spread antibiotic resistant pathogens, as well as insight into policy options to help contain the spread of resistance. Here the history of mathematical modeling of infectious diseases and a selection of its achievements and limitations is briefly examined, followed by a discussion of the need to develop models of disease spread that incorporate individual behavior with reference to how this can improve models of bacterial pathogens.

6.2   MATHEMATICAL MODELING OF INFECTIOUS DISEASES

Mathematical models of infectious disease have been used since the eighteenth century to guide interventions to control disease [11]. At the beginning of the twentieth century, models explaining the dynamics of measles [41] as well as malaria [71] had been developed. Kermack and McKendrick then established the modern mathematical theory of infectious disease [49], clarifying in deterministic settings the threshold nature of epidemics and the central idea of herd immunity. Important factors such as the impact of stochasticity and critical community size on sustaining epidemics were later introduced [9, 10], as were further refinements describing the invasion and persistence of human pathogens [2, 40]. Similar techniques have been applied to the study of the spread of animal and plant diseases, both in agricultural and natural landscapes [17, 36, 39, 48, 75]. This has led to theories of how epidemics spread spatially, and how control measures should be deployed in a wide range of pathogens, host populations, and environments. As the global epidemic of antibiotic resistance has increased in recent years, mathematical models of the spread of antibiotic resistance have also been developed [4,5,26,27,59,68,69]. These models have elucidated important aspects of transmission in hospital settings, nursing homes, and other inpatient facilities.

The models underlying most theories on epidemic spread have generally been formulated as "SIR" models [43] in which individuals of the same epidemiological status are lumped together in homogeneous pools: S is the size of the susceptible class, I is the infectious, and R is the removed (e.g., dead or recovered). Control treatments are then modeled either through quantitative changes to model parameters, notably transmission rates and infectious periods, or by introducing additional transitions, such as when a vaccination program switches individuals from a susceptible to a removed class [2]. Like central idealizations in other fields—the simple harmonic oscillator or ideal gas in physics, or the Lotka–Volterra equations in ecology—the classical "SIR" differential equations are an elegant, tractable, and important foundation. And while they have been instrumental in understanding fundamental aspects of disease spread, just as in these other fields, they are not fully equal to the complexity of real-world settings, which amalgamate pathogen evolution, risk behavior, spatial dynamics, and policy. Classical modeling thus generally ignores heterogeneities between individuals, and simply cannot represent the direct interactions between individuals, which ultimately generate the patterns that emerge at the population level.

Despite these limitations, computational models have begun increasingly to include realistic descriptions of human behavior in understanding how diseases spread. The catalyst for this was largely sexually transmitted diseases where behavior (and modification of behavior) is most pronounced. Thus, factors such as heterogeneous sexual mixing rates [12, 44], rational responses to economic incentives [67], more realistic descriptions of risk-taking behavior [13–15, 29, 45], and the spread of fear and its impact on disease transmission [32] have all been examined within this classical modeling framework. More recently, models in this framework have examined the impact of behavioral responses of individuals on the spread of antibiotic resistance. For example, including heterogeneities in age-related mixing can dramatically affect the spread of community-associated strains of methicillin-resistant *Staphylococcus aureus* (CA-MRSA) [55], which has important implications for controlling the disease.

Computational power and speed has been one of the major limitations to modeling individual behavior that in recent years has been removed. This has allowed for the development of agent-based models that can include realistic contact networks [33, 62, 63] and examine how behaviors such as policy resistance to vaccination [83] impacts the spread of a disease. Individual-based models have also been developed to understand the complexity of bacterial strain interaction [21] as well as heterogenties in malaria transmission and the impact this has on emergence and spread of drug resistance [52]. Despite these advances, there still exists room to innovate and integrate, particularly as individuals can behave in unexpected ways during epidemics. Epidemics are contexts fraught with fear, distrust of health authorities, and poor information. In such settings, understanding how people behave and how that behavior may affect transmission can guide decision making and policy options. While epidemiological modeling has come a long way in including behavior, many models posit that individuals behave rationally. The model of "rational behavior" can be misguiding in contexts, such as healthcare, where decision making must cope with great uncertainty in emotionally charged contexts. Systematic deviations from rationality have been repeatedly observed in both laboratory and field experiments (see, e.g., Refs. [42, 74]). These deviations are explained by decision heuristics that can both approximate rational choices as well as yield irrational biases [37], for example:

1. *Representativeness Heuristic.* People may rely on this heuristic to make judgments about an uncertain object, event, or process. For example, what is the probability that object x belongs to set A? Tversky and Kahneman [47] have demonstrated that when people use the representativeness heuristic to make these judgments, they estimate the subjective probability of x belonging to set A by using the perceived similarity of x to the other objects in set A as the probability. The perceived similarity is the extent to which x resembles the other objects in set A, either in "essential characteristics to its parent population" or "reflects the salient features of the process by which it is generated." If the perceived similarity is high, the probability estimate is high. If the perceived similarity is low, the probability estimate is low. Extensive study of the representativeness heuristic has identified many systematic errors in intuitive

likelihood estimates, including a central finding of pervasive insensitivity to sample size [81], and base-rate neglect (Example 2).

In an epidemic context, the representativeness heuristic can have enormous importance. Individuals, including experts, may use the representativeness of the individuals they interact with and their personal social network, instead of accurate probability estimates, to produce false conclusions about the prevalence, contact risk, and morbidity of a disease.

2. *Base-Rate Neglect.* Experimental psychology has demonstrated that people tend to ignore prior probabilities, or base rates, when making probability judgments about uncertain events, particularly when presented with individuating information [1, 7, 46]. Vaccination is an important example where base rate neglect can affect the spread of disease. For example, in determining whether individuals will vaccinate or not, rational actor models may use Bayes Theorem to estimate the probability of having an adverse reaction (A) given a vaccination (V),

$$P(A|V) = \frac{P(V|A)P(A)}{P(V)} = \frac{P(V|A)P(A)}{P(V|A)P(A) + P(V|A^c)P(A^c)}$$

where $P(A|V)$ equals the probability of an adverse reaction given vaccination, and $P(V|A)$ is the probability of having been vaccinated given an adverse reaction. These are multiplied by the prior probabilities, or base rate $P(A)/P(V)$. When base rates are excluded, the inverse probabilities are equated,

$$P(A|V) = P(V|A)$$

Media attention typically focuses on the person who had an adverse reaction after receiving a vaccine. In the presence of this individuating information, individuals may ignore base rates or see them as irrelevant. As a result, overestimation of the likelihood of an adverse reaction, and undervaccination of the public, may occur.

3. *Hyperbolic Discounting.* Individual's discount rates are often inconsistent across time spans, tending toward far greater impatience in the short run [57]. This is relevant when modeling diseases that have long incubation periods or for which the risk of infection changes over long periods.

4. *Conditional Expected Utility (or the Illusion of Control).* Individuals frequently overweigh how much their own actions matter when estimating the probability of an outcome [58]. Thus, while proper hygiene may reduce the probability of contracting a disease, it is possible that individuals will systematically overstate their own control over disease outcomes.

These are just a few selected departures from the rational actor model that explain some of the systematic ways in which people fail to make optimal decisions or maximize their expected utility. Future models must incorporate the psychology of risk [35] and take account of behavioral aspects that might influence the spread of disease.

6.3   ANTIBIOTIC   RESISTANCE,   BEHAVIOR,   AND   MATHEMATICAL MODELING

6.3.1   Why an integrated approach?

Ecologically the dynamics of bacterial pathogens are complicated by the fact that they are embedded in the microbiome. Thus, these pathogens must compete with other species of microorganism—some of which are closely related-both directly, for resources, and indirectly, as the hosts immune response alters the ecological landscape. The ecology of the environment, including agricultural practices, may also affect disease dynamics by promoting increased resistance or the long-distance spread of certain pathogens. Epidemiologically many bacterial pathogens can be difficult to track as they do not always cause disease. Thus, individuals can remain colonized and infectious without ever experiencing clinical symptoms, making it difficult to understand the transmission pathways of a disease. In addition, the mechanisms by which colonized individuals progress to disease are not well understood. However, the ecology and epidemiology are also affected by individual/institutional behavioral decisions. For instance, there is a strong link between increased antibiotic use and resistance [6, 78]; thus, individual decisions on antibiotic usage strongly influence the rate that resistance will emerge and spread. Prescription rates for antibiotics are shaped by patients' expectations for antibiotic therapy, and studies show that physicians often prescribe based on their beliefs about what patients expect [22, 66, 73], even though doctors are actually not particularly good at divining patients' expectations [66]. This communication gap leads to injudicious use of antibiotics. On the other side of the prescription decision, patients often will not complete a course of antibiotic therapy, which can allow for the development of resistance. Transmission of antibiotic resistance is also affected by contact networks, which are age dependent [65], as well as the effectiveness of interventions. Infection control in hospitals is one of the primary interventions aimed at reducing the spread of resistance. Even so, hospitals may base their level of investment in infection control on the amount other hospitals that share the same population invest in infection control [76]. At the individual level, rates of hand-washing compliance, which is seen as paramount in reducing the probability of transmission within the hospital, have been notoriously difficult to increase [38, 82].

Usage of antibiotics is also likely to be subject to feedbacks as in other diseases (i.e., people are likely to respond to the prevalence of resistance by changing their usage patterns). A central example is methicillin-resistant *S. aureus* (MRSA), a multiply resistant pathogen that is a scourge of hospitals and has become resistant to nearly every known antibiotic. An October 2007 report estimated that MRSA killed more people annually than HIV [54]. This finding led to an enormous surge in interest in MRSA that then waned exponentially toward baseline levels over time

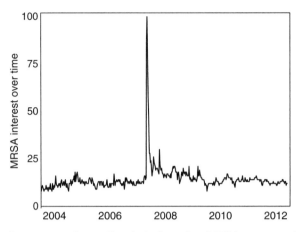

FIGURE 6.1   Data are searches on Google for how often MRSA was entered as a search term relative to the total search volume across the world. The horizontal axis of the main graph represents time (starting from 2004) and the vertical is how often a term is searched for relative to the total number of searches, globally. Data Source: Google Trends (www.google.com/trends).

(see Fig. 6.1). The surge and decline in MRSA interest raises important questions. What sort of impact does such a surge in concern about a resistant bacterial pathogen that is increasing in frequency have on the transmission of that pathogen and the usage of antibiotics? Along with the primary article were numerous reports discussing ways to mitigate transmission, including efforts to change behavior in places such as locker-rooms. What sort of lasting effect did this have? Do these types of behavioral changes explain why rates of MRSA have largely stagnated (or even gone down) in recent years [53], or is its stagnation the result of primarily ecological factors? These are important open questions. Addressing them adequately will require integrated approaches that encompass the roles of economics, sociology (network dynamics), the psychology of risk, computer science, and the ecology and epidemiology of disease.

Over the past several decades, significant advances have been made in understanding disease transmission, individual behavior, and social structures. For instance, one of the biggest advances in the area of epidemic modeling is use of Bayesian inference in conjunction with Markov chain Monte Carlo methods to impute unobserved data [23]. This methodology has been used to estimate epidemic trees of bacterial pathogens where many of the infections may be asymptomatic and thus unobserved [24, 56]. Nonetheless, much of the research has remained "balkanized" with few of the advances within each discipline spreading to others. This is particularly unfortunate in that understanding the complex interplay between behavior, transmission, and disease evolution requires an approach synthesizing insights from each discipline. For instance, exploring how the announcement of an emerging disease affects the future transmission of that disease requires an understanding of how people respond to different types of information, how the information and associated emotions such as fear of the disease spread, and how the resulting behavioral responses (e.g., fear-driven self isolation, long-range flight) alter social contact patterns, treatment seeking (or refusing) behavior, and the transmission of the disease.

Adding to the complexity of the problem, these behavioral responses and their effects unfold against a backdrop of evolving pathogens that may change based on transmission itself. For example, ecological evidence and theory suggest that when drug-resistant pathogens first emerge they may be more virulent than drug-sensitive pathogens [16, 19, 20, 25, 51, 72, 79, 86]. At the same time, awareness of this increased virulence can change individuals' behavior (improved hygiene, increased hospital visitation, modification of contact networks, etc.). These individual behavioral adaptations can in turn alter the future ecological and evolutionary trajectory of the pathogen, to an extent as great as, if not greater than, the changes in the pathogens intrinsic virulence.

In the face of so much complexity, mathematical and computational models are extremely valuable research and policy tools. They can help strip away extraneous detail, revealing the core generators of complex dynamics. They can offer predictions, forecast the effectiveness of interventions, and even prioritize empirical research [31, 84]. This is particularly important as individual behavior is a significant driver of the emergence and spread of antibacterial resistance. While there are more than 260 million prescriptions written for antibiotics annually, it has been estimated that as much as 60% of antibiotic use is inappropriate or unwarranted [78]. Inappropriate antimicrobial use is the result of individual behavior on the part of both patients and physicians (who prescribe antibiotics). Overuse of antibiotics accelerates the emergence and the transmission of antibiotic resistance in the population, and in some cases (e.g., carbapenem-resistant enterococci) all known antibiotics may soon be exhausted. All of this makes antibiotic resistance, and the diverse behaviors driving it, a particularly timely area for integrated modeling.

6.3.2   The role of symptomology

One important area of modeling bacterial pathogens that can be instructive in devising new policy options is how symptoms (or lack thereof) of infection affect behavior and disease spread. While clinical symptoms are a clear signal to which an individual can respond (stay at home, go to the doctor, etc.) or to which others can respond (e.g., avoid unnecessary contact or sharing of food), many bacterial pathogens can also colonize people for long periods of time. These asymptomatic infections allow individuals to go about their daily routine unfettered. This has the potential to greatly impact the dynamics of disease. To capture behavior appropriately, models of bacterial resistance should account for infection status, distinguishing between (1) clinically ill, (2) asymptomatic, and (3) well individuals.

1. *The clinically ill.* Clinically diseased individuals can have multiple behavioral responses, but the most important include increased probability of contact with the healthcare system and decreased contact with large parts of their "normal" network. These two factors can either increase spread or decrease it, depending

on the protective response of infected individuals and others in their network. For instance, an individual that goes to the hospital could spread the disease more widely than they would have had they only maintained normal contact patterns. In the modern European history of smallpox, for example, half of all transmissions took place in hospitals [30,61].

2. *The Asymptomatically Infected.* This category of individuals is often ignored or subsumed in the infected class in determining transmission patterns, particularly in bacterial resistance [77], but can in fact be more influential than individuals expressing symptoms. An infection can be asymptomatic at the beginning of its natural history (due to genetic or immune factors for example), or at the end, after the resolution of symptoms (spontaneously or therapeutically) without clearance of the pathogen. Asymptomatic infection can be momentous as exemplified by Typhoid Mary, who remained infectious for typhoid throughout her life despite never becoming symptomatic. This phenomenon has been recognized as a problem for enteric diseases such as cholera [50] and should also be included in models of bacterial pathogens. The behavior of asymptomatically infected individuals depends as well on awareness of their status. As seen with HIV, people may act very differently if they know they are infectious [80], all of which invites integrated modeling.

3. *"Well" individuals.* Uninfected individuals can also respond to disease, of course. The "worried well," as they are sometimes called, have been known to adopt extreme behaviors of prevention, such as self-sequestration (e.g., locking oneself in the basement), and spatial flight. Both can be very consequential for disease spread and can be driven by contagious fear, the extreme forms of which qualify as a mass sociogenic illness [8]. This behavior at its extreme can inundate emergency departments, which may be understaffed due to absenteeism among "worried well" health care personnel, degrading vaccination programs, delaying care for the truly ill, and hampering other control measures. In sum, fear contagion among the perfectly well can sharply inflate the demand for emergency resources while at the same time depressing the supply, a vicious spiral to be sure. Worried well are also likely to overuse antibiotics, making them more susceptible to infection. In addition, every individual that takes antibiotics produces some resistant bacteria, though this does not necessarily mean pathogenic bacteria. Still, the use of antibiotics will produce some resistant bacteria (at least transiently). These resistant bacteria are transmitted to other individuals, or are excreted from the body and enter the environment, spreading resistant genes to other bacteria, including pathogenic ones.

6.4   CONCLUSION

Understanding the interaction of epidemiological, evolutionary, and behavioral factors is crucial if we are to design innovative public health strategies against bacterial pathogens. Such strategies can significantly improve the health and well-being of millions in the United States and internationally. Understanding of this multifaceted problem requires an integrated approach that incorporates insights from multiple disciplines, and focuses on specific pathogens as bacterial pathogens can have very different routes of transmission (airborne vs. environmental vs. person-to-person) as well as different networks (community vs. hospital) and different ecological niches (gut vs. skin). Future models of antibacterial resistance must also take account of human behavior, taking particular care to develop behavioral considerations for different disease classifications of individuals, specifically addressing how asymptomatically infected individuals behave relative to sick individuals and how this might impact the dynamics of the disease. Cross-discipline cooperation that can incorporate the particulars of a disease into models that include social structure and behavior are more likely to provide useful policy recommendations.

Acknowledgments

We thank Dr. Joshua M. Epstein for his comments and support on the drafting of the manuscript. This work was supported in part by The National Center for the Study of Preparedness and Catastrophic Event Response (PACER) at the Johns Hopkins University, through the U.S. Department of Homeland Security Grant N00014-06-1-0991, through the National Institute of General Medical Sciences Models of Infectious Disease Agent Study (MIDAS) through grant 5U54GM088491, and through the Center for Public Health Practice by the Centers for Disease Control and Prevention (cooperative agreement 1P01TP000304). This project was also supported by Pioneer Award Number DP1OD003874 awarded to J.M. Epstein by the Office of the Director, National Institutes of Health. The content is solely the responsibility of the authors and does not necessarily represent the official views of the Office of the Director, National Institutes of Health or the National Institutes of Health, or the Office of Naval Research or the official policy or position of the Department of Defense or the U.S. Government. The funders had no role in study design, data collection and analysis, decision to publish, or preparation of the manuscript.

REFERENCES

1. I. Ajzen. Intuitive theories of events and the effects of base-rate information on prediction. *Journal of Personality and Social Psychology*, **35**(5):303 (1977).

2. R. M. Anderson and R. M. May. *Infectious Diseases of Humans*. Oxford University Press, Oxford, 1991.

3. D. I. Andersson and D. Hughes. Antibiotic resistance and its cost: is it possible to reverse resistance? *Nat Rev Micro*, **8**(4):260–271 (2010).

4. E. Armeanu, M. J. M. Bonten, and R. A. Weinstein. Control of vancomycin-resistant Enterococci: one size fits all? *Clinical Infectious Diseases*, **41**(2):210–216 (2005).

5. D. J. Austin, M. J. M. Bonten, R. A. Weinstein, S. Slaughter, and R. M. Anderson. Vancomycin-resistant enterococci in intensive-care hospital settings: transmission

dynamics, persistence, and the impact of infection control programs. *Proceedings of the National Academy of Sciences*, **96**(12):6908–6913 (1999).

6. D. J. Austin, K. G. Kristinsson, and R. M. Anderson. The relationship between the volume of antimicrobial consumption in human communities and the frequency of resistance. *PNAS*, **96**:1152–1156 (1999).

7. M. Bar-Hillel. The base-rate fallacy in probability judgments. *Acta Psychologica*, **44**(3):211–233 (1980).

8. R. E. Bartholomew and S. Wessely. Protean nature of mass sociogenic illness. *The British Journal of Psychiatry*, **180**(4):300–306 (2002).

9. M. Bartlett. "Deterministic and stochastic models for recurrent epidemics" in *Proceedings of the Third Berkley Symposium on Mathematical Statistics and Probability*, vol. 4, p. 81–109. University of California Press, Berkley, 1956.

10. M. Bartlett. Measles periodicity and community size. *Journal of the Royal Statistical Society, Series A (General)*, **120**(1):48–70 (1957).

11. D. Bernoulli and S. Blower. An attempt at a new analysis of the mortality caused by smallpox and of the advantages of inoculation to prevent it. *Reviews in Medical Virology*, **14**(5):275–288 (2004).

12. S. M. Blower and A. R. McLean. Mixing ecology and epidemiology. *Proceedings: Biological Sciences*, **245**(1314):187–192 (1991).

13. S. M. Blower and A. R. McLean. Prophylactic vaccines, risk behavior change, and the probability of eradicating HIV in San Francisco. *Science*, **265**(5177): 1451–1454 (1994).

14. S. M. Blower, H. B. Gershengorn, and R. M. Grant. A tale of two futures: HIV and antiretroviral therapy in San Francisco. *Science*, **287**(5453):650–654 (2000).

15. S.M Blower, K. Koelle, and J. Mills. "Health policy modeling: epidemic control, HIV caccines, and risky behavior" in (E. H. Kaplan and R. Brookmeyer, editors), *Quantitative Evaluation of HIV Prevention Programs*, pp. 260–289. Yale University Press, New Haven, CT, 2002.

16. R. Bødker, W. Kisinza, R. Malima, H. Msangeni, and S. Lindsay. Resurgence of malaria in the Usambara Mountains, Tanzania, an epidemic of drug-resistant parasites. *Global Change & Human Health*, **1**(2):134–153 (2000).

17. U. Carlsson-Granr and P. H. Thrall. The spatial distribution of plant populations, disease dynamics and evolution of resistance. *Oikos*, **97**(1):97–110 (2002).

18. Centers for Disease Control and Prevention (CDC). Antibiotic resistance threats in the united states, 2013. Technical report, Centers for Disease Control and Prevention, 2013.

19. H. F. Chambers. Community-associated MRSA resistance and virulence converge. *New England Journal of Medicine*, **352**(14):1485–1487 (2005).

20. H. F. Chambers and F. R. DeLeo. Waves of resistance: *Staphylococcus aureus* in the antibiotic era. *Nature Reviews: Microbiology*, **7**(9):629–641 (2009).

21. S. Cobey and M. Lipsitch. Niche and neutral effects of acquired immunity permit coexistence of pneumococcal serotypes. *Science*, **335**(6074):1376–1380 (2012).

22. J. Cockburn and S. Pit. Prescribing behaviour in clinical practice: patients' expectations and doctors' perceptions of patients' expectations–a questionnaire study. *British Medical Journal*, **315**(7107):520–523 (1997).

23. B. S. Cooper, G. F. Medley, S. J. Bradley and G. M. Scott. An augmented data method for the analysis of nosocomial infection data. *American Journal of Epidemiology*, **168**(5): 548–557 (2008).

24. B. S. Cooper, T. Kypraios, R. Batra, D. Wyncoll, O. Tosas, and J. D. Edgeworth. Quantifying type-specific reproduction numbers for nosocomial pathogens: evidence for heightened transmission of an asian sequence type 239 MRSA clone. *PLoS Computational Biology*, **8**(4):e1002454 (2012).

25. M. H. Craig, I. Kleinschmidt, D. Le Sueur, and B. L. Sharp. Exploring 30 years of malaria case data in KwaZulu-Natal, South Africa: part II. The impact of non-climatic factors. *Tropical Medicine & International Health*, **9**(12):1258–1266 (2004).

26. E. M. C. D'Agata, G. Webb, and M. Horn. A mathematical model quantifying the impact of antibiotic exposure and other interventions on the endemic prevalence of vancomycin-resistant enterococci. *Journal of Infectious Diseases*, **192**(11):2004–2011 (2005).

27. E. M. C. D'Agata, G. F. Webb, M. A. Horn, R. C. Moellering, and S. Ruan. Modeling the invasion of community-acquired methicillin-resistant *Staphylococcus aureus* into hospitals. *Clinical Infectious Diseases*, **48**(3):274–284 (2009).

28. J. Davies and D. Davies. Origins and evolution of antibiotic resistance. *Microbiology and Molecular Biology Reviews*, **74**(3):417–33 (2010).

29. S. Del Valle, H. Hethcote, J. M. Hyman, and C. Castillo-Chavez. Effects of behavioral changes in a smallpox attack model. *Mathematical Biosciences*, **195**(2):228–251 (2005).

30. J. M. Epstein. *Toward a Containment Strategy for Smallpox Bioterror: An Individual-Based Computational Approach*. Brookings Institution Press, Washington, DC, 2004.

31. J. M. Epstein. Why model? *Journal of Artificial Societies and Social Simulation*, **11**(4):12 (2008).

32. J. M. Epstein, J. Parker, D. Cummings, and R. A. Hammond. Coupled contagion dynamics of fear and disease: mathematical and computational explorations. *PLoS ONE*, **3**(12):e3955 (2008).

33. S. Eubank, H. Guclu, V. A. Kumar, M. V. Marathe, A. Srinivasan, Z. Toroczkai, and N. Wang. Modelling disease outbreaks in realistic urban social networks. *Nature*, **429**(6988):180–184 (2004).

34. G. A. Filice, J. A. Nyman, C. Lexau, C. H. Lees, L. A. Bockstedt, K. Como-Sabetti, L. J. Lesher, and R. Lynfield. Excess costs and utilization associated with methicillin resistance for patients with *Staphylococcus aureus* infection. *Infection Control and Hospital Epidemiology*, **31**(4):365–373 (2010).

35. B. Fischhoff. *Risk Analysis and Human Behavior*. Earthscan, New York, 2013.

36. C. A. Gilligan. An epidemiological framework for disease management. *Advances in Botanical Research*, **38**:1–64 (2002).

37. T. Gilovich, D. Griffin, and D. Kahneman. *Heuristics and Biases: The Psychology of Intuitive Judgment*. Cambridge University Press, Cambridge, 2002.

38. D. A. Goldmann, R. A. Weinstein, R. P. Wenzel, O. Tablan, R. Duma, R. Gaynes, J. Schlosser, and W. Martone. Strategies to prevent and control the emergence and spread of antimicrobial-resistant microorganisms in hospitals: a challenge to hospital leadership. *Journal of American Medical Association*, **275**(3):234–240 (1996).

39. B. T. Grenfell and A. P. Dobson. *Ecology of Infectious Diseases in Natural Populations*. Cambridge University Press, Cambridge, 1995.

40. B. T. Grenfell, O. N. Bjornstad, and B. F. Finkenstadt. Dynamics of measles epidemics: scaling noise, determinism, and predictability with the TSIR model. *Ecological Monographs*, **72**(2):185–232 (2002).

41. W. H. Hamer. The Milroy Lectures on epidemic disease in England the evidence of variability and of persistency of type. *The Lancet*, **167**:733–739 (1906).

42. R. Hastie and R. M. Dawes. *Rational Choice in an Uncertain World: The Psychology of Judgment and Decision Making*. Sage, London, 2010.

43. H. W. Hethcote. The mathematics of infectious diseases. *SIAM Review*, **42**(4):599–653 (2000).

44. H. W. Hethcote and J. A. Yorke. Gonorrhea transmission dynamics and control. *Lecture Notes in Biomathematics*, **56** (1984).

45. J. M. Hyman and J. Li. Behavior changes in SIS STD models with selective mixing. *SIAM Journal on Applied Mathematics*, **57**(4):1082–1094 (1997).

46. D. Kahneman and A. Tversky. On the psychology of prediction. *Psychological Review*, **80**(4):237 (1973).

47. D. Kahneman and A. Tversky. "Subjective probability: a judgment of representativeness" in (C.A.S staël von Holstein, editor), *The Concept of Probability in Psychological Experiments*, pp. 25–48. Springer, New York, 1974.

48. M. J. Keeling, O. N. Bjrnstad, and B. T. Grenfell. *Metapopulation Dynamics of Infectious Disease*. Elsevier, Amsterdam, 2004.

49. W. O. Kermack and A. G. McKendrick. A contribution to the mathematical theory of epidemics. *Proceedings of the Royal Society of London. Series A*, **115**:700–721 (1927).

50. A. A. King, E. L. Ionides, M. Pascual, and M. J. Bouma. Inapparent infections and cholera dynamics. *Nature*, **454**(7206):877–880 (2008).

51. E. Klein. *Anti-Malarial Drug Resistance*. Princeton University, Princeton, NJ, 2012.

52. E. Klein. The impact of heterogeneous transmission on the establishment and spread of antimalarial drug resistance. *Journal of Theoretical Biology*, **340**:177–185 (2014).

53. E. Y. Klein, L. Sun, D. L. Smith, and R. Laximarayan. The changing epidemiology of methicillin-resistant *Staphylococcus aureus* in the United States: a national observational study. *American Journal of Epidemiology*, **177**(7):666–674 (2013).

54. R. M. Klevens, M. A. Morrison, J. Nadle, S. Petit, K. Gershman, S. Ray, L. H. Harrison, R. Lynfield, G. Dumyati, and J. M. Townes. Invasive methicillin-resistant *Staphylococcus aureus* infections in the United States. *Journal of the American Medical Association*, **298**(15):1763–1771 (2007).

55. R. Kouyos, E. Y. Klein, and B. Grenfell. Hospital-community interactions foster coexistence between methicillin-resistant strains of *Staphylococcus aureus*. *PLoS Pathogens*, **9**(2):e1003134 (2013).

56. T. Kypraios, P. O'Neill, S. Huang, S. Rifas-Shiman, and B. Cooper. Assessing the role of undetected colonization and isolation precautions in reducing methicillin-resistant *Staphylococcus aureus* transmission in intensive care units. *BMC Infectious Diseases*, **10**(1):29 (2010).

57. D. Laibson. Golden eggs and hyperbolic discounting. *The Quarterly Journal of Economics*, 112: 443–477 (1997).

58. E. J. Langer. The illusion of control. *Journal of Personality and Social Psychology*, **32**(2):311 (1975).

59. B. Y. Lee, S. M. McGlone, K. F. Wong, S. L. Yilmaz, T. R. Avery, Y. Song, R. Christie, S. Eubank, S. T. Brown, J. M. Epstein, J. I. Parker, D. S. Burke, R. M. Platt, and S. S. Huang. Modeling the spread of methicillin-resistant *Staphylococcus aureus* (MRSA) Outbreaks throughout the hospitals in Orange County, California. *Infection Control and Hospital Epidemiology*, **32**(6):562–572 (2011).

60. W. Lew, M. Pai, O. Oxlade, D. Martin and D. Menzies. Initial drug resistance and Tuberculosis treatment outcomes: systematic review and meta-analysis. *Annals of Internal Medicine*, **149**(2):123–134 (2008).

61. I. M. Longini Jr, M. Elizabeth Halloran, A. Nizam, Y. Yang, S. Xu, D. S. Burke, D. A. T. Cummings, and J. M. Epstein. Containing a large bioterrorist smallpox attack: a computer simulation approach. *International Journal of Infectious Diseases*, **11**(2):98–108 (2007).

62. A. Marathe, B. Lewis, J. Chen, and S. Eubank. Sensitivity of household transmission to household contact structure and size. *PLoS ONE*, **6**(8):e22461 (2011).

63. M. Marathe and A. K. S. Vullikanti. Computational epidemiology. *Communications of the ACM*, **56**(7):88–96 (2013).

64. P. D. Mauldin, C. D. Salgado, I. S. Hansen, D. T. Durup, and J. A. Bosso. Attributable hospital cost and length of stay associated with health care-associated infections caused by antibiotic-resistant gram-negative bacteria. *Antimicrobial Agents and Chemotherapy*, **54**(1):109–115 (2010).

65. J. Mossong, N. Hens, M. Jit, P. Beutels, K. Auranen, R. Mikolajczyk, M. Massari, S. Salmaso, G. S. Tomba, J. Wallinga, J. Heijne, M. Sadkowska-Todys, M. Rosinska, and W. J. Edmunds. Social contacts and mixing patterns relevant to the spread of infectious diseases. *PLoS Medicine*, **5**(3):e74 (2008).

66. S. Ong, J. Nakase, G. J. Moran, D. J. Karras, M. J. Kuehnert, and D. A. Talan. Antibiotic use for emergency department patients with upper respiratory infections: prescribing practices, patient expectations, and patient satisfaction. *Annals of Emergency Medicine*, **50**(3):213–220 (2007).

67. T. J. Philipson and R. A. Posner. *Private Choices and Public Health: The AIDS Epidemic in an Economic Perspective*. Harvard University Press, Cambridge, MA, 1993.

68. D. Pittet, B. Allegranzi, H. Sax, S. Dharan, C. L. Pessoa-Silva, L. Donaldson, and J. M. Boyce. Evidence-based model for hand transmission during patient care and the role of improved practices. *The Lancet Infectious Diseases*, **6**(10):641–652 (2006).

69. J. P. Raboud, R. M. Saskin, A. M. D. Simor, M. M. D. M. Loeb, K. R. N. Green, M. D. Don E. Low, and M. D. M. Allison McGeer. Modeling transmission of methicillin resistant *Staphylococcus aureus* among patients admitted to a hospital. *Infection Control and Hospital Epidemiology*, **26**(7):607–615 (2005).

70. R. R. Roberts, B. Hota, I. Ahmad, R. D. Scott, S. D. Foster, F. Abbasi, S. Schabowski, L. M. Kampe, G. G. Ciavarella, and M. Supino. Hospital and societal costs of antimicrobial-resistant infections in a Chicago teaching hospital: implications for antibiotic stewardship. *Clinical Infectious Diseases*, **49**(8):1175–1184 (2009).

71. R. Ross. *The Prevention of Malaria, 2nd ed.* Murray, London, 1911.

72. G. Shanks, K. Biomndo, S. Hay, and R. Snow. Changing patterns of clinical malaria since 1965 among a tea estate population located in the Kenyan highlands. *Transactions of the Royal Society of Tropical Medicine and Hygiene*, **94**(3):253–255 (2000).

73. E. Shapiro. Injudicious antibiotic use: an unforeseen consequence of the emphasis on patient satisfaction? *Clinical Therapeutics*, **24**(1):197–204 (2002).

74. H. A. Simon. *Models of Man; Social and Rational*. Wiley, Oxford, 1957.

75. D. L. Smith, B. T. Lucey, L. A. Waller, J. E. Childs, and L. A. Real. Predicting the spatial dynamics of rabies epidemics on heterogeneous landscapes. *Proceedings of the National Academy of Sciences of the United States of America*, **99**:3668–3672 (2002).

76. D. L. Smith, S. A. Levin, and R. Laxminarayan. Strategic interactions in multi-institutional epidemics of antibiotic resistance. *Proceedings of the National Academy of Sciences of the United States of America*, **102**(8):3153–3158 (2005).

77. I. H. Spicknall, B. Foxman, C. F. Marrs, and J. N. Eisenberg. A modeling framework for the evolution and spread of antibiotic resistance: literature review and model categorization. *American Journal of Epidemiology*, **178**(4):508–520 (2013).

78. L. Sun, E. Y. Klein, and R. Laxminarayan. Seasonality and temporal correlation between community antibiotic use and resistance in the United States. *Clinical Infectious Diseases*, **55**(5):687–694 (2012).

79. J. F. Trape, G. Pison, M. P. Preziosi, C. Enel, A. Desgrees du Lou, V. Delaunay, B. Samb, E. Lagarde, J. F. Molez, and F. Simondon. Impact of chloroquine resistance on malaria mortality. *Comptes Rendus de l'Academie des Sciences. Serie III, Sciences de la Vie*, **321**(8):689–697 (1998).

80. H. M. Truong, T. Kellogg, J. Klausner, M. Katz, J. Dilley, K. Knapper, S. Chen, R. Prabhu, R. Grant, and B. Louie. Increases in sexually transmitted infections and sexual risk behaviour without a concurrent increase in HIV incidence among men who have sex with men in San Francisco: a suggestion of HIV serosorting? *Sexually Transmitted Infections*, **82**(6):461–466 (2006).

81. A. Tversky and D. Kahneman. Belief in the law of small numbers. *Psychological Bulletin*, **76**(2):105–110 (1971).

82. R. A. Weinstein, M. J. M. Bonten, D. J. Austin, and M. Lipsitch. Understanding the spread of antibiotic resistant pathogens in hospitals: mathematical models as tools for control. *Clinical Infectious Diseases*, **33**(10):1739–1746 (2001).

83. C. R. Wells, E. Y. Klein, and C. T. Bauch. Policy resistance undermines superspreader vaccination strategies for influenza. *PLoS Computational Biololgy*, **9**(3):e1002945 (2013).

84. W. C. Wimsatt. "False models as means to truer theories" in (M. Nitecki and A. Holfmen, editors), *Neutral Models in Biology*, p. 23–55. Oxford University Press, New York, 1987.

85. N. Woodford and D. M. Livermore. Infections caused by Gram-positive bacteria: a review of the global challenge. *Journal of Infection*, **59** (Supplement 1):S4–S16 (2009).

86. J. R. Zucker, E. M. Lackritz, T. K. Ruebush, II, A. W. Hightower, J. E. Adungosi, J. B. O. Were, B. Metchock, E. Patrick, and C. C. Campbell. Childhood mortality during and after hospitalization in western Kenya: effect of malaria treatment Regimens. *American Journal of Tropical Medicine and Hygiene*, **55**(6):655–660 (1996).

SECTION 4

MATHEMATICAL MODELS AND ANALYSIS FOR SCIENCE AND ENGINEERING

7

DATA-DRIVEN METHODS FOR DYNAMICAL SYSTEMS: QUANTIFYING PREDICTABILITY AND EXTRACTING SPATIOTEMPORAL PATTERNS

DIMITRIOS GIANNAKIS AND ANDREW J. MAJDA

Courant Institute of Mathematical Sciences, New York University, New York, NY, USA

Large-scale datasets generated by dynamical systems arise in a diverse range of disciplines in science and engineering, including fluid dynamics [50, 96], materials science [58, 59], molecular dynamics [27, 84], and geophysics [31, 75]. A major challenge in these domains is to utilize the vast of data that are being collected by observational networks or output by large-scale numerical models to advance scientific understanding of the operating physical processes and reveal their predictability. For instance, in climate atmosphere ocean science (CAOS), the dynamics takes place in an infinite-dimensional phase space where the coupled nonlinear partial differential equations for fluid flow and thermodynamics are defined, and the observed data correspond to functions of that phase space, such as temperature or circulation measured over a set of spatial points. There exists a strong need for data analysis algorithms to extract and create reduced representations of the large-scale coherent patterns which are an outcome of these dynamics, including the El Niño Southern Oscillation (ENSO) in the ocean [100] and the Madden-Julian Oscillation (MJO) in the atmosphere [70]. Advances in the scientific understanding and forecasting capability of these phenomena have potentially high socioeconomic impact.

Mathematical and Computational Modeling: With Applications in Natural and Social Sciences, Engineering, and the Arts, First Edition. Roderick Melnik.

In this chapter, we review the work of the authors and their collaborators on data-driven methods for dynamical systems to address these objectives. In particular, in Sections 7.1 and 7.2, respectively, we present (i) methods based on data clustering and information theory to reveal predictability in high-dimensional dynamical systems and quantify the fidelity of forecasts made with imperfect models [42, 43, 46]; and (ii) nonlinear Laplacian spectral analysis (NLSA) algorithms [14, 40, 41, 44, 45, 103] for decomposition of spatiotemporal data. The common theme in these topics is that aspects of the coarse-grained geometry of the data in phase space play a role. Specifically, in (i) we discuss how the affiliation of the system state to a discrete partition of phase space can be used as a surrogate variable replacing a high-dimensional vector of initial data in relative entropy functionals measuring predictability and model error. In (ii), we discuss about the coarse-grained geometry of the data that will enter through discrete diffusion operators constructed using dynamics-adapted kernels to provide basis functions for temporal modes of variability analogous to linear-projection coordinates in principal components analysis (PCA, e.g., [50]).

Throughout, we illustrate these techniques with applications to CAOS. In particular, in (i) we study long-range predictability in a simple model [81] of ocean circulation in an idealized basin featuring a current analogous to the Gulf Stream and the Kuroshio Current in the Atlantic and Pacific Oceans, respectively. Such currents are known to undergo changes in configuration affecting continental-scale climate patterns on timescales spanning several months to years. Revealing the predictability of these circulation regimes is important for making skillful initial-value decadal forecasts [82]. In (ii) we present an application of NLSA to a complex spatiotemporal signal of infrared brightness temperature (T_b) acquired through satellites (the CLAUS archive [49]). Because T_b is a good proxy variable for atmospheric convection (deep-penetrating clouds are cold, and therefore produce a strong T_b signal against the emission background from the Earth's surface), an objective decomposition of such data can provide important information about a plethora of climatic processes, including ENSO, the MJO, as well as diurnal-scale processes. This application of NLSA to two-dimensional CLAUS data has not been previously published.

We conclude in Section 7.3 with a synthesis discussion of open problems and possible connections between the two topics.

7.1   QUANTIFYING LONG-RANGE PREDICTABILITY AND MODEL ERROR THROUGH DATA CLUSTERING AND INFORMATION THEORY

7.1.1   Background

Since the classical work of Lorenz [68] and Epstein [33], predictability within dynamical systems has been the focus of extensive study. In the applications outlined in the chapter's preamble, the dynamics span multiple spatial and temporal scales, take place in phase spaces of large dimension, and are strongly mixing. Yet, despite the complex underlying dynamics, several phenomena of interest are organized around a relatively small number of persistent states (so-called regimes), which are predictable over timescales significantly longer than suggested by decorrelation times or Lyapunov exponents. Such phenomena often occur in these applications in variables with nearly Gaussian equilibrium statistics [7, 79] and with dynamics that is very different [34] from the more familiar gradient flows (arising, e.g., in molecular dynamics), where long-range predictability also often occurs [27, 28]. In certain cases, such as CAOS [54, 76] and econometrics [101], seasonal effects play an important role, resulting in time-periodic statistics. In either case, revealing predictability in these systems is important from both a practical and a theoretical standpoint.

Another issue of key importance is to quantify the fidelity of predictions made with imperfect models when (as is usually the case) the true dynamics of nature cannot be feasibly integrated or are simply not known [72, 88]. Prominent techniques for building imperfect predictive models of regime behavior include finite-state methods, such as hidden Markov models [34, 79] and cluster-weighted models [12], as well as continuous models based on approximate equations of motion, for example, linear inverse models [86, 99] and stochastic mode elimination [78]. Other methods blend aspects of finite-state and continuous models, employing clustering algorithms to derive a continuous local model for each regime, together with a finite-state process describing the transitions between regimes [36, 52–54].

The fundamental perspective adopted here is that predictions in dynamical systems correspond to transfer of information, specifically, transfer of information between the initial data (which in general do not suffice to completely determine the state of the system) and a target variable to be forecast. This opens up the possibility of using the mathematical framework of information theory to characterize predictability and model error [23, 24, 64–66, 72, 77, 88, 90, 94, 99], as well as to perform uncertainty quantification [71].

The prototypical problem we wish to address here is illustrated in Figure 7.1. There, our prior knowledge about an observable a_t to be predicted at time t is represented by a distribution $p(a_t)$, which in general may be time-dependent. For instance, if a_t is the temperature measured over a geographical region, then time-dependence

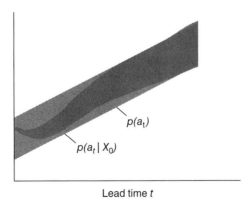

Lead time t

FIGURE 7.1   Illustration of a statistical forecasting problem. Figure adopted from Ref. [10].

of $p(a_t)$ would be due to the seasonality of the Earth's climate or an underlying slow trend occurring in a climate change scenario. Contrasted with $p(a_t)$ is the posterior distribution $p(a_t \mid X_0)$ representing our knowledge about a_t given that we have observed initial data X_0 at time $t = 0$. In the temperature forecasting example, atmospheric variables, such as wind fields, pressure, and moisture, as well oceanic circulation, would all be employed to make an initial-value forecast about a_t. At short times, one would expect $p(a_t \mid X_0)$ to depend very strongly on the initial data and have mass concentrated in a significantly narrower range of a_t values than the prior distribution. This situation, where the availability of highly-resolved initial data plays a crucial role, has been termed by Lorenz [69] as a predictability problem of the first kind. On the other hand, due to mixing dynamics, the predictive informa-tion contributed by X_0 is expected to decay at late times, and eventually $p(a_t \mid X_0)$ will converge to $p(a_t)$. In this second-kind predictability problem, knowledge of the "boundary conditions" is important. In climate science, boundary conditions would include anthropogenic and volcanic emissions, changes in solar insolation, and so on. At the interface between these two types of predictability problems lie long-range initial-value forecasts (e.g., [82]), which will be the focus of the work presented here. Here, the forecast lead time is short enough so that $p(a_t \mid X_0)$ differs significantly from $p(a_t)$, but long enough so that fine-grained aspects of the initial data contribute little predictive information beyond forecasts with coarse-grained initial data.

In all of the cases discussed, a common challenge is that the initial data X_0 are generally high-dimensional even if the target observable a_t is a scalar. Indeed, in present-day numerical weather and climate forecasting, one has to assimilate a very comprehensive set of initial data if only to produce a point forecast about a quan-tity of interest. Here, we advocate that the information-theoretic framework can be combined with data clustering algorithms to produce lower bounds to intrinsic pre-dictability and model error, which are practically computable for high-dimensional initial data. These bounds are derived by replacing the high-dimensional space of ini-tial data by the integer-valued affiliation function to a coarse-grained partition of that space, constructed by clustering a training dataset generated by a potentially imper-fect model. The algorithm for building the partition can be general and designed according to the problem at hand in the following text, we describe a concrete scheme based on K-means clustering augmented with a notion of temporal persistence in clus-ter affiliation through running averages of initial data. We apply this scheme to study long-range predictability in an equivalent barotropic, double-gyre model of ocean circulation, and the fidelity of coarse-grained Markov models for ocean circulation regime transitions.

7.1.2 Information theory, predictability, and model error

7.1.2.1 Predictability in a perfect-model environment We consider the general setting of a stochastic dynamical system

$$dz = F(z,t)\,dt + G(z,t)\,dW \quad \text{with } z \in \mathbb{R}^m, \tag{7.1}$$

which is observed through (typically, incomplete) measurements

$$x(t) = H(z(t)), \quad x(t) \in \mathbb{R}^n, \quad n \leq m. \tag{7.2}$$

As reflected by the explicit dependence of the deterministic and stochastic coefficients in (7.1) on time and the state vector, the dynamics of $z(t)$ may be nonstationary and forced by nonadditive noise. However, in the application of Section 7.1.5, the dynamics will be deterministic with time-independent equilibrium statistics. In particular, $z(t)$ is given by the streamfunction ψ of an equivalent-barotropic ocean model, and H will be a projection operator from z to the leading 20 principal components (PCs) of ψ. We refer the reader to Ref. [46] for applications involving non-stationary stochastic dynamics.

Let $a_t = a(z(t))$ be a target variable for prediction which can be expressed as a function of the state vector. Let also

$$X_t = (x(t), x(t - \delta t), \ldots, x(t - (q-1)\delta t)), \quad X_t \in \mathbb{R}^N, \quad N = qn, \tag{7.3}$$

with $x(t_i)$ given from (7.2) and q a positive integer, be a history of observations collected over a time window $\Delta t = (q-1)\delta t$. Hereafter, we refer to the observations X_0 at time $t = 0$ as initial data. Broadly speaking, the question of dynamical predictability in the setting of (7.1) and (7.2) may be posed as follows: Given the initial data, how much information have we gained about a_t at time $t > 0$ in the future? Here, uncertainty in a_t arises because of both the incomplete nature of the measurements in (7.2) and the stochastic component of the dynamical system in (7.1). Thus, it is appropriate to describe a_t via some time-dependent probability distribution $p(a_t \mid X_0)$ conditioned on the initial data. Predictability of a_t is understood in this context as the additional information contained in $p(a_t \mid X_0)$ relative to the prior distribution [23,41,64], $p(a_t) = \mathbb{E}_{X_0} p(a_t \mid X_0)$, which we now specify.

Throughout, we consider that our knowledge of the system before the observations becomes available is described by a statistical equilibrium state $p_{eq}(z(t))$, which is time dependent (e.g., time-periodic [46]). An assumption made here when $p_{eq}(z(t))$ is time independent is that $z(t)$ is ergodic, with

$$\mathbb{E}_{p_{eq}} a_t \approx \frac{1}{s} \sum_{i=0}^{s-1} a(z(t - i\delta t)) \tag{7.4}$$

for a large-enough number of samples s. In all of these cases, the prior distributions for a_t and X_t are the distributions $p_{eq}(a_t)$ and $p_{eq}(X_t)$ induced on these variables by $p_{eq}(z(t))$, that is,

$$p(a_t) = p_{eq}(a_t), \quad p(X_t) = p_{eq}(X_t). \tag{7.5}$$

As the forecast lead time grows, $p(a_t \mid X_0)$ converges to $p_{eq}(a_t)$, at which point X_0 contributes no additional information about a_t beyond equilibrium.

The natural mathematical framework to quantify predictability in this setting is information theory [19] and, in particular, the concept of relative entropy. The latter is defined as the functional

$$P(p'(a_t), p(a_t)) = \mathbb{E}_{p'} \left(\frac{\log p'(a_t)}{p(a_t)} \right)$$

between two probability measures, $p'(a_t)$ and $p(a_t)$, and has the attractive properties that (i) it vanishes if and only if $p = p'$ and is positive if $p \neq p'$ and (ii) is invariant under general invertible transformations of a_t. For our purposes, the so-called Bayesian-update interpretation of relative entropy is also of key importance. This states that if $p'(a_t) = p(a_t \mid X_0)$ is the posterior distribution of a_t conditioned on some variable X_0 and p is the corresponding prior distribution, then $P(p'(a_t), p(a_t))$ measures the additional information beyond p about a_t gained by having observed X_0. This interpretation stems from the fact that

$$P(p(a_t \mid X_0), p(a_t)) = \mathbb{E}_{a_t \mid X_0} \log \left(\frac{p(a_t \mid X_0)}{p(a_t)} \right) \tag{7.6}$$

is a nonnegative quantity (by Jensen's inequality), measuring the expected reduction in ignorance about a_t relative to the prior distribution $p(a_t)$ when X_0 has become available [19, 88]. It is therefore crucial that $p(a_t \mid X_0)$ is inserted in the first argument of $P(\cdot, \cdot)$ for a correct assessment of predictability.

The natural information-theoretic measure of predictability compatible with the prior distribution $p(a_t)$ in (7.5) is

$$\mathcal{D}(a_t \mid X_0) = P(p(a_t \mid X_0), p(a_t)). \tag{7.7}$$

As one may explicitly verify, the expectation value of $\mathcal{D}(a_t \mid X_0)$ with respect to the prior distribution for X_0,

$$\mathcal{D}(a_t, X_0) = \mathbb{E}_{X_0} \mathcal{D}(a_t \mid X_0)$$
$$= \mathbb{E}_{a_t, X_0} \log \left(\frac{p(a_t \mid X_0)}{p(a_t)} \right) = P(p(a_t, X_0), p(a_t)p(X_0)) \tag{7.8}$$

is also a relative entropy, here, between the joint distribution of the target variable and the initial data and the product of their marginal distributions. This quantity is known as the mutual information between a_t and X_0, measuring the expected predictability of the target variable over the initial data [23, 41, 90].

One of the classical results in information theory is that the mutual information between the source and output of a channel measures the rate of information flow across the channel [19]. The maximum mutual information over the possible source distributions corresponds to the channel capacity. In this regard, an interesting parallel between prediction in dynamical systems and communication across channels is that the combination of dynamical system and observation apparatus (represented here by (7.1) and (7.2)) can be thought of as an abstract communication channel with the initial data X_0 as input and the target a_t as output.

7.1.2.2 Quantifying the error of imperfect models

The analysis in Section 7.1.2.1 was performed in a perfect-model environment. Frequently, however, instead of the true forecast distributions $p(a_t \mid X_0)$, one has access to distributions $p^M(a_t \mid X_0)$ generated by an imperfect model,

$$dz(t) = F^M(z,t)\,dt + G^M(z,t)\,dW. \tag{7.9}$$

Such situations arise, for instance, when one cannot afford to feasibly integrate the full dynamical system in (7.1) (e.g., simulations of biomolecules dissolved in a large number of water molecules) or when the laws governing $z(t)$ are simply not known (e.g., condensation mechanisms in atmospheric clouds). In other cases, the objective is to develop reliable reduced models for $z(t)$ to be used as components of coupled models (e.g., parameterization schemes in climate models [55]). In this context, assessments of the error in the model prediction distributions are of key importance but frequently not carried out in an objective manner that takes into account both the mean and variance [72].

Relative entropy again emerges as the natural information-theoretic functional for quantifying model error. Now, the analog between dynamical systems and coding theory is with suboptimal coding schemes. In coding theory, the expected penalty in the number of bits needed to encode a string assuming that it is drawn from a probability distribution q, when in reality the source probability distribution is p', is given by $\mathcal{P}(p',q)$ (evaluated in this case with base-2 logarithms). Similarly, $\mathcal{P}(p',q)$ with p' and q equal to the distributions of a_t conditioned on X_0 in the perfect and imperfect model, respectively, leads to the error measure

$$\mathcal{E}(a_t \mid X_0) = \mathcal{P}(p(a_t \mid X_0), p^M(a_t \mid X_0)). \tag{7.10}$$

By direct analogy with (7.6), $\mathcal{E}(a_t \mid X_0)$ is a nonnegative quantity measuring the expected increase in ignorance about a_t incurred by using the imperfect model distribution $p^M(a_t \mid X_0)$ when the true state of the system is given by $p(a_t \mid X_0)$ [72,77,88]. As with (7.7), $p(a_t \mid X_0)$ must appear in the first argument of $\mathcal{P}(\cdot, \cdot)$ for a correct assessment of model error. Moreover, $\mathcal{E}(a_t \mid X_0)$ may be aggregated into an expected model error over the initial data,

$$\mathcal{E}(a_t, X_0) = \mathbb{E}_{X_0} \mathcal{E}(a_t \mid X_0) = \mathbb{E}_{a_t, X_0} \log\left(\frac{p(a_t \mid X_0)}{p^M(a_t \mid X_0)}\right). \tag{7.11}$$

However, unlike $\mathcal{D}(a_t, X_0)$ in (7.8), $\mathcal{E}(a_t, X_0)$ does not correspond to a mutual information between random variables.

Note that by writing down (7.10) and (7.11), we have tacitly assumed that the target variable can be simultaneously defined in the perfect and imperfect models, that is, a_t can be expressed as a function of either $z(t)$ or $z^M(t)$. Even though z and z^M may lie in completely different phase spaces, in practice one is typically interested in large-scale coarse-grained target variables (e.g., the mean temperature over a geographical region of interest), which are well-defined in both the perfect and imperfect model.

7.1.3   Coarse-graining phase space to reveal long-range predictability

Despite their theoretical appeal, the predictability and model error measures $\mathcal{D}(a_t \mid X_0)$ and $\mathcal{E}(a_t \mid X_0)$ are frequently infeasible to evaluate in practice, the reason being that both of these measures require the evaluation of an expectation value over the initial data X_0. As stated in Section 7.1.1, the spaces of initial data used for making predictions in complex systems are generally high dimensional, even if the target observable a_t is a scalar. Operationally, computing the expectation $\mathbb{E}_{X_0}$ requires evaluation of an integral over X_0 that rapidly becomes intractable as the dimension of X_0 grows. Here, we address this "curse of dimension" issue by replacing X_0 with an integer-valued surrogate variable S_0 representing the affiliation of X_0 in a partition of the initial-data space. By the data-processing inequality in information theory [19], the coarse-grained predictability and model error metrics $\mathcal{D}(a_t, S_0)$ and $\mathcal{E}(a_t, S_0)$, respectively, provide lower bounds to $\mathcal{D}(a_t, X_0)$ and $\mathcal{E}(a_t, X_0)$ which are practically computable for high-dimensional initial data.

7.1.3.1   *Perfect-model scenario*   Our method of partitioning the space of initial data, described also in Refs. [41, 46, 46], proceeds in two stages: a training stage and prediction stage. The training stage involves taking a dataset

$$\mathcal{X} = \{x(0), x(\delta t), \ldots, x((s-1)\,\delta t)\}, \tag{7.12}$$

of s observation samples $x(t) \in \mathbb{R}^n$ and computing via data clustering a collection

$$\Theta = \{\theta_1, \ldots, \theta_K\}, \quad \theta_k \in \mathbb{R}^p \tag{7.13}$$

of parameter vectors θ_k characterizing the clusters. Used in conjunction with a rule (e.g., (7.36) ahead), for determining the integer-valued affiliation $S(X_0)$ of a vector X_0 from (7.3), the cluster parameters lead to a mutually-disjoint partition of the set of initial data , viz.

$$\Xi = \{\xi_1, \ldots, \xi_K\}, \quad \xi_k \subset \mathbb{R}^N, \tag{7.14}$$

such that $S(X_0) = S_0$ indicates that the membership of X_0 is with cluster $\xi_{S_0} \in \Xi$. Thus, a dynamical regime is understood here as an element ξ_k of Ξ and coarse-graining as a projection $X_0 \mapsto S_0$ from the (generally, high dimensional) space of initial data to the integer-valued membership S_0 in the partition. It is important to note that $\mathcal{X}$ may consist of either observations $x(t)$ of the perfect model from (7.2), or data generated by an imperfect model (which does not have to be the same as the model in (7.9) used for prediction). In the latter case, the error in the training data influences the amount of information loss by coarse graining but does not introduce biases that would lead one to overestimate predictability.

Because $S_0 = S(X_0)$ is uniquely determined from X_0, it follows that

$$p(a_t \mid X_0, S_0) = p(a_t \mid X_0). \tag{7.15}$$

The above expresses the fact that no additional information about the target variable a_t is gained through knowledge of S_0 if X_0 is known. Moreover, (7.15) leads to a Markov property between the random variables a_t, X_0, and S_0, viz.

$$p(a_t, X_0, S_0) = p(a_t \mid X_0, S_0)p(X_0 \mid S_0)p(S_0) = p(a_t \mid X_0)p(X_0 \mid S_0)p(S_0). \tag{7.16}$$

The latter is a necessary condition for the predictability and model error bounds discussed below.

Equation (7.15) also implies that the forecasting scheme based on X_0 is statistically sufficient [11, 22] for the scheme based on S_0. That is, the predictive distribution $p(a_t \mid S_0)$ conditioned on the coarse-grained initial data can be expressed as an expectation value

$$p(a_t \mid S_0) = \mathbb{E}_{X_0 \mid S_0} p(a_t \mid X_0) \tag{7.17}$$

of $p(a_t \mid X_0)$ with respect to the distribution $p(X_0 \mid S_0)$ of the fine-grained initial data X_0 given S_0. We use the shorthand notation

$$p_t^k = p(a_t \mid S_0 = k), \tag{7.18}$$

for the forecast distribution for a_t conditioned on the kth cluster.

In the prediction stage, the p_t^k are estimated for each $k \in \{1, \dots, K\}$ by bin-counting joint realizations of a_t and S_0, using data that are independent from the dataset $\mathcal{X}$ employed in the training stage (details about the bin-counting procedure are provided in Section 7.1.4). The predictive information content in the partition is then measured via coarse-grained analogs of the relative-entropy metrics in (7.7) and (7.8), viz.,

$$\mathcal{D}(a_t \mid S_0) = \mathcal{P}(p(a_t \mid S_0), p(a_t)) \quad \text{and} \quad \mathcal{D}(a_t, S_0) = \mathbb{E}_{S_0} \mathcal{D}(a_t \mid S_0). \tag{7.19}$$

By the same arguments used to derive (7.8), it follows that the expected predictability measure $\mathcal{D}(a_t, S_0)$ is equal to the mutual information between the target variable a_t at time $t \geq 0$ and the membership S_0 of the initial data in the partition at time $t = 0$. Note the formula

$$\mathcal{D}(a_t, S) = \sum_{k=1}^{K} \pi_k \mathcal{D}_t^k, \quad \text{with } \mathcal{D}_t^k = \mathcal{P}(p_t^k, p_{\text{eq}}), \quad \pi_k = p(S = k). \tag{7.20}$$

Two key properties of $\mathcal{D}(a_t, S)$ are as follows:

1. It provides a lower bound to the predictability measure $\mathcal{D}(a_t, X_0)$ in (7.8) determined from the fine-grained initial data X_0, that is,

$$\mathcal{D}(a_t, X_0) \geq \mathcal{D}(a_t, S_0) \tag{7.21}$$

2. Unlike $\mathcal{D}(a_t, X_0)$, which requires evaluation of an integral over X_0 that rapidly becomes intractable as the dimension of X_0 grows (even if the target variable is scalar), $\mathcal{D}(a_t, S_0)$ requires only the evaluation of a discrete sum over S_0.

Equation (7.21), which is known in information theory as data-processing inequality [24, 41], expresses the fact that coarse-graining, $X_0 \mapsto S(X_0)$, can only lead to conservation or loss of information. In particular, it can be shown [13] that the Markov property in (7.16) leads to the relation

$$\mathcal{D}(a_t, X_0) = \mathcal{D}(a_t, S_0) + \mathcal{I}, \tag{7.22}$$

where

$$\mathcal{I} = \mathbb{E}_{a_t, X_0, S_0} \log \left(\frac{p(X_0 \mid a_t, S_0)}{p(X_0 \mid S_0)} \right) \tag{7.23}$$

is a nonnegative term measuring the loss of predictive information due to coarse-graining of the initial data. Because the nonnegativity of $\mathcal{I}$ relies only on the existence of a coarse-graining function meeting the condition (7.15), and not on the properties of the training data $\mathcal{X}$ used to construct that function, there is no danger of over-estimating predictability through $\mathcal{D}(a_t, S_0)$, even if an imperfect model is employed to generate $\mathcal{X}$. Thus, $\mathcal{D}(a_t, S_0)$ can be used practically as a sufficient condition for predictability, irrespective of model error in $\mathcal{X}$ and/or suboptimality of the clustering algorithm.

In general, the information loss $\mathcal{I}$ will be large at short lead times, but in many applications involving strongly-mixing dynamical systems, the predictive information in the fine-grained aspects of the initial data will rapidly decay as t grows. In such scenarios, $\mathcal{D}(a_t, S_0)$ provides a tight bound to $\mathcal{D}(a_t, X_0)$, with the crucial advantage of being feasibly computable with high-dimensional initial data. Of course, failure to establish predictability on the basis of $\mathcal{D}(a_t, S_0)$ does not imply absence of intrinsic predictability, for it could be that $\mathcal{D}(a_t, S_0)$ is small because $\mathcal{I}$ is comparable to $\mathcal{D}(a_t, X_0)$.

Since relative entropy is unbounded from above, it is useful to convert $\mathcal{D}(a_t, S_0)$ into a predictability score lying in the unit interval

$$\delta_t = 1 - \exp(-2\mathcal{D}(a_t, S_0)). \tag{7.24}$$

Joe [57] shows that the above definition for δ_t is equivalent to a squared correlation measure, at least in problems involving Gaussian random variables.

7.1.3.2 *Quantifying the model error in long-range forecasts* Consider now an imperfect model that, as described in Section 7.1.2.2, produces prediction distributions

$$p_t^{Mk} = p^M(a_t \mid S_0 = k) \tag{7.25}$$

which may be systematically biased away from p_t^k in (7.18). Similarly to Section 7.1.3.1, we consider that the random variables a_t, X_0, and S_0 in the imperfect model have a Markov property,

$$\begin{aligned} p^M(a_t, X_0, S) &= p^M(a_t \mid X_0, S_0)p(X_0 \mid S_0)p(S_0) \\ &= p^M(a_t \mid X_0)p(X_0 \mid S_0)p(S_0), \end{aligned}$$

where we have also assumed that the same initial data and cluster affiliation function are employed to compare the perfect and imperfect models (i.e., $p^M(X_0 \mid S_0) = p(X_0 \mid S_0)$ and $p^M(S_0) = p(S_0)$). As a result, the coarse-grained forecast distributions in (7.25) can be determined via (cf. (7.17))

$$p^M(a_t \mid S_0) = \mathbb{E}_{X_0 \mid S_0} p^M(a_t \mid X_0). \tag{7.26}$$

In this setup, an obvious candidate measure for predictability follows by writing down (7.19) with p_t^k replaced by p_t^{Mk}, that is,

$$\mathcal{D}^M(a_t, S_0) = \mathbb{E}_{S_0} \mathcal{D}^M(a_t \mid S_0) = \sum_{k=1}^{K} \pi_k \mathcal{D}_t^{Mk}, \quad \text{with} \quad \mathcal{D}_t^{Mk} = \mathcal{P}(p_t^{Mk}, p_{\mathrm{eq}}^M). \tag{7.27}$$

By direct analogy with (7.21), $\mathcal{D}^M(a_t, S_0)$ provides a nonnegative lower bound of $\mathcal{D}^M(a_t, X_0)$. Clearly, an important deficiency of this measure is that by being based solely on forecast distributions internal to the model, it fails to take into account model error or "ignorance" of the imperfect model in (7.9) relative to the perfect model in (7.1) [40,72,88]. Nevertheless, $\mathcal{D}^M(a_t, S_0)$ provides an additional metric to discriminate between imperfect models with similar $\mathcal{E}(a_t, X_0)$ scores from (7.11) and estimate how far a given imperfect forecast is from the model's equilibrium distribution. For the latter reasons, we include $\mathcal{D}^M(a_t, S_0)$ as part of our model assessment framework. Following (7.24), we introduce for convenience a unit-interval normalized score

$$\delta_t^M = 1 - \exp(-2\mathcal{D}^M(a_t, S_0)). \tag{7.28}$$

Next, note the distinguished role that the imperfect-model equilibrium distribution plays in (7.27): If $p_{\mathrm{eq}}^M(a_t)$ differs systematically from the equilibrium distribution $p_{\mathrm{eq}}(a_t)$ in the perfect model, then $\mathcal{D}^M(a_t, S_0)$ conveys false predictability at *all* times (including $t = 0$), irrespective of the fidelity of $p^M(a_t \mid S_0)$ at finite times. This observation leads naturally to the requirement that long-range forecasting models must reproduce the equilibrium statistics of the perfect model with high fidelity. In the information-theoretic framework of Section 7.1.2.2, this is expressed as

$$\varepsilon_{\mathrm{eq}} \ll 1, \quad \mathrm{with} \quad \varepsilon_{\mathrm{eq}} = 1 - \exp(-2\mathcal{E}_{\mathrm{eq}}(a_t)) \tag{7.29}$$

and

$$\mathcal{E}_{\mathrm{eq}}(a_t) = \mathcal{P}(p_{\mathrm{eq}}(a_t), p_{\mathrm{eq}}^M(a_t)).$$

Here, we refer to the criterion in (7.29) as equilibrium consistency; an equivalent condition is called fidelity [25] or climate consistency [40] in CAOS work.

Even though equilibrium consistency is a necessary condition for skillful long-range forecasts, it is not a sufficient condition. In particular, the model error $\mathcal{E}(a_t, X_0)$ at finite lead time t may be large, despite eventually decaying to a small value at asymptotic times. The expected error in the coarse-grained forecast distributions is expressed in direct analogy with (7.11) as

$$\mathcal{E}(a_t, S_0) = \mathbb{E}_{S_0} \mathcal{E}(a_t \mid S_0) = \sum_{k=1}^{K} \pi_k \mathcal{E}_t^k, \quad \mathrm{with} \quad \mathcal{E}_t^k = \mathcal{P}(p_t^k, p_t^{Mk}), \tag{7.30}$$

and corresponding error score

$$\varepsilon_t = 1 - \exp(-2\mathcal{E}_t^K), \quad \varepsilon_t \in [0, 1). \tag{7.31}$$

Similar arguments to those used to derive (7.22) lead to a decomposition [46]

$$\mathcal{E}(a_t, X_0) = \mathcal{E}(a_t, S_0) + \mathcal{I} - \mathcal{J} \tag{7.32}$$

of the model error $\mathcal{E}(a_t, X_0)$ into the coarse-grained measure $\mathcal{E}(a_t, S_0)$, the information loss term $\mathcal{I}$ due to coarse graining in (7.23), and a term

$$\mathcal{J} = \mathbb{E}_{a_t, X_0, S_0} \log \left(\frac{p^M(a_t \mid X_0)}{p^M(a_t \mid S_0)} \right)$$

reflecting the relative ignorance of the fine-grained and coarse-grained forecast distributions in the imperfect model. The important point about $\mathcal{J}$ is that it obeys the bound [46]

$$\mathcal{J} \le \mathcal{I}.$$

As a result, $\mathcal{E}(a_t, S_0)$ is a lower bound of the fine-grained error measure $\mathcal{E}(a_t, X_0)$ in (7.11), that is,

$$\mathcal{E}(a_t, X_0) \ge \mathcal{E}(a_t, S_0). \tag{7.33}$$

Because of (7.33), a detection of a significant $\mathcal{E}(a_t, S_0)$ is sufficient to reject a forecasting scheme based on the fined-grained distributions $p^M(a_t \mid X_0)$. The reverse statement, however, is generally not true. In particular, the error measure $\mathcal{E}(a_t, X_0)$ may be significantly larger than $\mathcal{E}(a_t, S_0)$, even if the information loss $\mathcal{I}$ due to coarse-graining is small. Indeed, unlike $\mathcal{I}$, the $\mathcal{J}$ term in (7.32) is not bounded from below and can take arbitrarily large negative values. This is because the coarse-grained forecast distributions $p^M(a_t \mid S_0)$ are determined through (7.26) by averaging the fine-grained distributions $p^M(a_t \mid X_0)$, and averaging can lead to cancellation of model error. Such a situation with negative $\mathcal{J}$ cannot arise with the forecast distributions of the perfect model, where, as manifested by the nonnegativity of $\mathcal{I}$, coarse-graining can at most preserve information.

In summary, our framework for assessing long-range coarse-grained forecasts with imperfect models takes into consideration all of $\varepsilon_{\mathrm{eq}}$, ε_t, and δ_t^M as follows:

- $\varepsilon_{\mathrm{eq}}$ must be small, that is, the imperfect model should be able to reproduce with high fidelity the distribution of the target variable a_t at asymptotic times (the prior distribution, relative to which long-range predictability is measured).

- The imperfect model must have correct statistical behavior at finite times, that is, ε_t must be small at the forecast lead time of interest.

- At the forecast lead time of interest, the additional information beyond equilibrium δ_t^M must be large, otherwise the model has no utility compared with a trivial forecast drawn for the equilibrium distribution.

In order to evaluate these metrics in practice, the following two ingredients are needed: (i) the training dataset $\mathcal{X}$ in (7.12) to compute the cluster parameters Θ from (7.13). and (ii) simultaneous realizations of a_t (in both the perfect and imperfect models) and $x(t)$ (which must be statistically independent from the data in (i)), to evaluate the cluster-conditional distributions p_t^k and p_t^{Mk}. Note that neither access to the full state vectors $z(t)$ and $z^M(t)$ of the perfect and imperfect models nor knowledge of the equations of motions is required to evaluate the predictability and model error scores proposed here. Moreover, the training dataset $\mathcal{X}$ can be generated by an imperfect model. The resulting partition in that case will generally be less informative in the sense of the $\mathcal{D}(a_t, S_0)$ and $\mathcal{E}(a_t, S_0)$ metrics, but, so long as (ii) can be carried out with small sampling error, $\mathcal{D}(a_t, S_0)$ and $\mathcal{E}(a_t, S_0)$ will still be lower bounds of $\mathcal{D}(a_t, X_0)$ and $\mathcal{E}(a_t, X_0)$, respectively. See Ref. [46] for an example where $\mathcal{D}(a_t, S_0)$ and $\mathcal{E}(a_t, S_0)$ reveal long-range predictability and model error despite substantial model error in the training data.

7.1.4   *K*-means clustering with persistence

We now describe a method based on K-means clustering and running-average smoothing of training and initial data that are able to reveal predictability beyond decorrelation time in the ocean model in Section 7.1.5, as well as in stochastic models with nonlinearities [41]. Besides the number of clusters (regimes) K, our algorithm has two additional free parameters. These are temporal windows, $\Delta t'$ and Δt, used to take running averages of $x(t)$ in the training and prediction stages, respectively. This procedure, which is reminiscent of kernel density estimation methods [92], leads to a two-parameter family of partitions as follows.

First, set an integer $q' \geq 1$ and replace $x(t)$ in (7.12) with the averages over a time window $\Delta t' = (q' - 1)\delta t$, that is,

$$x^{\Delta t'}(t) = \sum_{i=1}^{q'} x(t - (i-1)\delta t)/q'. \tag{7.34}$$

Next, apply K-means clustering [30] to the above coarse-grained training data. This leads to a set of parameters Θ from (7.13) that minimize the sum-of-squares error functional,

$$L(\Theta) = \sum_{k=1}^{K} \sum_{i=q'-1}^{s-1} \gamma_k(i\delta t) \|x^{\Delta t'}(i\delta t) - \theta_k^{\Delta t'}\|_2^2,$$

where

$$\gamma_k(t) = \begin{cases} 1, & k = \Gamma(t), \\ 0, & \text{otherwise}, \end{cases} \qquad \Gamma(t) = \arg\min_j \|x^{\Delta t'}(t) - \theta_j^{\Delta t'}\|_2 \tag{7.35}$$

is the weight of the kth cluster at time $t = i\delta t$, and $\|v\|_2 = \left(\sum_{i=1}^{n} v_i^2\right)^{1/2}$ denotes the Euclidean norm. Note that temporal persistence of $\Gamma(t)$ is an outcome of running-average smoothing of the training data.

In the second (prediction) stage of the procedure, data

$$X_0 = (x(0), x(-\delta t), \dots, x(-(q-1)\delta t))$$

of the form (7.3) are collected over an interval $[-\Delta t, 0]$ with $\Delta t = (q-1)\delta t$, and their average $x^{\Delta t}(0)$ is computed via an analogous formula to (7.34). It is important to note that the initial data X_0 used in the prediction stage are independent of the training dataset. The affiliation function S is then given by

$$S(X_0) = \arg\min_k (\|x^{\Delta t}(0) - \theta_k^{\Delta t'}\|_2); \tag{7.36}$$

that is, $S(X_0)$ depends on both Δt and $\Delta t'$. Because $x^{\Delta t}$ can be uniquely determined from the initial-data vector X_0, (7.36) provides a mapping from X_0 to $\{1, \dots, K\}$, defining the elements of the partition in (7.14) through

$$\xi_k = \{X_t : S(X_t) = k\}.$$

Physically, the width of Δt controls the influence of the past history of the system relative to its current state in assigning cluster affiliation. If the target variable exhibits significant memory effects, taking the running average over a window comparable to the memory time scale should lead to gains of predictive information $\mathcal{D}(a_t, S_0)$, at least for lead times of order Δt or less. We provide an example of this behavior in Section 7.1.5.

For ergodic dynamical systems satisfying (7.4), the cluster-conditional densities p_t^k in (7.18) may be estimated as follows. First, obtain a sequence of observations $x(t')$ (independent of the training dataset $\mathcal{X}$ in (7.12)) and the corresponding time series $a_{t'}$ of the target variable. Second, using (7.36), compute the membership sequence $S_{t'} = S(X_{t'})$ for every time t'. For given lead time t, and for each $k \in \{1, \ldots, K\}$, collect the values

$$\mathcal{A}_t^k = \{a_{t+t'} : S_{t'} = k\}.$$

Then, set distribution bin boundaries $A_0 < A_1 < \ldots$ and compute the occurrence frequencies

$$\hat{p}_t^k(i) = \frac{N_i}{N},$$

where N_i is the number of elements of $\mathcal{A}_t^k$ lying in $[a_{i-1}, a_i]$ and $N = \sum_i N_i$. Note that the A_i are vector-valued if a_t is multivariate. By ergodicity, in the limit of an infinite number of bins and samples, the estimators $\hat{p}_t^k(i)$ converge to the continuous densities p_t^k in (7.18). The equilibrium distribution $p_{\text{eq}}(a_t)$ and the cluster affiliation probabilities π_k in (7.20) may be evaluated in a similar manner. Together, the estimates for p_t^k, p_{eq}, and π_k are sufficient to determine the predictability metrics $\mathcal{D}_t^k$ from (7.19). In particular, if a_t is a scalar variable (as will be the case below), the relative-entropy integrals in (7.19) can be carried out by standard one-dimensional quadrature, for example, the trapezoidal rule. This simple procedure is sufficient to estimate the cluster-conditional densities with little sampling error for the univariate target variables in Section 7.1.5. For non-ergodic systems and/or lack of availability of long realizations, more elaborate methods (e.g., [61]) may be required to produce reliable estimates of $\mathcal{D}(a_t, S_0)$.

We close this section with an important point about the forecast distributions from (7.18): Because p_t^k are evaluated independently for each pair $\Delta \mathcal{T} = (\Delta t, \Delta t')$ of running-average intervals, there is no reason why one should use the same $p_t^k|_{\Delta \mathcal{T}}$ for all lead times. In particular, given a collection $\{\Delta \mathcal{T}_1, \Delta \mathcal{T}_2, \ldots\}$ of coarse-graining parameters, the natural forecast distribution to use are the ones that maximize the expected predictability (7.20), viz.

$$p_t^{*k} = p_t^k|_{\Delta \mathcal{T}_i}, \quad i = \arg\max_j \mathcal{D}(a_t, S_0)|_{\Delta \mathcal{T}_j},$$

with corresponding predictability score

$$\mathcal{D}^*(a_t, S_0) = \mathcal{D}(a_t, S_0)|_{\Delta \mathcal{T}_i}, \quad \delta_t^* = 1 - \exp(-2\mathcal{D}^*(a_t, S_0)). \tag{7.37}$$

We will see in Section 7.1.5 that the p_t^{*k} can contain significantly more information than the individual forecast distributions p_t^k.

7.1.5 Demonstration in a double-gyre ocean model

The so-called 1.5-layer model [81] describes the dynamics of wind-driven ocean circulation as the motion of two immiscible, vertically-averaged layers of fluid of different density under the influence of wind-induced shear, Coriolis force (in the β-plane approximation), and subgrid-scale diffusion. The lower layer is assumed to be infinitely deep and at rest, whereas the upper layer is governed by a quasi-geostrophic equation for the streamfunction $\psi(\mathbf{r},t)$ (which, in this case is equal to the interface displacement) at position $\mathbf{r} = (x,y)$ and time t, giving the velocity vector $\mathbf{v} = (\partial_y \psi, -\partial_x \psi)$. The kinetic and potential energies, respectively, given by $E_{\text{kin}} = C_{\text{kin}} \int d\mathbf{r} \, \|\mathbf{v}(\mathbf{r},t)\|^2$ and $E_{\text{pot}} = C_{\text{pot}} \int d\mathbf{r} \, \psi^2(\mathbf{r},t)$ with C_{kin}, C_{pot} constants, make up the total energy, $E = E_{\text{kin}} + E_{\text{pot}}$. The latter will be one of our main prediction observables.

We adopt throughout the model parameter values in section 2 of Ref. [81], as well as their canonical asymmetric double-gyre wind forcing. With this forcing, the 1.5-layer model develops an eastward-flowing separating jet configuration analogous to the Gulf Stream in the North Atlantic. Moreover, the model features the essential dynamical mechanisms of equivalent barotropic Rossby waves, lateral shear instability, and damping. The model was integrated by R. Abramov using a pseudospectral code on a 180×140 uniform grid of size $\Delta r = 20$ km and fourth-order Runge-Kutta timestepping of size $t^+ = 3$ h. The resulting time-averaged streamfunction and its standard deviation, $\Psi(\mathbf{r}) = \langle \psi(\mathbf{r},t) \rangle$ and $\sigma(\mathbf{r}) = \langle \psi^2(\mathbf{r},t) - \Psi^2(\mathbf{r}) \rangle^{1/2}$, where $\langle f(t) \rangle = \int_0^T dt f(t)/T$ denotes empirical temporal averaging, are shown in Figure 7.2. In that, the eastward jet is seen to separate from the western boundary $x = 0$ approximately at meridional coordinate $y = 0$ and to follow a characteristic sinusoid path as it penetrates into the basin. The meridional asymmetry of the wind forcing is manifested in the somewhat stronger anticyclonic gyre in the southern portion of the domain.

The phenomenological study of McCalpin and Haidvogel [81] has determined that in this parameter regime the time of viscous decay of westward-propagating eddies can be small, comparable, or large relative to the drift time taken for the eddies to reach the western boundary current (the drift time increases with the eastward position in the domain where an eddy forms). The eddies that survive long enough to reach the western meridional boundary perturb the eastward current, resulting in a meander-like pattern. Otherwise, in the absence of eddy interaction, the current penetrates deeply into the basin. As shown in Figure 7.2, most of the variance of the time-averaged state is concentrated in the portion of the domain occupied by the jet.

Because of the intermittent nature of the current-eddy interaction, the model exhibits interesting low-frequency variability, characterized by infrequent transitions between a small number of metastable states. These metastable states may be differentiated by their distinct ranges of energy content (e.g., Fig. 7.7). Empirically, three metastable states have been identified, consisting of high, middle, and low-energy

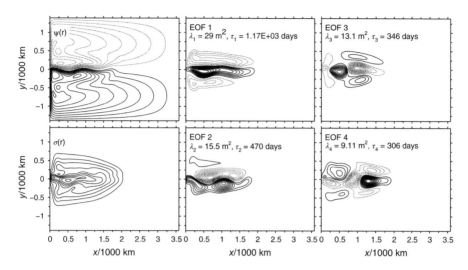

FIGURE 7.2 The time-averaged state, $\Psi(\mathbf{r})$, its standard deviation, $\sigma(\mathbf{r})$, and the leading four streamfunction-metric EOFs, evaluated using an equilibrated realization of the 1.5-layer model of length $T = 10{,}000$ years sampled every $\delta t = 20$ days. The contour levels in the panels for $\Psi(\mathbf{r})$ and $\sigma(\mathbf{r})$ are spaced by 12.5 m, spanning the interval $[-150, 150]$ m. Contours are drawn every 12.5 arbitrary units in the panels for $\text{EOF}_i(\mathbf{r})$, which also indicate the corresponding eigenvalues and correlation times, respectively, λ_i and τ_i. Solid and dotted lines correspond to positive and negative contour levels, respectively. The separation point of the eastward jet is located near the coordinate origin, $\mathbf{r} = (x, y) = (0, 0)$. The eigenvalues and EOFs are the solutions of the eigenproblem $\int d\mathbf{r}' \, C(\mathbf{r}, \mathbf{r}') \text{EOF}_i(\mathbf{r}') = \lambda_i \text{EOF}_i(\mathbf{r})$ associated with the covariance matrix $C(\mathbf{r}, \mathbf{r}') = \int_0^T dt \, \psi'(\mathbf{r}, t) \psi'(\mathbf{r}', t)/T$, where $\psi'(\mathbf{r}, t)$ is the streamfunction anomaly. With this definition, the physical dimension of the λ_i is $(\text{length})^2$. The correlation times are given by $\tau_i = \int_0^T dt \, |\rho_i(t)|$, where ρ_i is the autocorrelation function of the corresponding PC (see Fig. 7.3).

configurations [6, 81]. As illustrated in Figure 7.6, the high-energy state is dominated by a strong elongated jet, which penetrates deep into the basin. On the other hand, the jet is significantly weakened in the low-energy state, where the most prominent features are meandering flow structures. The middle-energy state is characterized by a moderately-penetrating jet that correlates strongly with the spatial configuration of the mean state. Yet, in spite of the prominent regime behavior, the equilibrium distributions of the leading PCs and the energy are unimodal (Fig. 7.3). Note that regime behavior accompanied by unimodality in the equilibrium statistics arises more generally in geophysical flows [79].

In what follows, we view the solution of the 1.5-layer model as the true signal (7.1) from nature, that is, we set $z(t) = \psi(\mathbf{r}, t)$. Moreover, we consider that $z(t)$ is observed through the leading 20 PCs of the streamfunction,

$$\text{PC}_i(t) = \int d\mathbf{r} \, \text{EOF}_i(\mathbf{r}) \psi'(\mathbf{r}, t),$$

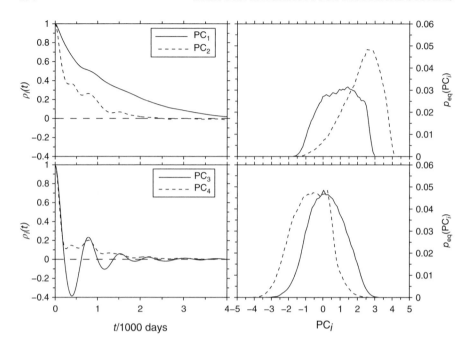

FIGURE 7.3 Empirical autocorrelation functions, $\rho_i(t) = \int_0^T dt'\, \mathrm{PC}_i(t')\mathrm{PC}_i(t'+t)/T$, and equilibrium densities, $p_{\mathrm{eq}}(\mathrm{PC}_i)$, of the leading four streamfunction PCs. Among these PCs, only PC_3 has significantly negative values in $\rho_i(t)$. All the autocorrelation functions of PC_i with $i \in [5,20]$ (not shown here) take negative values. Note that $p_{\mathrm{eq}}(\mathrm{PC}_i)$ are all unimodal, yet the system exhibits long-lived affiliations to regimes (see Fig. 7.7).

where $\mathrm{EOF}_i(\mathbf{r})$ is the ith empirical orthogonal function (EOF) in the streamfunction metric (see the caption of Figure 7.3 for a definition) and $\psi'(\mathbf{r},t) = \psi(\mathbf{r},t) - \Psi(\mathbf{r})$ is the streamfunction anomaly. Thus, the observation vector from (7.2), $x(t) = H(z(t)) = (\mathrm{PC}_1(t),\dots,\mathrm{PC}_{20}(t))$ is 20-dimensional.

7.1.5.1 Predictability bounds for coarse-grained observables

For our clustering and forecast distribution calculations, we took a time series $x(t)$ consisting of a total of $s = 1.6 \times 10^5$ samples taken uniformly every $\delta t = 20$ days. That is, the total observation time span is $s\,\delta t = 3.2 \times 10^6$ days ≈ 8767 years. Our training dataset $\mathcal{X}$ (7.12) is the first half of that time series, that is, $t \in [0,T]$, with $T = 1.6 \times 10^6$ days.

The prediction observables a_t considered in this study are the energy E and the leading-two streamfunction PCs. In light of the conventional low-, middle-, and high-energy phenomenology of the 1.5-layer model [6,81], energy is a natural observable to consider for long-range forecasting. Moreover, the time-averaged spatial features of the circulation regimes are well captured by the leading PCs. We used the portion of the time series with $t > T$ to compute the cluster-conditional time-dependent distributions p_t^k (7.18) for these observables via the procedure described in Section 7.1.4.

Thus, the data used to estimate p_t^k are independent of the input data to the algorithm for evaluating the cluster coordinates θ_k in (7.13).

All forecast densities were estimated by binning the $s/2$ prediction samples in $n_B = 100$ bins of uniform width. The entropy integrals in the predictability metric $\mathcal{D}(a_t, S_0)$ in (7.20) were evaluated via the standard trapezoidal rule. We verified the robustness of our results against sampling errors by halving the length of the prediction time series and repeating the calculation of p_t^k for each half. Quadrature errors were assessed by halving the number n_B of distribution bins and reevaluating $\mathcal{D}(a_t, S_0)$. In all cases, the predictability scores in Figure 7.4 did not change significantly. Moreover, we tested for robustness of the computed cluster-coordinates θ_k in (7.13) by using either of the two halves of our training data. This did not impart significant changes in the spatial structure of the regimes in Figure 7.6.

Following the strategy laid out in Section 7.1.4, we vary the running-average time intervals $\Delta t'$ and Δt, used respectively to coarse-grain $\mathcal{X}$ and the time series (7.3) of

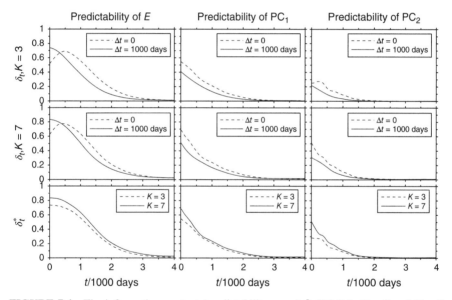

FIGURE 7.4 The information content (predictability score) δ_t (7.24) in $K = 3$ and $K = 7$ partitions (7.14) as a function of prediction horizon t for the energy E and the leading two PCs of the streamfunction. Two values for the running-average interval for initial cluster affiliation are displayed ($\Delta t = 0$ and 1000 days), as well as the optimal score δ_t^* for various values of Δt in the interval $[0, 1000]$ days. In all cases, the running-average interval for coarse-graining the training dataset is $\Delta t' = 1000$ days. The δ_t lines for energy with $\Delta t = 0$ illustrate that the decay of relative entropy to equilibrium may be non-monotonic, a behavior that cannot be replicated by Markov models (see Fig. 7.8). The $K = 7$ partitions have higher information content than the $K = 3$ ones in the leading PCs (i.e., the large-scale structures in the flow) for $t \lesssim 600$ days or about half the decorrelation time of the leading PC (see Fig. 7.3). However, $K = 7$ contributes essentially no additional predictive information beyond $K = 3$ for decadal forecasts.

initial data, seeking to maximize (for the given choice of observable and forecast lead time t) the information content $\mathcal{D}(a_t, S_0)$ from (7.20) beyond equilibrium (or, equivalently, the predictability score δ_t in (7.24)) in the resulting partition from (7.14). In Figure 7.4, we display a sample of the δ_t results for fixed $\Delta t' = 1000$ days (i.e., a value comparable to the decorrelation time, $t_1 = 1165$ days, of PC$_1$) and representative values of short and long initial-data windows, respectively $\Delta t = 0$ and $\Delta t = \Delta t' = 1000$ days. For the time being, we consider models with either $K = 3$ or $K = 7$ clusters and subsequently (in Section 7.1.5.2) study in more detail the relevance of these choices from a physical standpoint.

There are a number of important points to be made about Figure 7.4. First, for the chosen observables, the predictability score δ_t^* (7.37) of the optimal partitions is significant for prediction horizons that exceed the longest decorrelation time in the X_t components used for clustering by a large margin. The fact that decorrelation times are poor indicators of intrinsic long-range predictability has been noted in other CAOS applications [99]. Here, the decay in the δ_t^* score for energy over one e-folding time corresponding to t_1 is $\delta_{t_1}^*/\delta_0^* \simeq 0.7$ or a factor of five weaker decay than $e^{-2} \simeq 0.14$ expected for a purely exponential decay (the comparison is with e^{-2} rather than e^{-1} because δ_t^* is associated with squared correlations). Predictability of energy remains significant up to $t \simeq 3000$ days ($\delta_{3000}^*/\delta_0^* \simeq 0.07$), or three times the decorrelation time of PC$_1$. This means that predictions approaching the decadal scale are possible for E, given knowledge at time $t = 0$ of the system's affiliation to the regimes associated with partition Ξ in (7.14). Note that no fine-grained information about the initial conditions is needed to make these forecasts. Uncertainty in initial conditions is a well-known obstacle in long-range forecasts [56, 60, 82, 95].

Second, as illustrated by the discrepancy between the δ_t scores evaluated for $\Delta t = 0$ and 1000 days, the time window Δt that maximizes the information beyond equilibrium in the partition depends on both the observable and the forecast lead time. More specifically, in the calculations used to produce the δ_t^* versus t lines in Figure 7.4, the optimal Δt for mid-term prediction ($t \lesssim 500$ days) of the energy is around 500–1000 days, but that value rapidly decreases to essentially no coarse-graining ($\Delta t = 0$) when t extends beyond the 2-year horizon. On the other hand, $\Delta t = 0$ is optimal for all values of the prediction lead time t in the case of the PCs. The fact that the optimal Δt for long-range forecasting is small is beneficial from a practical standpoint, since it alleviates the need of collecting initial data over long periods.

Third, as alluded to in the beginning of this section, the $K = 7$ partitions carry significantly higher predictive information than the $K = 3$ ones for mid-range forecasts (up to 3 years), but that additional information is lost in the large lead-time regime. In particular, the δ_t^* scores of the $K = 3$ and $K = 7$ models meet at approximately $t = 2000$ days for E, 500 days for PC$_1$, and 1000 days for PC$_2$.

A final point about Figure 7.4 pertains to the non-monotonicity of δ_t (equivalently, $\mathcal{D}(a_t, S_0)$) for E. It is a general result, sometimes referred to as the generalized second law of thermodynamics, that if the dynamics of an observable are Markovian, then the corresponding relative entropy $\mathcal{D}(a_t, S_0)$ decreases monotonically with t [19, 64, 65]. Thus, the increasing portion of the $\delta_t(E)$ versus t curve for $\Delta t = 0$ and

$t \lesssim 500$ days is a direct evidence of non-Markovianity of the energy observable. As discussed in Section 7.1.5.3, this has important implications for model error when the corresponding cluster affiliation sequence is approximated by a Markov process.

Non-Markovianity of energy is consistent with the fact that longer running-average windows are favored for optimal predictions of this observable for moderate lead times. Physically, as follows from (7.36), the width of Δt controls the influence of the past history of the system relative to its current state in assigning cluster affiliation. If a prediction observable exhibits significant memory effects, taking the running average over a window comparable to the memory time scale should lead to gains of predictive information, at least for lead times of order Δt or less. This is reflected in the δ_t results for energy in Figure 7.4, where forecasts made using a 1000-day averaging window are more skillful than the corresponding forecasts with $\Delta t = 0$, provided that the lead time does not exceed 500 days or so.

7.1.5.2 The physical properties of the regimes We now study the spatial and temporal properties of the regimes associated with the coarse-grained partitions of Section 7.1.5.1. For concreteness, we focus on a $K = 3$ partition with running-average windows $(\Delta t, \Delta t') = (1000, 1000)$ days; see Ref. [41] for results with $K = 7$ partitions and other running-average windows. The $K = 3$ partition was motivated by the analyses in Refs. [6,81], which associate the meandering, mean-flow resembling, and extensional circulation regimes of the 1.5-layer model with bands of low, moderate, and high values of the energy observable. More specifically, the chosen Δt value is a reasonable compromise for simultaneously maximizing the predictability metrics in Figure 7.5 for energy and the leading two PCs.

The key objects facilitating our study of the physical properties of the regimes are the cluster-conditional mean and standard deviation of the streamfunction anomaly, $\psi'_k(\mathbf{r}) = \langle \psi'(\mathbf{r},t) \rangle_k$ and $\sigma_k(\mathbf{r}) = \langle (\psi'(\mathbf{r},t) - \psi'_k(\mathbf{r}))^2 \rangle_k^{1/2}$, which are shown in Figure 7.6. Here, $\langle \cdot \rangle_k$ denotes expectation value with respect to p_t^k from (7.18) at $t = 0$, which, by ergodicity (7.4), can be evaluated by taking temporal averages conditioned on $S(X_t) = k$. First, it is clear from Figure 7.6 that the circulation regimes identified by the K-means clustering algorithm with $K = 3$ and running averaging are in good agreement with the semi-empirical phenomenology established for 1.5-layer double-gyre ocean models [6, 81]. Specifically, state 1, which has a low expected value of energy, $E_1 = \langle E(t) \rangle_1 = 3.5 \times 10^{17}$ J, features a meandering jet pattern; state 2, with $E_2 = 3.9 \times 10^{17}$ J resembles the time-averaged state $\Psi(\mathbf{r})$; and state 3 is dominated by a strong, deeply-penetrating jet and has large mean energy $E_3 = 4.2 \times 10^{17}$ J. As one might expect from the corresponding relative increase in information content (see Fig. 7.5), the basic spatial features of the $K = 3$ regimes are captured with significantly higher fidelity by $K = 7$ partitions (see Ref. [42]).

Turning to the temporal aspects of the switching process between the regimes, Figure 7.7 illustrates that the cluster affiliation sequence $S_t = S(X_t)$ from (7.36) of the $K = 3$ partition of observation space with $\Delta t = 1000$ days leads to a natural splitting of the energy time series into persistent regimes (with decadal mean lifetimes), as expected from the high information content of that partition about energy. As remarked in Section 7.1.4, imposing temporal regularity in S_t is frequently a challenge

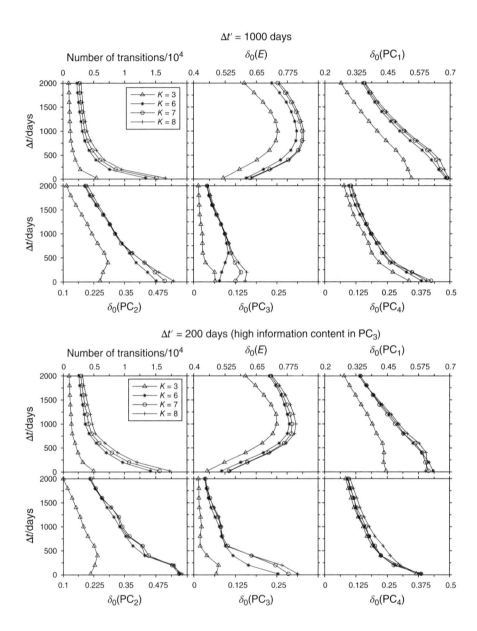

FIGURE 7.5 The dependence of the number of transitions and the relative-entropy predictability score δ_t (7.24) on the running-average interval Δt (initial data for prediction), evaluated at time $t = 0$ for the energy E and leading four PCs for partitions with $K \in \{3, 6, 7, 8\}$ clusters. The running-average interval for coarse-graining the training data is $\Delta t' = 1000$ days and 200 days, respectively, in the top and bottom set of panels.

Circulation regimes for $K = 3$, clustering optimized for energy

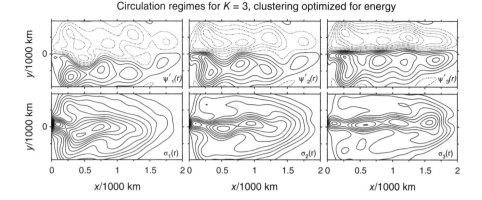

FIGURE 7.6 Mean streamfunction anomaly, $\psi_k'(\mathbf{r})$, and its standard deviation, $\sigma_k(\mathbf{r})$, conditioned on the clusters of a $K = 3$ partition. The contour-level spacing for $\psi_k'(\mathbf{r})$ and $\sigma_k(\mathbf{r})$ is 25 and 10 m, respectively. Solid and dotted lines respectively represent positive and negative contour levels. This partition of observation space has been evaluated using running-average windows of duration $\Delta t = \Delta t' = 1000$ days and is optimized for maximal information content beyond equilibrium about energy (see Fig. 7.5). The spatial features of the circulation regimes identified here via running-average K-means clustering are in good agreement with the meandering (ψ_1'), mean-flow resembling (ψ_2'), and extensional (ψ_3') phases of the jet in Ref. [81], with correspondingly low, moderate, and high values of mean energy (see Fig. 7.7).

in CAOS applications (e.g., standard K-means analysis of this dataset results in unphysical, high-frequency transitions between the regimes), but it emerges here automatically by virtue of coarse-graining the training dataset and the interval Δt for initial cluster affiliation. It is important to emphasize, however, that persistence is not synonymous with skill. For instance, the δ_0 score for PC_1 in Figure 7.5 is a decreasing function of Δt, even though the persistence of the regimes exhibits a corresponding increase (as indicated by the drop in the number of transitions with Δt). Information theory allows one to tell when a persistent cluster affiliation sequence actually carries information for prediction (or classification, as is the case for the $t = 0$ examples considered here) or is too crude of a description of the intrinsic low-frequency dynamics.

7.1.5.3 *Markov models of regime behavior in the 1.5-layer ocean model* We now apply the tools developed in Section 7.1.2.2 to assess the model error in Markov models of regime behavior in the 1.5-layer model. Throughout, we treat the output of the 1.5-layer model as the perfect model and Markov models of the switching process between the regimes identified in Section 7.1.5.1 and 7.1.5.2 as imperfect reduced models with dynamical model error. In particular, we introduce model error by evaluating the forecast distributions p_t^{Mk} in (7.25) under the assumption that the affiliation sequence $\Gamma(t)$ in (7.35) is a Markov process. The Markov assumption for $\Gamma(t)$ is made frequently in cluster analyses of time series in atmosphere–ocean

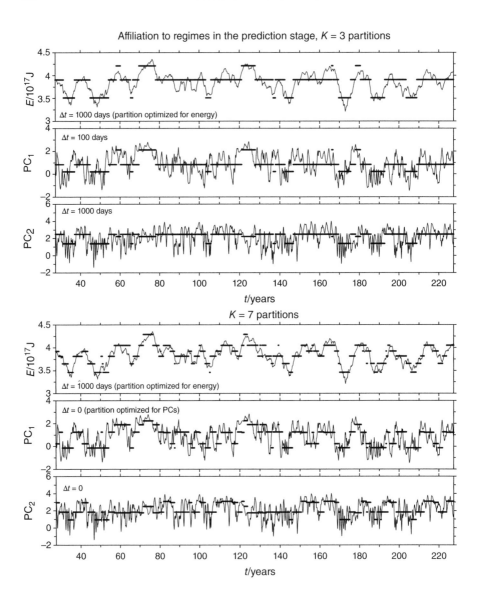

FIGURE 7.7 Time series of the energy E and the leading two PCs spanning a 200-year interval. The thick horizontal lines show the discrete energy and PC affiliation sequences, $\langle E \rangle_{S(t)}$ and $\langle \mathrm{PC}_i \rangle_{S(t)}$, where $\langle \cdot \rangle_k$ denotes cluster-conditional expectation value and $S(t)$ is the cluster affiliation sequence in (7.36). Throughout, the running-average window in the training stage is $\Delta t' = 1000$ days, and regimes are ordered in order of increasing $\langle E \rangle_k$. In the $K = 7$ panels, the running-average window Δt is chosen so that the corresponding partitions of observation space contain high information about energy ($\Delta t = 1000$ days) or the leading PCs ($\Delta t = 0$; see Fig. 7.5).

science [35–37, 51, 52, 54, 79], but as we demonstrate in Section 7.1.5.4, it can lead to false predictability, as measured by the $\mathcal{D}^M(a_t, S_0)$ metric in (7.27). The benefit of the scheme presented here is that false predictability can be detected directly through the measures of model error $\mathcal{E}_{eq}$ and $\mathcal{E}(a_t, S_0)$.

The fundamental assumption in the Markov models studied here is that there exists a $K \times K$ generator matrix L, such that

$$P(t)_{ij} = p(\Gamma(t) = j \mid \Gamma(0) = i) = \exp(tL)_{ij}, \tag{7.38}$$

where $\Gamma(t)$ is defined in (7.35) and t is an integer multiple of the sampling interval δt. In general, the existence of L is not guaranteed, even if $\Gamma(t)$ is indeed Markovian. Nevertheless, one may always try to estimate a Markov generator that is consistent with the given realization $\Gamma(t)$ using one of the available algorithms in the literature [21,83] and verify a posteriori its consistency by computing $\mathcal{E}(a_t, S_0)$ from (7.30) for prediction observables a_t of interest. Here, the cluster-conditional probabilities (7.25) in the Markov model are given (via Bayes' theorem) by

$$p_t^{Mk} = \sum_{j=1}^{K} \exp(tL)_{kj} \phi^j, \tag{7.39}$$

where

$$\phi^k = p(a_t | \Gamma(t) = k)$$

are the distributions for a_t conditioned on the value of $\Gamma(t)$ for the training data. These distributions can be estimated by cluster-conditional bin counting of simultaneous realizations of a_t and $\Gamma(t)$, as described in Section 7.1.4. As with Section 7.1.5.1, our primary observables of interest are the total energy in the flow, E, and the leading two PCs of the streamfunction.

Since for sufficiently long training time series the ϕ^k are equivalent to the p_t^k distributions in (7.18) evaluated at $t = 0$ with equal running-average windows in the training and prediction stages (i.e., $\Delta t = \Delta t'$), the model probabilities in (7.39) have no error at time $t = 0$; that is, $\mathcal{E}_0^k$ in (7.30) is zero by construction. Moreover, $\mathcal{E}_t^k$ will vanish as $t \to \infty$ for models that meet the equilibrium consistency condition in (7.29) for

$$p_{eq}^M = \sum_{k=1}^{K} \pi_k^M \phi^k,$$

where $\pi^M = (\pi_1^M, \ldots, \pi_K^M)$ is the equilibrium distribution of the Markov model, defined by the requirement for all t,

$$\sum_{i=1}^{K} \pi_i^M P(t)_{ij} = \pi_j^M.$$

However, due to dynamical model error, $\mathcal{E}(a_t, S_0)$ will generally be nonzero for finite and nonzero t.

7.1.5.4 The model error in long-range predictions with coarse-grained Markov models

To construct our Markov models, we took the same training data used in Section 7.1.5.1 consisting of the leading 20 PCs of the streamfunction in the 1.5-layer model, $x(t) = (\mathrm{PC}_1(t), \ldots, \mathrm{PC}_{20}(t))$ and computed affiliation sequences $\Gamma(t)$ from (7.35), applying the procedure described in Section 7.1.4 for various choices of K and running-average windows $\Delta t'$. In each case, we determined P by fitting the generator matrix L in (7.38) to $\Gamma(t)$ using the Bayesian algorithm of Metzner et al. [83]. We checked for robustness against sampling errors by repeating our calculations using either of the two halves of the training time series. This resulted to relative changes of order 10^{-3} in L, as measured by the ratio of Frobenius norms $\|\delta\mathrm{L}\|/\|\mathrm{L}\|$. Likewise, the changes in the results in Figure 7.8 were not significant.

The two main risks with the assumption that $\Gamma(t)$ has the Markov property are that (1) the Markov model fails to meet the equilibrium consistency condition in (7.29), that is, the Markov equilibrium distribution deviates systematically from

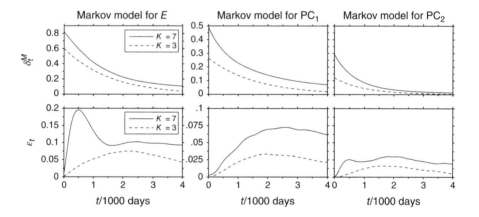

FIGURE 7.8 Internal predictability score, δ_t^M, and model error, ϵ_t, of $K = 3$ and $K = 7$ Markov models as a function of the forecast lead time t. The observables under consideration are the energy E and the leading two PCs. The coarse-graining interval $\Delta t'$ for creating the partition in the training stage is $\Delta t' = 2000$ and 1000 days, respectively, for $K = 3$ and $K = 7$, with corresponding model error in the energy equilibrium $\epsilon_{\mathrm{eq}} \sim 10^{-5}$ and 0.068; that is, the three-state Markov model meets equilibrium consistency (7.29), but the seven-state model does not. At finite t, small values of ϵ_t mean that the relative-entropy score δ_t^M in (7.28) is an appropriate surrogate for the true predictive skill of the Markov models. On the other hand, if ϵ_t and δ_t^M are both large, then δ_t^M is biased and measures false predictive skill. The equilibration time of the Markov models (given by $-1/\mu_1$, where μ_1 is the first nontrivial eigenvalue of the generator of the Markov process) is 3000 and 4260 days, respectively, for $K = 3$ and $K = 7$. Thus, in this example, the most erroneous Markov model has the largest false skill and is also the most persistent.

the truth and (2) the discrepancy $\mathcal{D}^M(a_t, S_0)$ in (7.27) from equilibrium of the Markov model measures false predictability, for example, for a Markov model that relaxes to equilibrium unjustifiably slowly. The latter two pitfalls may arise independently of one another, since the discrepancy $\mathcal{E}(a_t, S_0)$ in (7.11) of a model from the truth as it relaxes to equilibrium can be large for some forecast lead time t, even if the model error $\mathcal{E}_{eq}$ in equilibrium is small. Nevertheless, the $\mathcal{E}(a_t, S_0)$ and $\mathcal{E}_{eq}(a_t)$ metrics (and the corresponding normalized error scores ϵ_t and ϵ_{eq} in (7.31) and (7.29), respectively) allow one to detect these types of error a posteriori, given the Markov matrix P fitted to the data.

In Figure 7.8 we illustrate these issues through a comparison between a $K = 7$ and a $K = 3$ Markov model. The seven-state model was constructed using the $\Gamma(t)$ affiliation sequence of the $K = 7$ partitions of Section 7.1.5.1; that is, the training time series $x(t)$ was coarse-grained using a running-average window of $\Delta t' = 1000$ days. The same training time series was used to evaluate the generator of the $K = 3$ Markov model, but in this case $\Delta t'$ was increased to 2000 days. First, it is clear from the graphs for ϵ_t that the seven-state Markov model asymptotes to an incorrect equilibrium distribution. For this reason, the relative-entropy score δ_t^M measures false predictability for all forecast horizons, including $t = 0$. On the other hand, the $K = 3$ model of Figure 7.8 does meet climate equilibrium consistency (with $\epsilon_{eq} \sim 10^{-5}$), which means that for this model, δ_0^M is a true measure of classification skill beyond equilibrium. That model, however, experiences a gradual ramp-up of ϵ_t, peaking at around $t = 2000$ days, and as a result, its predictions cannot be deemed accurate beyond, say, a horizon $t \gtrsim 1000$ days.

Note now how the second pitfall might lead one to believe that the seven-state Markov model is more skillful than the three-state one: The smallest nontrivial eigenvalue, $\mu_1 = (\log \lambda_1)/\delta t \simeq -1/(4000 \text{ days})$, of the generator matrix of the $K = 7$ model has smaller absolute value than the corresponding eigenvalue, $\mu_1 \simeq -1/(3000 \text{ days})$, of the $K = 3$ model. That is, for long-enough prediction horizons, the seven-state model relaxes more slowly to equilibrium than the three-state model, that is, it is more persistent. By monitoring ϵ_{eq} and ϵ_t, it is possible to identify models with false persistence as illustrated above.

7.2 NLSA ALGORITHMS FOR DECOMPOSITION OF SPATIOTEMPORAL DATA

7.2.1 Background

In a wide range of disciplines in science and engineering, including those outlined in the chapter's preamble, there exists a strong interest in extracting physically meaningful information about the spatial and temporal variabilities of data from models, experiments, or observations with the goal of enhancing the understanding of the underlying dynamics. Frequently, observations of the system under study are incomplete, that is, only a subset of the full phase space is accessible.

A classical way of attacking this problem is through singular spectrum analysis (SSA) or one of its variants [1, 13, 38, 47, 105, 108]. Here, a low-rank approximation of a dynamic process is constructed by first embedding a time series of a scalar or multivariate observable in a high-dimensional vector space $\mathbb{R}^N$ (here referred to as lagged-embedding space) using the method of delays [29, 85, 89, 97], and then performing a truncated singular value decomposition (SVD) of the matrix X containing the embedded data. In this manner, information about the dynamical process is extracted from the left and right singular vectors of X with the l largest singular values. The left (spatiotemporal) singular vectors form a set of so-called extended empirical orthogonal functions (EEOFs) in $\mathbb{R}^N$, which, at each instance of time, are weighted by the corresponding principal components (PCs) determined from the right (temporal) singular vectors to yield a rank-l approximation of X.

A potential drawback of this approach is that it is based on minimizing an operator norm that may be unsuitable for signals generated by nonlinear dynamical systems. Specifically, the PCs are computed by projecting onto the principal axes of the l-dimensional ellipsoid that best fits the covariance of the data in lagged-embedding space in the least-squares sense. This construction is optimal when the data lie on a linear subspace of $\mathbb{R}^N$, but nonlinear processes and/or observation functions will in general produce data lying on a nonlinear submanifold $\mathcal{M} \subset \mathbb{R}^N$ with non-Gaussian distributions departing significantly from the ellipsoid defined by the covariance operator of the data. Physically, a prominent manifestation of this phenomenon is failure to capture via SSA the intermittent patterns arising in turbulent dynamical systems, that is, temporal processes that carry low variance but play an important dynamical role [2, 20].

Despite their inherently nonlinear character, such datasets possess natural linear structures, namely Hilbert spaces $\Lambda^0 \mathcal{M}$ of square-integrable functions on $\mathcal{M}$ with inner product inherited from the volume form of a Riemannian metric induced on the data by lagged embedding. Moreover, intrinsically discrete analogs $\Lambda^0 M$ of $\Lambda^0 \mathcal{M}$ can be constructed empirically for the set $M \subset \mathcal{M}$ of observed data using techniques from discrete exterior calculus (DEC, e.g., Refs. [26, 48, 112]). These spaces may be thought of as the collection of all possible weights that can be assigned to the data samples when making a low-rank reconstruction, that is, they are analogous to the temporal spaces of SSA. Based on these observations, it is reasonable to develop algorithms for data decomposition which are based on suitably-defined maps from $\Lambda^0 M$ to lagged-embedding space $\mathbb{R}^N$. Such maps, denoted here by A, have the advantage of being simultaneously linear and compatible with the nonlinear geometry of the data.

Here, we advocate that this approach, implemented via algorithms developed in machine learning [3, 17], can reveal important aspects of complex, high-dimensional signals, which are not accessible to classical SSA. In this framework, which we refer to as nonlinear Laplacian spectral analysis (NLSA), an orthonormal basis for $\Lambda^0 M$ is constructed through eigenfunctions of a diffusion operator associated with a kernel in lagged-embedding space with explicit dependence on the dynamical vector field on $\mathcal{M}$ generating the data. Projecting the data from embedding space onto these eigenfunctions then gives a matrix representation of A leading, via SVD, to a decomposition of the dataset into a biorthonormal set of spatiotemporal patterns.

7.2.2 Mathematical framework

We consider a scenario where the dynamics is described by a flow $F_t: \mathcal{F} \mapsto \mathcal{F}$ operating in a phase space $\mathcal{F}$ and evolving on an attractor $\mathcal{M} \subseteq \mathcal{F}$. Moreover, observations are taken uniformly in time with a timestep $\delta t > 0$ on the attractor via a smooth vector-valued function $G: \mathcal{F} \mapsto \mathbb{R}^n$, forming a dataset $x = (x_1, \ldots, x_s)$ with

$$x_i = G(z_i), \quad z_i = F_{t_i} z_0, \quad t_i = i\delta t, \quad z_0 \in \mathcal{M}. \tag{7.40}$$

In general, we are interested in cases where the observation function is incomplete, that is, the x_i alone is not sufficient to uniquely determine the state of the system in $\mathcal{M}$. Geometrically, this means that the image manifold $G(\mathcal{M}) \subset \mathbb{R}^n$ is not diffeomorphic to $\mathcal{M}$. Our objective is to produce a decomposition of x_i into a set of l spatiotemporal patterns $\hat{x}_i^k$,

$$x_i \approx \tilde{x}_i = \sum_{k=1}^{l} \hat{x}_i^k, \tag{7.41}$$

taking into account the underlying dynamical system operating on $\mathcal{M}$. That is, compared to generic point clouds of data, an additional structure that we have at our disposal here is the time ordering of the observations, which carries meaningful information about F_t. Therefore, we seek that the decomposition (7.41) depends on that time ordering.

The methodology employed here to address this objective consists of four basic steps: (1) Embed the observed data in a vector space $\mathbb{R}^N$ of dimension greater than n via the method of delays; (2) construct a linear map A_l taking a Hilbert space of scalar functions on $\mathcal{M}$ representing temporal patterns to the spatiotemporal patterns in $\mathbb{R}^N$; (3) perform an SVD in a basis of orthonormal diffusion eigenfunctions to extract the spatial and temporal modes associated with A_l; (4) project the modes from $\mathbb{R}^N$ to data space $\mathbb{R}^n$ to obtain the spatiotemporal patterns $\hat{x}_t^k$ in (7.41). Below, we provide a description of each step, as well as an outline of SSA algorithms to draw connections between the two approaches. Further details of the NLSA framework, as well as pseudocode, are presented in Refs. [41,44]. A Matlab implementation is available upon request from the corresponding author.

Hereafter, we shall consider that $\mathcal{M}$ has integer dimension m and is compact and smooth, so that a well-defined continuous spectral theory exists [5]. Moreover, we assume that the dynamical vector field $\dot{F}$ induced on $\mathcal{M}$, given by

$$\dot{F}(f) = \lim_{t \to 0} (f(F_t z) - f(z))/t, \quad \text{with} \quad z \in \mathcal{M}, \quad f \in C^\infty \mathcal{M}, \tag{7.42}$$

is also smooth. We emphasize, however, that the smoothness assumptions for $\mathcal{M}$ and $\dot{F}$ are to be viewed as "Platonic ideals" serving as a guidelines for algorithm design and analysis but seldom encountered in practice (e.g., due to finite number of samples and/or non-smoothness of the attractor). Operationally, one works in the intrinsically discrete framework of spectral graph theory [16] and DEC [26,48], which exist independently of the continuous theory, even if the latter was used as a means of gaining insight.

7.2.2.1 Time-lagged embedding

This step is familiar from the qualitative theory of dynamical systems [29,85,89,97]. Under generic conditions, the image of $z_i \in \mathcal{M}$ in embedding space, $\mathbb{R}^N$, under the delay-coordinate mapping,

$$H(z_i) = X_i = (G(z_i), G(z_{i-1}), \dots, G(z_{i-(q-1)})), \quad X_i \in \mathbb{R}^N, \quad N = qn, \quad (7.43)$$

lies on a manifold $H(\mathcal{M})$ which is diffeomorphic to $\mathcal{M}$, provided that the dimension N is sufficiently large. Thus, given a sufficiently-long embedding window $\Delta t = (q-1)\delta t$, we obtain a representation of the attractor underlying our incomplete observations.

Broadly speaking, preprocessing the data by time lagged-embedding produces both topological and geometrical changes. In particular, the topology of the embedded dataset $H(\mathcal{M})$ will be different from that of the original $G(\mathcal{M})$ data if the observation map is incomplete, recovering the manifold structure of the attractor lost through partial observations. An implication of this is that the time series X_i in (7.43) becomes Markovian, or, equivalently, the dynamical vector field $\dot{F}$ on $\mathcal{M}$ from (7.42) is carried along by means of the derivative map DH to a smooth vector field $\dot{F}_* = DH\dot{F}$ on $H(\mathcal{M})$.

Besides topological properties, time-lagged embedding also influences the geometry of the data, in the sense that the Riemannian metric h induced on $\mathcal{M}$ by pulling back the canonical inner product of $\mathbb{R}^N$ depends explicitly on the dynamical flow generating the data. To see this, let $(u^1, \dots, u^m)$ be a coordinate system in a neighborhood of $z_i \in \mathcal{M}$. In this coordinate system, the induced metric h at z_i has components

$$h_{\mu\nu}|_i = \sum_{\alpha=1}^{N} \frac{\partial X_i^\alpha}{\partial u^\mu} \frac{\partial X_i^\alpha}{\partial u^\nu} = \sum_{\alpha=1}^{n} \sum_{j=0}^{q-1} \frac{\partial x_{i-j}^\alpha}{\partial u^\mu} \frac{\partial x_{i-j}^\alpha}{\partial u^\nu} = \sum_{j=0}^{q-1} g_{\mu\nu}|_{i-j}, \quad (7.44)$$

where g is the induced metric on the original dataset $G(\mathcal{M})$ and X_i^α (x_i^α) the components of X_i (x_i) in an orthonormal basis of $\mathbb{R}^N$ ($\mathbb{R}^n$). It therefore follows that the induced metric on $\mathcal{M}$ is a "running-averaged" version of the induced metric on $G(\mathcal{M})$ along orbits of the dynamical system in (7.40) of temporal extent Δt. In other words, the results of a data analysis algorithm operating in $\mathbb{R}^N$ that processes the data based on distance-based affinity metrics will depend on the dynamical flow F_t.

Note that this effect takes place even if the observation map is complete (i.e., $G(\mathcal{M})$ is a diffeomorphic copy of $\mathcal{M}$), which suggests that time-lagged embedding may be used as a tool to control the dataset geometry even in fully observed dynamical systems. This question has been studied in detail in recent work by Berry and collaborators [9], who have established a correspondence between the Lyapunov metric of a dynamical system acting along the most stable Osedelets subspace and the induced metric in a suitable lagged-embedding space.

7.2.2.2   Overview of singular spectrum analysis   The classical linear framework for creating low-rank approximations of a dataset in lagged-embedding space is essentially identical to PCA or proper orthogonal decomposition (POD) algorithms [1, 50], apart from the fact that one obtains a biorthonormal basis of temporal–spatiotemporal patterns rather than the standard temporal–spatial basis. Algorithms of this type are interchangeably called SSA [38, 105], singular system analysis [13], and EEOFs [108]. We use the term SSA to refer to this family of algorithms.

Let $X = (X_1, \ldots, X_S)$ with $S = s - q + 1$ be the data matrix in lagged embedding space, dimensioned $N \times S$. In SSA, the biorthonormal basis that optimally fits the data is constructed through the SVD of X,

$$X = U\Sigma V^T,$$

$$U = (U_1, \ldots, U_N), \quad \Sigma = \mathrm{diag}(\sigma_1, \ldots, \sigma_{\min\{N,S\}}), \quad V = (V_1, \ldots, V_S),$$
$$U_i \in \mathbb{R}^N, \quad \sigma_i \geq 0, \quad V_i \in \mathbb{R}^S, \tag{7.45}$$
$$U_i^T U_j = V_i^T V_j = \delta_{ij}.$$

Here, U and V are orthogonal matrices of dimension $N \times N$ and $S \times S$, respectively, and Σ a diagonal matrix with nonnegative diagonal entries ordered in decreasing order. In (7.45), the jth column of V gives rise to a function of time (a PC)

$$\tilde{v}_j(t_i) = V_{ij}. \tag{7.46}$$

Moreover, the corresponding column U_j of U represents a spatiotemporal process $u_j(\tau_i)$ in time duration Δt, viz.

$$u_j(\tau_i) = (U_{n(i-1)+1,j}, \ldots, U_{ni,j}) \in \mathbb{R}^n, \quad \tau_i = (i-1)\delta t. \tag{7.47}$$

A rank-l approximation X_l of the dataset in lagged embedding space is then constructed through the leading l singular vectors from (7.45),

$$X_l = U_l \Sigma_l V_l = X V_l^T V_l,$$
$$U_l = (U_1, \ldots, U_l), \quad \Sigma_l = \mathrm{diag}(\sigma_1, \ldots, \sigma_l), \quad V_l = (V_1, \ldots, V_l). \tag{7.48}$$

It is a standard result from linear algebra that X_l is the optimal rank-l approximation of X with respect to the Frobenius norm of linear operators.

More abstractly, implicit to (7.45) is the notion that the dataset induces a linear map $X : \mathbb{R}^S \mapsto \mathbb{R}^N$ taking the so-called "chronos" space of temporal patterns, $\mathbb{R}^S$, to the "topos" space of spatiotemporal patterns, $\mathbb{R}^N$, via matrix multiplication [1], that is, $f \mapsto Xf$ with $f \in \mathbb{R}^S$. This picture will be useful in the development of NLSA algorithms ahead.

7.2.2.3   Spaces of temporal patterns   Another useful way of interpreting the chronos modes is to view them as scalar functions on the data manifold. In particular, we think of the components of each $(f_1, \ldots, f_S) \in \mathbb{R}^S$, as the values of a function

$f : \mathcal{M} \mapsto \mathbb{R}$ sampled at the states z_i in (7.40), that is, $f(z_i) = f_i$. In particular, to each V_j from (7.45) we associate a scalar function $v_j(z_i) = V_{ij}$. The main tenet in NLSA algorithms is that the extracted temporal modes should belong to low-dimensional families of "well-behaved" functions on $\mathcal{M}$. The function space in question is spanned by the leading eigenfunctions of diffusion operators on $\mathcal{M}$, as we now discuss.

Let $\Lambda^p \mathcal{M}$ denote the vector space of smooth p-form fields on $\mathcal{M}$. For our purposes, a diffusion operator $\mathcal{L}$ will be an elliptic, second-order differential operator acting on scalar functions in $\Lambda^0 \mathcal{M}$, which annihilates constant functions, that is,

$$f = \text{const.} \quad \Longrightarrow \quad \mathcal{L}f = 0.$$

An important theoretical result (e.g., [32, 87]) is that every diffusion operator $\mathcal{L}$ induces a Riemannian geometry on $\mathcal{M}$, in the sense that one can associate to $\mathcal{L}$ a unique metric tensor k. More specifically, to every $\mathcal{L}$, there corresponds a unique codifferential operator $\delta : \Lambda^1 \mathcal{M} \mapsto \Lambda^0 \mathcal{M}$ which produces the factorization

$$\mathcal{L} = \delta d, \tag{7.49}$$

and gives the metric implicitly through the relation

$$\langle \delta\omega, f \rangle_k = \int_{\mathcal{M}} \delta\omega f \, d\mu = \langle \omega, df \rangle_k = \int_{\mathcal{M}} \sum_{\alpha,\beta=1}^{m} k^{\alpha\beta} w_\alpha \, d_\beta f \, d\mu. \tag{7.50}$$

Here, f and ω are arbitrary smooth scalar functions and one-forms, $\langle \cdot, \cdot \rangle_k$ the Hodge inner product between p-forms, $d : \Lambda^p \mathcal{M} \mapsto \Lambda^{p+1} \mathcal{M}$ the exterior derivative, $d\mu = \sqrt{\det k} \, du^1 \wedge \cdots \wedge du^m$ the volume form of k, and $k^{\alpha\beta}$ the components of the "inverse metric" associated with k in the u^μ coordinates. A local expression for the codifferential in terms of the metric is

$$\delta(\omega) = -\frac{1}{\sqrt{\det k}} \sum_{\alpha,\beta=1}^{m} \frac{\partial}{\partial u^\alpha} \left(k^{\alpha\beta} \sqrt{\det k} \omega_\beta \right). \tag{7.51}$$

It follows from (7.51) that the Riemannian metric k associated with $\mathcal{L}$ has the property that the corresponding codifferential δ is the adjoint of d with respect to the Hodge inner product. This construction leads naturally to a normalized Dirichlet form

$$E_k(f) = \frac{\langle f, \mathcal{L}f \rangle_k}{\|f\|_k^2} = \frac{\langle df, df \rangle_k}{\|f\|_k^2} \geq 0, \quad \|f\|_k = \langle f, f \rangle_k^{1/2}, \tag{7.52}$$

which characterizes how strongly oscillatory a scalar function f is. Note that $E_k(f)$ depends significantly on k.

Let $\phi_0, \phi_1, \ldots$ be normalized eigenfunctions of $\mathcal{L}$ with corresponding eigenvalues $\lambda_0, \lambda_1, \ldots,$

$$\mathcal{L}\phi_i = \lambda_i \phi_i, \quad \langle \phi_i, \phi_j \rangle_k = \delta_{ij}, \quad 0 = \lambda_0 < \lambda_1 \leq \lambda_2 \ldots.$$

The basic requirement in NLSA is that the recovered patterns $v_j(z_i)$ of temporal variability should have bounded Dirichlet form with respect to a Riemannian metric (see (7.65) ahead) constructed in lagged embedding space with an explicit dependence on the dynamical vector field $\vec{F}$. Specifically, for a function $f = \sum_i c_i \phi_i$ we require that $c_i = 0$ for $i > l$, or, equivalently, $E_k(f) \leq \lambda_l$. Operationally, this criterion is enforced by introducing the l-dimensional space of functions spanned by the leading l eigenfunctions of $\mathcal{L}$,

$$\Phi_l = \text{span}\{\phi_0, \ldots, \phi_{l-1}\}, \quad \dim \Phi_l = l, \tag{7.53}$$

and replacing the linear map X in (7.45) by a linear map A_l whose domain is Φ_l. We will return to this point in Section 7.2.2.6.

We remark that this viewpoint is fundamentally different from SSA and related variance-optimizing algorithms. In those algorithms, the unimportant features of the data are spanned by vectors in embedding space onto which the dataset projects weakly. On the other hand, in NLSA, the unimportant features are those that require large-λ_i basis functions on the data manifold to be described. In particular, there may be temporal modes of variability that carry a small portion of the variance of the total signal but are "large-scale" on $\mathcal{M}$ in the sense of small E_k. Such modes will generally not be accessible to SSA algorithms.

7.2.2.4 *Discrete formulation*

In practical applications, one has seldom access to a densely sampled smooth manifold $\mathcal{M}$. However, using the machinery of DEC [26,48] and spectral graph theory [3, 17], it is possible to design an algorithm that has the same fundamental properties as the continuous formulation in Section 7.2.2.3 but is intrinsically discrete (i.e., not a discretization of a continuous theory). In this regard, the main role of the continuous picture is to provide a guideline for building a discrete algorithm.

Let $M = \{z_q, z_{q+1}, \ldots, z_{S+q-1}\} \subset \mathcal{M}$ be the discrete set of states on the attractor which are available for data analysis after time-lagged embedding (the initial $q-1$ states, $z_1, \ldots, z_{q-1}$, must be discarded in order to apply (7.43)). The first step of the classical procedure for building a discrete diffusion operator L analogous to $\mathcal{L}$ in (7.49) is to identify the spaces $\Lambda^0 M$ and $\Lambda^1 M$ of scalar-valued functions and 1-forms on the discrete dataset M. These spaces consist of functions defined on the vertices M and edges $M \times M$, respectively, of an undirected graph, whose nodes $1, \ldots, S$ correspond to the states $z_q, \ldots, z_{S+q-1}$ of the dynamical system at which observations are taken. That graph is further equipped with an ergodic, reversible[1] Markov chain (typically constructed through a kernel, as described in Section 7.2.2.5)

---

[1]Reversibility of the Markov chain is not strictly necessary, but it simplifies the expression for the codifferential in (7.56). Moreover, Markov chains derived from kernels are reversible by construction. See Ref. [112] for more general expressions applicable to nonreversible Markov chains, as well as higher-order forms.

whose state space is M. That is, we have

$$\sum_{i=1}^{S} \pi_i p_{ij} = \pi_j, \quad \pi_i p_{ij} = \pi_j p_{ji}, \tag{7.54}$$

where p and π are the transition probability matrix and invariant distribution of the Markov chain, respectively. The latter are used to construct the inner products $\langle \cdot, \cdot \rangle_p$ of $\Lambda^0 M$ and $\Lambda^1 M$ via the formulas

$$\langle f, f' \rangle_p = \sum_{i=1}^{S} \pi_i f(i) f'(i), \quad \langle \omega, \omega' \rangle_p = \sum_{i,j=1}^{S} \pi_i p_{ij} \omega([ij]) \omega'([ij]), \tag{7.55}$$

where $f, f' \in \Lambda^0 M$, $\omega, \omega' \in \Lambda^1 M$, and $[ij]$ is the edge connecting vertices i and j. Introducing the discrete exterior derivative $d : \Lambda^0 M \mapsto \Lambda^1 M$ with $df([ij]) = f(j) - f(i)$, the codifferential $\delta : \Lambda^1 M \mapsto \Lambda^0 M$ is defined as the adjoint of d with respect to the $\langle \cdot, \cdot \rangle_p$ inner product. That is, for any $f \in \Lambda^0 M$ and $\omega \in \Lambda^1 M$ we have

$$\langle \omega, df \rangle_p = \langle \delta \omega, f \rangle_p.$$

An explicit formula for δ (which must be modified if p is not reversible) is

$$\delta \omega(i) = \sum_{j=1}^{S} p_{ij} (\omega([ji]) - \omega([ij])). \tag{7.56}$$

Equations 7.55 and 7.56 are the discrete counterparts of (7.50) and (7.51), respectively.

With these definitions, the discrete diffusion operator is constructed in direct analogy to (7.49), viz.

$$L = \delta d, \quad Lf(i) = 2 \sum_{i,j=1}^{S} p_{ij} (f(i) - f(j)), \quad f \in \Lambda^0 M. \tag{7.57}$$

This operator provides a tool for computing orthonormal basis functions of $\Lambda^0 M$ through its associated eigenfunctions,

$$L\phi_i = \lambda_i \phi_i, \quad \langle \phi_i, \phi_j \rangle_p = \delta_{ij}, \quad 0 = \lambda_0 < \lambda_1 \leq \lambda_2 \leq \cdots. \tag{7.58}$$

Moreover, it provides a measure for the oscillatory character of a function $f \in \Lambda^0 M$ through the associated Dirichlet form [cf. (7.52)], $E_p(f) = \langle f, Lf \rangle_p / \|f\|_p^2$. The spaces of admissible temporal patterns in NLSA,

$$\Phi_l = \mathrm{span}\{\phi_0, \ldots, \phi_{l-1}\}, \tag{7.59}$$

are modeled after (7.53) and have the property

$$E_p(f) \leq \lambda_l \quad \text{for every } f \in \Phi_l. \tag{7.60}$$

7.2.2.5 Dynamics-adapted kernels In order to turn the framework of Section 7.2.2.4 into a complete algorithm, we must specify the Markov transition probability matrix p associated with the discrete diffusion operator L in (7.57). Here, we follow the widely adopted approach in the literature (e.g., [3, 4, 8, 9, 17, 18, 39]), whereby p is constructed through a suitable local kernel whose asymptotic properties provide a connection between L and a diffusion operator $\mathcal{L}$ in the continuum limit. In other words, the design of the Markov matrix employed in the discrete algorithm is informed by the asymptotic properties of the kernel and the Riemannian geometry associated with $\mathcal{L}$ via (7.50). Because kernels can be computed using only observed quantities in data space without having to know a priori the structure of the underlying phase space $\mathcal{F}$ and flow F_t, this approach opens the possibility of model-free analysis of dynamical system data [8].

Recall that a kernel is a function that maps pairs of states in $\mathcal{M}$ to a positive number and decays exponentially fast away from the basepoint at which it is evaluated. A standard choice in this context is the isotropic Gaussian kernel [3, 4, 17, 18],

$$\bar{K}_\epsilon(z_i, z_j) = \exp\left(\frac{-\|H(z_i) - H(z_j)\|^2}{\epsilon}\right), \tag{7.61}$$

where ϵ is a positive parameter and $\|\cdot\|$ the canonical Euclidean norm. In writing down (7.61) using the delay-coordinate map H from (7.43), we have made explicit the fact that our focus is kernels defined lagged-embedding space. In NLSA, we work with a "locally scaled" kernel

$$K_{\delta t}(z_i, z_j) = \exp(-\|H(z_j) - H(z_i)\|^2 / (\|\xi_i\| \|\xi_j\|)), \quad \xi_i = X_i - X_{i-1}, \tag{7.62}$$

where ξ_i is the displacement vector between temporal nearest neighbors. One may explicitly verify that the quantities ξ_i and ξ_j are finite-difference approximations to the dynamical vector field carried along to lagged-embedding space [39], that is,

$$\dot{F}_*|_{z_i} = \xi_i / \delta t + O(\delta t). \tag{7.63}$$

Thus, the NLSA kernel depends on the dynamics implicitly through time-lagged embedding, and explicitly through ξ.

Given a choice of kernel such as the examples above, it is possible to construct a Markov transition probability p by performing suitable normalizations to convert $K_{ij} = K(z_i, z_j)$ to a row-stochastic matrix. Here, we adopt the normalization procedure developed by Coifman and Lafon [17] in the diffusion map (DM) family of algorithms. In DM, the Markov matrix p is constructed from K by introducing a scalar parameter α and performing the sequence of operations

$$Q(z_i) = \sum_{z_j \in M} K(z_i, z_j), \quad \tilde{K}(z_i, z_j) = \frac{K(z_i, z_j)}{(Q_i Q_j)^\alpha},$$

$$\tilde{Q}(z_i) = \sum_{z_j \in M} \tilde{K}(z_i, z_j), \quad p_{ij} = \frac{\tilde{K}(z_i, z_j)}{\tilde{Q}(z_i)}. \tag{7.64}$$

In Ref. [17] it was shown that with this definition of p and $\alpha = 1$, the discrete diffusion operator L associated with an isotropic, exponentially decaying kernel converges as $\epsilon \to 0$ (and the sample number S grows at least as fast as $\epsilon^{-m/2-1}$ [93]) to $\mathcal{L} = \Delta$, where Δ is the Laplace–Beltrami operator associated with the Riemannian metric induced on $\mathcal{M}$ through the embedding $\mathcal{M} \mapsto H(\mathcal{M})$. Importantly, the $L \to \Delta$ convergence holds even if the sampling density on $\mathcal{M}$ is nonuniform relative to the Riemannian volume form. The metric associated with the kernel in (7.61) is the induced metric h in (7.44) determined by the delay-coordinate mapping.

This result was extended to the case of anisotropic kernels by Berry [8], who showed that under relatively weak assumptions (which are met by both of the kernels in (7.61) and (7.62)) the asymptotic diffusion operator and the induced metric k are determined by the Hessian of K evaluated at $z_j = z_i$. In particular, in the limit $\delta t \to 0$ the locally-scaled kernel (7.62) leads to the induced metric [39]

$$k = \frac{h}{\|\dot{F}\|_h^2}, \quad \|\dot{F}\|_h^2 = h(\dot{F}, \dot{F}). \tag{7.65}$$

Motivated by this asymptotic result, we work throughout with the $\alpha = 1$ DM normalization in conjunction with the locally-scaled kernel of (7.62).

It follows from (7.65) that the outcome of the $\|\xi_i\|$ scaling factors in (7.62) is a conformal transformation of the metric h in lagged-embedding space with a conformal factor $\|\dot{F}\|_h^2$ given by the squared magnitude of the "phase-space velocity" $\dot{F}$. In Ref. [41], this feature was found to be crucial for successful dimensional reduction of a dynamical system with chaotic metastability. Additional properties of ξ_i, in particular, its angle relative to the relative displacement vector $H(z_i) - H(z_j)$, can be incorporated in so-called "cone" kernels with stronger invariance properties [39].

A further desirable outcome of the local scaling by the $\|\xi_i\|$ factors is that the diffusion operator $\mathcal{L}$ and the associated Φ_l spaces of temporal patterns from (7.53) become conformally invariant [39]. In particular, equivalence classes of datasets related by conformal transformations of the metric in lagged embedding space,

$$\tilde{h}|_z = r(z)h|_z, \quad r(z) > 0,$$

lead asymptotically to the same $\mathcal{L}$ operator as $\delta t \to 0$.

Scale invariance is also beneficial in situations where H is a composite map $H : \mathcal{M} \mapsto \mathbb{R}^{N_1} \oplus \mathbb{R}^{N_2}$ such that $H(z) = (H_1(z), H_2(z))$ where both H_1 and H_2

are embeddings of $\mathcal{M}$. This scenario arises in practice when one has access to multivariate observations with distinct physical units, but there is no natural way of choosing a norm for the product space $\mathbb{R}^{N_1} \oplus \mathbb{R}^{N_2}$. Because the ratios $\|H_\beta(z_i) - H_\beta(z_j)\|^2 / \|\xi_i^{(\beta)}\| \|\xi_j^{(\beta)}\|$, $\beta \in \{1,2\}$, are invariant under scaling of the data by a constant (including change of units), the kernels (7.62) computed individually for H_1 and H_2 can be combined into a single product kernel without having to introduce additional scaling parameters, namely

$$K_{\delta t}(z_i, z_j) = \exp\left(-\frac{\|H_1(z_i) - H_1(z_j)\|^2}{\|\xi_i^{(1)}\| \|\xi_j^{(1)}\|} - \frac{\|H_2(z_i) - H_2(z_j)\|^2}{\|\xi_i^{(2)}\| \|\xi_j^{(2)}\|}\right).$$

A climate science application of this technique can be found in Ref. [14].

7.2.2.6 Singular value decomposition Having established the procedure to obtain the temporal spaces Φ_l in (7.59), the next step in NLSA is to form a family of linear maps $A_l : \Phi_l \mapsto \mathbb{R}^N$, which are represented by $N \times l$ matrices with elements

$$A_j^\alpha = \langle X^\alpha, \phi_j \rangle_p = \sum_{i=1}^S \pi_i X^\alpha(i) \phi_j(i), \quad X^\alpha(i) = \langle e_\alpha, X_i \rangle_{\mathbb{R}^N}. \tag{7.66}$$

Here, X^α is the scalar-valued function in $\Lambda^0 M$ giving the αth component of the observed data in an orthonormal basis $e_1, \ldots, e_N$ of $\mathbb{R}^N$. That is, a function $f = \sum_{k=1}^l c_k \phi_{k-1}$ in Φ_l is mapped to $y = A_l(f)$ with $y = (y^1, \ldots, y^N)$ and $y^\alpha = \sum_{j=1}^l A_j^\alpha c_j$. These linear maps replace the corresponding map for SSA in (7.1), enforcing the condition (7.60) on the discrete Dirichlet form. The spatial and temporal patterns associated with A_l follow in analogy with (7.45) by performing the SVD

$$A_l = U_l \Sigma_l V_l^T, \tag{7.67}$$

where $U_l = (U_1, \ldots, U_N)$ and $V_l = (V_1, \ldots, V_l)$ are $N \times N$ and $l \times l$ orthogonal matrices, and $\Sigma = \mathrm{diag}(\sigma_1^{(l)}, \ldots, \sigma_{\min\{N,l\}}^{(l)})$ a diagonal matrix of nonnegative singular values. Here, the matrix elements of V_l are expansion coefficients in the ϕ_i basis of Φ_l. In particular, the jth column V_l of V_l corresponds to a function $v_j \in \Lambda^0 M$ and a function of time $\tilde{v}_j$ given by

$$v_j(i) = \tilde{v}_j(t_i) = \sum_{k=1}^l v_{kj} \phi_{k-1}(i), \tag{7.68}$$

The above are the NLSA analogs of the chronos modes (7.46) in classical SSA. By the orthogonality properties of the ϕ_i basis functions, the v_j are orthogonal with respect to the inner product in (7.55). Note that unlike the rank-l truncated U_l matrix from SSA in (7.48), the first l columns of U_l from NLSA are not equal to the first l columns of U_{l+1} (the same is true for Σ_l and V_l). Moreover, the temporal patterns in (7.68) are not linear projections of the data onto the corresponding spatiotemporal patterns U_i.

Consider now the rank-l approximation X_l of the signal X in lagged-embedding space obtained by using the SVD of A_l,

$$X_l = U_l \Sigma_l V_l^T \Phi_l^T = X \Pi \Phi_l \Phi_l^T. \qquad (7.69)$$

Here, $\Pi = \mathrm{diag}(\pi_1, \ldots, \pi_S)$ is a diagonal matrix containing the invariant distribution (Riemannian measure) in (7.54), and $\Phi_l = (\phi_0, \ldots, \phi_{l-1})$ an $S \times l$ matrix of eigenfunction values. It follows by comparing (7.69) with (7.48) that the rank-l approximations of the signal in NLSA and SSA differ in their filtering kernel. NLSA filters the data by the diffusion kernel, $\Pi \Phi_l \Phi_l$, whereas SSA by the covariance kernel, $V_l V_l^T$. Note that besides differences in X_l, the spatiotemporal patterns corresponding to individual singular-vector pairs (i.e., the $\hat{X}_j$ terms in (7.72)) may differ substantially between the two methods.

We also mention a connection between NLSA and kernel PCA (KPCA, e.g., [91]). In particular, because the right singular vectors V_i in (7.67) are eigenvectors of $A_l^T A_l$ with corresponding eigenvalue $(\sigma_i^{(l)})^2$, it follows that the temporal modes in (7.68) are solutions of the kernel eigenvalue problem

$$K_l v_i = (\sigma_i^{(l)})^2 v_i, \quad \text{with} \quad K_l = \Phi_l \Phi_l^T \Pi X^T X \Pi. \qquad (7.70)$$

Thus, NLSA can formally be interpreted as KPCA with the kernel K_l in (7.70). Note that K_l should not be confused with the kernels in (7.61) and (7.62) used to construct the continuous and discrete diffusion operators in (7.49) and (7.57), respectively.

7.2.2.7 Setting the truncation level

As discussed in Sections 7.2.2.3 and 7.2.2.4, the parameter l controls the "wavenumber" on the data manifold resolved by the diffusion eigenfunctions spanning Φ_l. On the one hand, working at a tight truncation level (small l) is desirable in order to produce a parsimonious description of the data with minimal risk of overfitting (the variance of the discrete eigenfunction ϕ_l increases with l for a fixed number of samples S [93]). At the same time, a too drastic truncation will inevitably lead to important features of the data being unexplained. A useful heuristic criterion for selecting l is to monitor a relative spectral entropy D_l, measuring changes in the energy distribution among the modes of A_l as l grows [40]. This measure is given by the formula

$$D_l = \sum_{i=1}^{l} p_i^{(l+1)} \log \left(\frac{p_i^{(l+1)}}{\hat{p}_i^{(l+1)}} \right),$$

$$\hat{p}_i^{(l)} = \frac{(\hat{\sigma}_i^{(l)})^2}{(\sum_i^l (\hat{\sigma}_i^{(l)})^2)}, \quad (\hat{\sigma}_1^{(l)}, \ldots, \hat{\sigma}_{l-1}^{(l)}, \hat{\sigma}_{(l)}^l) = (\sigma_1^{(l-1)}, \ldots, \sigma_{l-1}^{(l-1)}, \sigma_{l-1}^{(l-1)}). \qquad (7.71)$$

The appearance of qualitatively new features in the spectrum of A_l is accompanied by spikes in D_l (e.g., Fig. 7.10a), suggesting that a reasonable truncation level is the minimum l beyond which D_l settles to small values.

Note that the compressed representation of the data in the $N \times l$-sized A_l results in substantial gains in computational efficiency compared to the SVD of the full data matrix X in large-scale applications where the ambient space dimension N and the sample number S are both large (e.g., Section 7.2.3). Of course, in NLSA one has to perform the pairwise kernel evaluations to form the diffusion operator L, but this computation can be straightforwardly parallelized. Moreover, by virtue of the exponential decay of the kernel, the eigenvalue problem (7.59) can be carried out efficiently using sparse iterative solvers.

Our experience from applications ranging from low-dimensional models [41], to comprehensive numerical models [14,40,41], and real-world observations [45,104], has been that the locally-scaled kernel in (7.62) in conjunction with the $\alpha = 1$ DM normalization in (7.64) and the Φ_l-restricted SVD in (7.67) leads to superior timescale separation and ability to detect physically meaningful low-variance patterns that are not accessible to classical linear-projection techniques such as PCA and SSA. However, a complete theoretical understanding of the SVD procedure, as well as its potential limitations, is still lacking.

7.2.2.8 Projection to data space   The final step in the NLSA pipeline is to construct the spatiotemporal patterns $\hat{x}_i^j$ in n-dimensional data space associated with the corresponding singular vectors and values, $\{U_j, V_j, \sigma_j^{(l)}\}$, of the A_l map in (7.66). Because A_l is a linear map, this procedure is significantly more straightforward and unambiguous than in methods based on nonlinear mapping functions (e.g., [15,67]), and consists of two steps: (1) Compute the $N \times S$ matrix $\hat{X}_j$ containing the jth spatiotemporal pattern in lagged embedding space, $\hat{X}_j = U_j \sigma_j^l V_j^T \Phi_l^T$ and (2) decompose each column of $\hat{X}_j$ into q blocks $\hat{x}_{ij}$ of size n,

$$
\hat{X}_j = \begin{pmatrix} \uparrow & & \uparrow \\ \hat{X}_1^j & \cdots & \hat{X}_S^j \\ \downarrow & & \downarrow \end{pmatrix} = \begin{pmatrix} \hat{x}_{11} & \cdots & \hat{x}_{1s'} \\ \vdots & \ddots & \vdots \\ \hat{x}_{q1} & \cdots & \hat{x}_{qs'} \end{pmatrix}, \tag{7.72}
$$

and take the average over the lagged embedding window,

$$
x_j = (\hat{x}_1^j, \ldots, \hat{x}_s^j), \quad \hat{x}_i^j = \sum_{k=1}^{\min\{q,i\}} \frac{\hat{x}_{j,i-k+1}}{\min\{q,i\}}. \tag{7.73}
$$

This leads to s samples in n-dimensional data space, completing the decomposition in (7.41).

7.2.3 Analysis of infrared brightness temperature satellite data for tropical dynamics

Satellite imagery has been used to study convection-coupled tropical disturbances since the 1970s. Substantial advances in the understanding of tropical waves have been made through linear theories and diagnostics guided by these theories (e.g., [63]). However, convection-coupled tropical motions are highly nonlinear and multi-scale. Among the most notable examples is the Madden–Julian oscillation (MJO, e.g., [70]); an eastward-propagating, planetary-scale envelope of organized tropical convection. Originating in the Indian Ocean and propagating eastward over the Indonesian Maritime Continent until its decay in the Western Pacific, the MJO has gross scales in the 30–90-day intraseasonal time range and zonal wavenumber of order 1–4. It dominates the tropical predictability in subseasonal time scales, exerting global influences through tropical–extratropical interactions, affecting weather and climate variability, and fundamentally interfacing the short-term weather prediction and long-term climate projections [106].

Conventional methods for extracting MJO signals from observations and models are linear, including linear bandpass filtering, regression, and EOFs [107]. On the other hand, theory development has suggested that the MJO is a nonlinear oscillator [73, 74]. With a nonlinear temporal filter, the observed MJO appears to be a stochastically driven chaotic oscillator [102].

In this section, we apply NLSA to extract the spatiotemporal patterns of the MJO and other convective processes from satellite infrared brightness temperature over the tropical belt. This analysis is an extension of the work Ref. [45, 104], which considered one-dimensional (1D) latitudinally-averaged data instead of the two-dimensional (2D) infrared brightness temperature field studied here.

7.2.3.1 Dataset description
7.2.3.1 Dataset description   The Cloud Archive User Service (CLAUS) Version 4.7 multisatellite infrared brightness temperature (denoted T_b) [49] is used for this study. Brightness temperature is a measure of the Earth's infrared emission in terms of the temperature of a hypothesized blackbody emitting the same amount of radiation at the same wavelength ($\sim$ 10–11 µm in the CLAUS data). It is a highly correlated variable with the total longwave emission of the Earth. In the tropics, positive (negative) T_b anomalies are associated with reduced (increased) cloudiness. The global CLAUS T_b data are on a 0.5° longitude by 0.5° latitude fixed grid, with 3-h time resolution from 00 to 21 UTC, spanning July 1, 1983 to June 30, 2006. T_b values range from 170 to 340 K with approximately 0.67 K resolution.

The subset of the data in the global tropical belt between 15°S and 15°N was taken to create a spatiotemporal dataset sampled at a uniform longitude–latitude grid of $n_x \times n_y = 720 \times 60$ gridpoints (with data space dimension $n = n_x n_y = 43{,}200$) and $s = 67{,}208$ temporal snapshots. Prior to analysis, the missing gridpoint values ($< 1\%$ of ns) were filled via linear interpolation in time. A portion of the data for the period of the observational campaign Tropical Ocean Global Atmosphere Coupled Ocean Atmosphere Response Experiment (TOGA COARE, November 1992 to February 1993) [109], which is discussed in Section 7.2.3.3, is shown in Figure 7.9 in a time–longitude plot of T_b averaged about the equator.

7.2.3.2 Modes recovered by NLSA
7.2.3.2 Modes recovered by NLSA   We have applied the NLSA algorithm described in Section 7.2.2 using an embedding window spanning $\Delta t = 64$ days. This amounts to an embedding space dimension $N = qn \approx 2.2 \times 10^7$ for the $\delta t = 3$ h

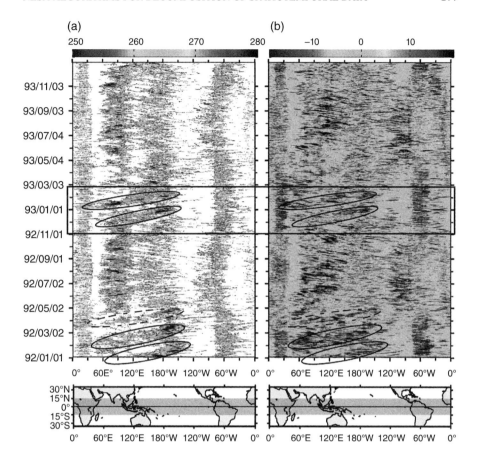

FIGURE 7.9 Time–longitude section of (a) symmetric and (b) antisymmetric brightness temperature T_b data (in K) from CLAUS for the period 1992–1993. In (a) only thresholded values <280 K are shown to emphasize convective activity. The bottom map in (a) indicates that the symmetric component was obtained via averaging over 15°S to 15°N. The antisymmetric component in (b) was obtained by subtracting the values at the northern latitudes from the corresponding southern latitudes. The boxed interval corresponds to the TOGA-CORE period. Ovals mark significant MJO events. Dates are expressed as YY-MM-DD.

sampling interval ($q = \Delta t / \delta t = 512$) and 0.5° resolution of our dataset. This choice of embedding window was motivated from our objective to resolve propagating structures such as the MJO with intraseasonal (30–90 days) characteristic timescales. In comparison, Kikuchi et al. [62] used a 10 days window with 5 days increments in their analysis of outgoing longwave radiation (OLR) data using EEOFs. Unlike conventional approaches [62, 80, 110], neither bandpass filtering nor seasonal detrending was applied prior to processing the data via NLSA.

For the calculation of the diffusion eigenfunctions in (7.58), we computed the pairwise kernel values from (7.62) in embedding space using brute force and evaluated

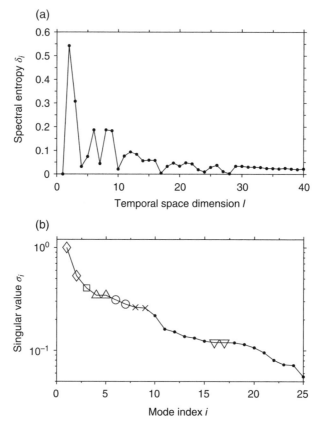

FIGURE 7.10   (a) Spectral entropy δ_l from (7.74) and (b) singular values $\sigma_i^{(l)}$ of the A_l linear map (7.66) with $l = 27$. The highlighted modes in (b) are the ($\Diamond$) annual; ($\Box$) interannual; ($\triangle$) symmetric diurnal; ($\bigcirc$) MJO; ($\times$) semiannual; ($\triangledown$) antisymmetric diurnal modes. See Figure 7.11 for the corresponding temporal patterns.

the Markov matrix retaining nonzero entries for 5000 nearest neighbors per data-point. The resulting spectral entropy D_l, computed via (7.71) and normalized to the unit interval by applying the transformation [57]

$$\delta_l = (1 - \exp(-2D_l))^{1/2} \tag{7.74}$$

is shown in Figure 7.10a. As described in Section 7.2.2.6, D_l exhibits a series of spikes as l increases from small to moderate values ($l \sim 20$), which correspond to qualitatively new spatiotemporal patterns entering in the spectrum of A_l. Eventually, D_l settles to small values for $l \gtrsim 25$. On the basis of the results in Figure 7.10a, hereafter we set the temporal space dimension in (7.59) to $l = 27$. The singular values $\sigma_i^{(l)}$ of the associated A_l linear map from (7.66) are displayed in Figure 7.10b.

With these NLSA parameter values, the recovered spatiotemporal modes describe several convective processes of interest operating in a wide range of spatial and temporal scales. Representative temporal patterns $\tilde{v}_j(t_i)$ from (7.68) are shown in Figure 7.11. Snapshots of the corresponding spatiotemporal patterns $\hat{x}_i^j$ from (7.73) are displayed in Figures 7.12 and 7.13. The properties of these modes are as follows.

- *Modes* $(1,2)$ *and* $(6,7)$. As manifested by the prominent lines at the once- and twice-per year frequencies in their temporal Fourier spectra, these modes respectively describe annual (Figs. 7.11a and b) and semiannual (Figs. 7.11f and g) periodic processes, which are expected to be prominent in tropical T_b signals. In the spatial domain, Modes $(1,2)$ are characterized by T_b anomalies of opposite sign in the Northern and Southern Hemispheres. The December 25 snapshot of Figure 7.12a corresponds to the dry season in the tropical North Hemisphere and wet season in the tropical Southern Hemisphere. The semiannual pair captures the march of intensified deep convection in the intratropical convergence zone (ITCZ), South Pacific convergence zone (SPCZ), Indo-Pacific warm pool, monsoons, and tropical storm tracks across the Southern and Northern Hemispheres. This pair of modes explains the apparent migration and amplitude modulation of convection signals due to the north–south asymmetry in land mass and bathymetry within the tropical belt.

- *Mode 3.* Due to the characteristic dipole pattern over the equatorial Pacific (Fig. 7.12c), and presence of spectral power on interannual timescales (Fig. 7.11c), this mode is interpreted as the El Niño Southern Oscillation (ENSO) [100] mode. The ENSO can enhance or inhibit MJO propagation by preconditioning the environment of the Western Pacific. For instance, the snapshot of Figure 7.12c coincides with an amplifying El Niño phase with warm sea surface temperature and enhanced convection (i.e., negative T_b anomaly), which is conducive to MJO propagation in the Western Pacific. On the other hand, MJO propagation is inhibited during La Niña years (not shown here).

- *Modes* $(8,9)$. This two-mode family corresponds the manifestation of the MJO in T_b data. Characterized by broad intraseasonal peaks (20–90 days) in their frequency spectra and phase-locked in quadrature (Figs. 7.11h and i), these modes describe a 5000 km-scale eastward-propagating envelope of organized convection (Figs. 7.12f and 7.13). This structure originates over the Indian Ocean and propagates eastward until it decays upon reaching the cold waters of the Central Pacific. The presence of semiannual lines in the frequency spectra of Modes $(8,9)$ is consistent with the fact that MJO events occur preferentially in boreal winter (November–March).

- *Modes* $(4,5)$ *and* $(16,17)$. Characterized by dominant peaks over the once-per-day frequency in their Fourier spectra (Figs. 7.11d, e, j, and k), these modes describe diurnal convective variability. The corresponding spatiotemporal patterns (Figs. 7.12b and g) are most prominent over land, where the diurnal cycle of convection is most active. The major difference between these modes is that

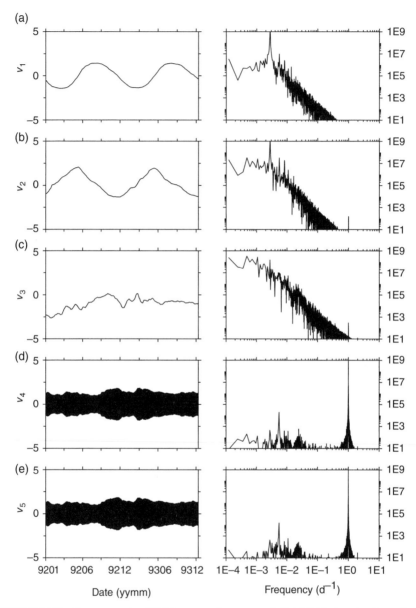

FIGURE 7.11   Representative NLSA temporal patterns of CLAUS T_b data for the period of January 1992 to December 1993 and their frequency power spectra. (a and b) Annual modes, (c) interannual (ENSO) mode, (d and e) symmetric diurnal pair, (f and g) semiannual modes (h and i) MJO pair, (j and k) antisymmetric diurnal pair.

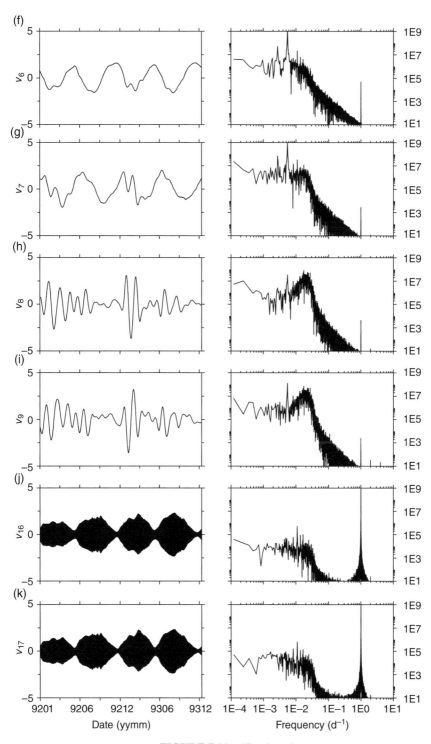

FIGURE 7.11   (Continued)

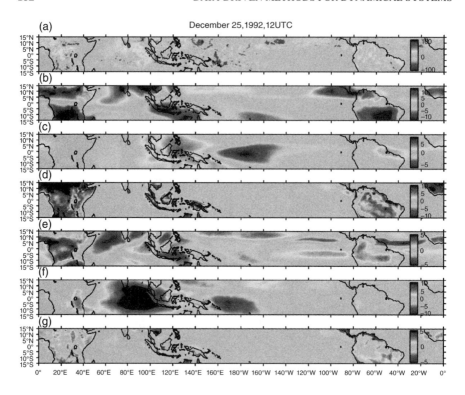

FIGURE 7.12 A snapshot of reconstructed CLAUS T_b (in K) data for December 25, 1992, 12UTC using the NLSA modes highlighted in Figures 7.10 and 7.11. Negative T_b anomalies indicate enhanced convection. Positive T_b anomalies indicate decreased cloudiness. (a) Raw data, (b) annual modes, $x_1 + x_2$, (c) interannual (ENSO) mode, x_3, (d) latitudinally symmetric diurnal pair, $x_4 + x_5$, (e) semiannual modes, $x_6 + x_7$, (f) MJO pair, $x_8 + x_9$, and (g) latitudinally antisymmetric diurnal pair, $x_{16} + x_{17}$. The prominent MJO event in (f) occurring over the Western Pacific was observed by the TOGA COARE field campaign [111].

(4, 5) are predominantly symmetric about the equator, whereas (16, 17) are predominantly antisymmetric. The symmetric pair is active year-round, but the antisymmetric pair is strongly modulated by the seasonal cycle.

The availability of this family of modes, determined by an objective algorithm requiring no preprocessing of the data, enables one to study interdependencies between modes of convection across multiple temporal and spatial scales. For instance, the interdependencies between ENSO, MJO, and the diurnal cycle are topics of significant current interest (e.g., [104] and references therein). Such analyses are outside the scope of this chapter, but we refer the reader to Refs. [103,104] and for a study involving NLSA modes from 1D CLAUS T_b data averaged about the equator, where a comparison between the NLSA and SSA modes is also made.

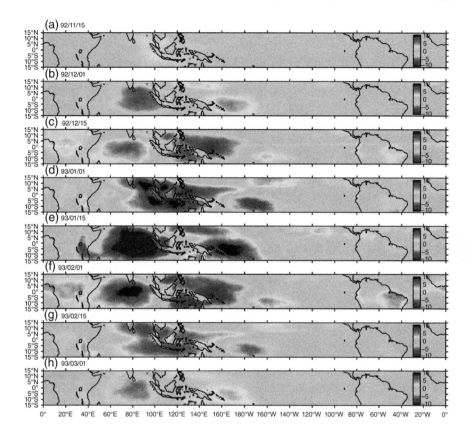

FIGURE 7.13 Reconstruction of the MJO waivetrain observed during the TOGA COARE intensive observing period (IOP) from November 1992 to March 1993. The color maps show temperature T_b anomalies (in K) obtained from the NLSA MJO modes of Figure 7.11h and i projected to data space via (7.73), that is, $x_8 + x_9$. Negative (positive) T_b anomalies indicate increased (decreased) convection. (a) No MJO activity is present; (b–d) the first MJO initiates over the Indian Ocean, propagates eastward over the Indonesian Maritime Continent, and decays after reaching the $180°$ dateline; (e–g) a second, stronger, MJO event with an initiation signal over East Africa; and (h) a weak third event starting at the end of the TOGA COARE IOP. See Figure 7.9 for the manifestation of these events in time–longitude section of the raw data.

7.2.3.3 Reconstruction of the TOGA COARE MJOs Two complete MJO events were observed during the TOGA COARE period (e.g., [111]). Figure 7.13 shows reconstructions of these events based on the NLSA modes. This reconstruction captures the salient features of the propagating envelope of deep convection associated with the MJO, including the initiation of enhanced deep convection (hence cold anomalies) over the Indian Ocean, the passage over the Maritime Continent, and the arrival and demise near the date line. The first event started near $75°E$ in late November, subsequently crossed the Maritime Continent around $100°–150°$E, then disappeared near $170°$W around January 10. The second event, being slightly faster than the first, started around January 5 and reached the central Pacific in early February. A third event started in March, after the end of the TOGA COARE period.

The TOGA COARE period was coincident with the amplifying phase of an El Niño event; therefore, the convective MJO superclusters propagated further east beyond the date line, where during normal years the cold sea surface temperature is not conducive to deep convection. The eastward-propagation speed of the reconstructed MJO events is consistent with the observed value of approximately $4–5$ ms^{-1}.

7.3 CONCLUSIONS

In this chapter, we have reviewed two examples of applied mathematics techniques for data analysis in dynamical systems: (1) Methods for quantifying predictability and model error based on data clustering and information theory (Section 7.1.2) and (2) NLSA algorithms for extracting spatiotemporal patterns from high-dimensional data (Section 7.2). We have highlighted these techniques with applications to CAOS, in particular, predictability assessment and Markov modeling of circulation regimes in a simple ocean model (Section 7.1.5) and extraction of modes of organized convection in the tropics from infrared brightness temperature (T_b) satellite data (Section 7.2.3).

A common theme in these methods has been the coarse-grained geometry of the data. In Section 7.1, we saw how a discrete partition of the space of initial data (constructed empirically through data clustering) can be used in conjunction with information theory to place practically computable lower bounds to the predictability of observables in dynamical systems and the error of forecasting these observables with imperfect models. In Section 7.2, the machinery of discrete exterior calculus and spectral graph theory was combined with delay-coordinate mappings of dynamical systems to extract spatiotemporal modes of variability which are describable in terms of low-dimensional sets of diffusion eigenfunctions, selected according to a "low-wavenumber" criterion on the data manifold formulated in an intrinsically discrete setting.

The techniques in (i) and (ii) can be naturally combined. In particular, recall that in Section 7.1.5, data space was spanned by the leading 20 PCs of the oceanic stream-function. One can consider replacing the PCs with the temporal patterns recovered by NLSA and seek predictable patterns in that space. A natural, but challenging application is to use the predictability framework of Section 7.1 to study MJO predictability (a problem of wide practical impact [106]) in the space of NLSA modes recovered from brightness temperature data and other relevant fields. In this case, it is likely that the complexity of the data in modal space compared to the 1.5-layer ocean model will require a modification of the K-means algorithm used in Section 7.1.5 in order to identify states with high MJO predictability. Likewise, we believe that it would be fruitful to explore alternative formulations to the locally-scaled NLSA kernel in (7.62) (which has been designed having a deterministic dynamical system in mind) to deal with stochastic dynamics. The recent work in Ref. [98] should be relevant in this context. Finally, as mentioned in Section 7.2.2.6, an open theoretical problem is the justification (and potential improvement) of the truncated SVD step in NLSA. We plan to pursue these topics in future work.

Acknowledgments

The authors would like to thank Rafail Abramov, Tyrus Berry, Grant Branstator, Mitch Bushuk, John Harlim, Illia Horenko, Eniko Szekely, Haiyan Teng, and Wen-wen Tung for their explicit and implicit contributions to this work. This work was supported by the Office of Naval Research, including ONR DRI grants N25-74200-F6607, N00014-10-1-0554, N00014-14-1-0150, and ONR MURI grant 25-74200-F7112. The results of Section 7.2.3 were obtained using the CLAUS archive held at the British Atmospheric Data Centre, produced using ISCCP source data distributed by the NASA Langley Data Center.

REFERENCES

1. N. Aubry, R. Guyonnet, and R. Lima. Spatiotemporal analysis of complex signals: Theory and applications. *J. Stat. Phys.*, 64:683–739, 1991.

2. N. Aubry, W.-Y. Lian, and E. S. Titi. Preserving symmetries in the proper orthogonal decomposition. *SIAM J. Sci. Comput.*, 14:483–505, 1993.

3. M. Belkin and P. Niyogi. Laplacian eigenmaps for dimensionality reduction and data representation. *Neural Comput.*, 15:1373–1396, 2003.

4. M. Belkin and P. Niyogi. Towards a theoretical foundation for Laplacian-based manifold methods. *J. Comput. Syst. Sci.*, 74(8):1289–1308, 2008.

5. P. H. Bérard. *Spectral Geometry: Direct and Inverse Problems*, volume 1207 of *Lecture Notes in Mathematics*. Springer-Verlag, Berlin, 1989.

6. P. S. Berloff and J. C. McWilliams. Large-scale, low-frequency variability in wind-driven ocean gyres. *J. Phys. Oceanogr.*, 29:1925–1949, 1999.

7. J. Berner and G. Branstator. Linear and nonlinear signatures in planetary wave dynamics of an AGCM: Probability density functions. *J. Atmos. Sci.*, 64:117–136, 2007.

8. T. Berry. *Model Free Techniques for Reduction of High-Dimensional Dynamics*. PhD thesis, George Mason University, 2013.

9. T. Berry, R. Cressman, Z. Greguric Ferencek, and T. Sauer. Time-scale separation from diffusion-mapped delay coordinates. *SIAM J. Appl. Dyn. Syst.*, 12:618–649, 2013.

10. G. Branstator and H. Teng. Two limits of initial-value decadal predictability in a CGCM. *J. Clim.*, 23:6292–6311, 2010.

11. J. Bröcker. Reliability, sufficiency and the decomposition of proper scores. *Q. J. R. Meteorol. Soc.*, 135:1512–1519, 2009.

12. J. Bröcker, D. Engster, and U. Parlitz. Probabilistic evaluation of time series models: A comparison of several approaches. *Chaos*, 19(4):04130, 2009.

13. D. S. Broomhead and G. P. King. Extracting qualitative dynamics from experimental data. *Phys. D*, 20(2–3):217–236, 1986.

14. M. Bushuk, D. Giannakis, and A. J. Majda. Reemergence mechanisms for North Pacific sea ice revealed through nonlinear Laplacian spectral analysis. *J. Climate*, 27: 6265–6287, 2013.

15. B. Christiansen. The shortcomings of nonlinear component analysis in identifying circulation regimes. *J. Clim.*, 18:4814–4823, 2005.

16. F. R. K. Chung. *Spectral Graph Theory*, volume 97 of *CBMS Regional Conference Series in Mathematics*. American Mathematical Society, Providence, 1997.

17. R. R. Coifman and S. Lafon. Diffusion maps. *Appl. Comput. Harmon. Anal.*, 21:5–30, 2006.

18. R R. Coifman, et al. Geometric diffusions as a tool for harmonic analysis and structure definition on data. *Proc. Natl. Acad. Sci.*, 102(21):7426–7431, 2005.

19. T. A. Cover and J. A. Thomas. *Elements of Information Theory*. Wiley-Interscience, Hoboken, 2nd edition, 2006.

20. D. T. Crommelin and A. J. Majda. Strategies for model reduction: Comparing different optimal bases. *J. Atmos. Sci.*, 61:2206–2217, 2004.

21. D. T. Crommelin and E. Vanden-Eijnden. Fitting timeseries by continuous-time Markov chains: A quadratic programming approach. *J. Comput. Phys.*, 217(2):782–805, 2006.

22. M. H. DeGroot and S. E. Fienberg. Assessing probability assessors: Calibration and refinements. In S. S. Gupta and J. O. Berger, editors, *Statistical Decision Theory and Related Topics III*, volume 1, pages 291–314. Academic Press, New York, 1982.

23. T. DelSole. Predictability and information theory. Part I: Measures of predictability. *J. Atmos. Sci.*, 61(20):2425–2440, 2004.

24. T. DelSole. Predictability and information theory. Part II: Imperfect models. *J. Atmos. Sci.*, 62(9):3368–3381, 2005.

25. T. DelSole and J. Shukla. Model fidelity versus skill in seasonal forecasting. *J. Clim.*, 23(18):4794–4806, 2010.

26. M. Desbrun, E. Kanso, and Y. Tong. Discrete differential forms for computational modeling. In *ACM SIGGRAPH 2005 Courses*, SIGGRAPH '05, Los Angeles, 2005.

27. P. Deuflhard, M. Dellnitz, O. Junge, and C. Schütte. Computation of essential molecular dynamics by subdivision techniques I: Basic concept. *Lect. Notes Comp. Sci. Eng.*, 4:98, 1999.

28. P. Deuflhard, W. Huisinga, A. Fischer, and C. Schütte. Identification of almost invariant aggregates in reversible nearly uncoupled Markov chains. *Linear Algebra Appl.*, 315:39, 2000.

29. E. R. Deyle and G. Sugihara. Generalized theorems for nonlinear state space reconstruction. *PLoS ONE*, 6(3):e18295, 2011.

30. R. O. Duda, P. E. Hart, and D. G. Stork. *Pattern Classification*. Wiley-Interscience, New York, 2nd edition, 2000.

31. V. P. Dymnikov and A. N. Filatov. *Mathematics of Climate Modeling*. Birkhäuser, Boston, 1997.

32. K. D. Elworthy, Y. Le Jan, and X.-M. Xi. *The Geometry of Filtering*. Frontiers in Mathematics. Birkhäuser, Basel, 2010.

33. E. S. Epstein. Stochastic dynamic prediction. *Tellus*, 21:739–759, 1969.

34. C. Franzke, A. J. Majda, and G. Branstator. The origin of nonlinear signatures of planetary wave dynamics: Mean phase space tendencies and contributions from non-Gaussianity. *J. Atmos. Sci.*, 64:3988, 2007.

35. C. Franzke, D. Crommelin, A. Fischer, and A. J. Majda. A hidden Markov model perspective on regimes and metastability in atmospheric flows. *J. Clim.*, 21(8):1740–1757, 2008.

36. C. Franzke, I. Horenko, A. J. Majda, and R. Klein. Systematic metastable regime identification in an AGCM. *J. Atmos. Sci.*, 66(9):1997–2012, 2009.

37. M. Ghil and A. W. Robertson. "Waves" vs. "particles" in the atmosphere's phase space: A pathway to long-range forecasting? *Proc. Natl. Acad. Sci*, 99(suppl. 1):2493–2500, 2002.

38. M. Ghil et al. Advanced spectral methods for climatic time series. *Rev. Geophys.*, 40: 3.1–3.41, 2002.

39. D. Giannakis. Dynamics-adapted cone kernels. *SIAM J. Appl. Dyn. Syst.*, In revision 2014.

40. D. Giannakis and A. J. Majda. Comparing low-frequency and intermittent variability in comprehensive climate models through nonlinear Laplacian spectral analysis. *Geophys. Res. Lett.*, 39:L10710, 2012.

41. D. Giannakis and A. J. Majda. Nonlinear Laplacian spectral analysis for time series with intermittency and low-frequency variability. *Proc. Natl. Acad. Sci.*, 109(7):2222–2227, 2012.

42. D. Giannakis and A. J. Majda. Quantifying the predictive skill in long-range forecasting. Part I: Coarse-grained predictions in a simple ocean model. *J. Clim.*, 25:1793–1813, 2012.

43. D. Giannakis and A. J. Majda. Quantifying the predictive skill in long-range forecasting. Part II: Model error in coarse-grained Markov models with application to ocean-circulation regimes. *J. Clim.*, 25:1814–1826, 2012.

44. D. Giannakis and A. J. Majda. Nonlinear Laplacian spectral analysis: Capturing intermittent and low-frequency spatiotemporal patterns in high-dimensional data. *Stat. Anal. Data Min.*, 6(3):180–194, 2013.

45. D. Giannakis, W.-w. Tung, and A. J. Majda. Hierarchical structure of the Madden-Julian oscillation in infrared brightness temperature revealed through nonlinear Laplacian spectral analysis. In *2012 Conference on Intelligent Data Understanding (CIDU)*, pages 55–62, Boulder, Colorado, 2012.

46. D. Giannakis, A. J. Majda, and I. Horenko. Information theory, model error, and predictive skill of stochastic models for complex nonlinear systems. *Phys. D.*, 241:1735–1752, 2012.

47. N. Golyandina, V. Nekrutkin, and A. Zhigljavsky. *Analysis of Time Series Structure: SSA and Related Techniques*. CRC Press, Boca Raton, 2001.

48. L. G. Grady and J. R. Polimeni. *Discrete Calculus*. Springer-Verlag, London, 2010.

49. K. Hodges, D.W. Chappell, G.J. Robinson, and G. Yang. An improved algorithm for generating global window brightness temperatures from multiple satellite infra-red imagery. *J. Atmos. Oceanic Technol.*, 17:1296–1312, 2000.

50. P. Holmes, J. L. Lumley, and G. Berkooz. *Turbulence, Coherent Structures, Dynamical Systems and Symmetry*. Cambridge University Press, Cambridge, 1996.

51. I. Horenko. On simultaneous data-based dimension reduction and hidden phase identification. *J. Atmos. Sci.*, 65:1941–1954, 2008.

52. I. Horenko. On robust estimation of low-frequency variability trends in discrete Markovian sequences of atmospheric circulation patterns. *J. Atmos. Sci.*, 66(7):2059–2072, 2009.

53. I. Horenko. On clustering of non-stationary meteorological time series. *Dyn. Atmos. Oceans*, 49:164–187, 2010.

54. I. Horenko. On the identification of nonstationary factor models and their application to atmospheric data analysis. *J. Atmos. Sci.*, 67(5):1559–1574, 2010.

55. I. Horenko. Nonstationarity in multifactor models of discrete jump processes, memory and application to cloud modeling. *J. Atmos. Sci.*, 68(7):1493–1506, 2011.

56. J. W. Hurrell et al. Decadal climate prediction: Opportunities and challenges. In *Proceedings of the OceanObs09 Conference: Sustained Ocean Observations and Information for Society*, pages 21–25, Venice, Italy, 2009.

57. H. Joe. Relative entropy measures of multivariate dependence. *J. Am. Stat. Assoc.*, 84(405):157–164, 1989.

58. M. A. Katsoulakis and D. G. Vlachos. Coarse-grained stochastic processes and kinetic Monte Carlo simulators for the diffusion of interacting particles. *J. Chem. Phys.*, 119(18):9412–9427, 2003.

59. M. A. Katsoulakis, A. J. Majda, and A. Sopasakis. Intermittency, metastability and coarse-graining for coupled deterministic–stochastic lattice systems. *Nonlinearity*, 19:1021–1047, 2006.

60. N. S. Keenlyside, M. Latif, J. Jungclaus, L. Kornblueh, and E. Roeckner. Advancing decadal-scale climate prediction in the North Atlantic sector. *Nature*, 453:84–88, 2008.

61. S. Khan, S. Bandyopadhyay, A. R. Ganguly, S. Saigal, D. J. Erickson III, V. Protopopescu, and G. Ostrouchov. Relative performance of mutual information estimation methods for quantifying the dependence among short and noisy data. *Phys. Rev. E*, 76:026209, 2007.

62. K. Kikuchi, B. Wang, and Y. Kajikawa. Bimodal representation of the tropical intraseasonal oscillation. *Clim. Dyn.*, 38:1989–2000, 2012.

63. G. N. Kiladis, M. C. Wheeler, P. T. Haertel, K. H. Straub, and P. E. Roundy. Convectively coupled equatorial waves. *Rev. Geophys.*, 47(2):RG2003, 2009.

64. R. Kleeman. Measuring dynamical prediction utility using relative entropy. *J. Atmos. Sci.*, 59(13):2057–2072, 2002.

65. R. Kleeman. Information theory and dynamical system predictability. *Entropy*, 13(3):612–649, 2011.

66. L.-Y. Leung and G. R. North. Information theory and climate prediction. *J. Clim.*, 3(1):5–14, 1990.

67. C. H. R. Lima, U. Lall, T. Jebara, and A. G. Barnston. Statistical prediction of ENSO from subsurface sea temperature using a nonlinear dimensionality reduction. *J. Clim.*, 22:4501–4519, 2009.

68. E. N. Lorenz. The predictability of a flow which possesses many scales of motion. *Tellus*, 21(3):289–307, 1969.

69. E. Lorenz. Climate predictability. In *The Physical Basis of Climate and Climate Modelling*, volume 16 of *GARP Publications Series*, pages 132–136. World Meteorological Organization, 1975.

70. R. A. Madden and P. R. Julian. Description of global-scale circulation cells in the tropics with a 40–50 day period. *J. Atmos. Sci.*, 29(6):1109–1123, 1972.

71. A. J. Majda and M. Branicki. Lessons in uncertainty quantification for turbulent dynamical systems. *Discrete Continuous Dyn. Syst.*, 32(9):3133–3221, 2012.

72. A. J. Majda and B. Gershgorin. Quantifying uncertainty in climate change science through empirical information theory. *Proc. Natl. Acad. Sci.*, 107(34):14958–14963, 2010.

73. A. J. Majda and S. N. Stechmann. The skeleton of tropical intraseasonal oscillations. *Proc. Natl. Acad. Sci.*, 106:8417–8422, 2009.

74. A. J. Majda and S. N. Stechmann. Nonlinear dynamics and regional variations in the MJO skeleton. *J. Atmos. Sci.*, 68:3053–3071, 2011.

75. A. J. Majda and X. Wang. *Nonlinear Dynamics and Statistical Theories for Basic Geophysical Flows*. Cambridge University Press, Cambridge, 2006.

76. A. J. Majda and X. Wang. Linear response theory for statistical ensembles in complex systems with time-periodic forcing. *Comm. Math. Sci.*, 8(1):145–172, 2010.

77. A. J. Majda, R. Kleeman, and D. Cai. A mathematical framework for quantifying predictability through relative entropy. *Methods Appl. Anal.*, 9(3):425–444, 2002.

78. A. J. Majda, I. I. Timofeyev, and E. Vanden Eijnden. Systematic strategies for stochastic mode reduction in climate. *J. Atmos. Sci.*, 60:1705, 2003.

79. A. J. Majda, C. Franzke, A. Fischer, and D. T. Crommelin. Distinct metastable atmospheric regimes despite nearly Gaussian statistics: A paradigm model. *Proc. Natl. Acad. Sci.*, 103(22):8309–8314, 2006.

80. E. D. Maloney and D. L. Hartmann. Frictional moisture convergence in a composite life cycle of the Madden-Julian oscillation. *J. Clim.*, 11:2387–2403, 1998.

81. J. D. McCalpin and D. B. Haidvogel. Phenomenology of the low-frequency variability in a reduced-gravity quasigeostrophic double-gyre model. *J. Phys. Oceanogr.*, 26(5):739–752, 1996.

82. G. A. Meehl, et al. Decadal prediction. Can it be skillful? *Bull. Am. Meteor. Soc.*, 90(10):1467–1485, 2009.

83. P. Metzner, E. Dittmer, T. Jahnke, and C. Schütte. Generator estimation of Markov jump processes based on incomplete observations equidistant in time. *J. Comput. Phys.*, 227(1):353–375, 2007.

84. B. Nadler, S. Lafon, R. R. Coifman, and I. Kevrikedes. Diffusion maps, spectral clustering, and reaction coordinates of dynamical systems. *Appl. Comput. Harmon. Anal.*, 21:113–127, 2006.

85. N. H. Packard et al. Geometry from a time series. *Phys. Rev. Lett.*, 45:712–716, 1980.

86. C. Penland. Random forcing and forecasting using principal oscillation pattern analysis. *Mon. Weather Rev.*, 117(10):2165–2185, 1989.

87. S. Rosenberg. *The Laplacian on a Riemannian Manifold*, volume 31 of *London Mathematical Society Student Texts*. Cambridge University Press, Cambridge, 1997.

88. M. S. Roulston and L. A. Smith. Evaluating probabilistic forecasts using information theory. *Mon. Weather Rev.*, 130(6):1653–1660, 2002.

89. T. Sauer, J. A. Yorke, and M. Casdagli. Embedology. *J. Stat. Phys.*, 65(3–4):579–616, 1991.

90. T. Schneider and S. M. Griffies. A conceptual framework for predictability studies. *J. Clim.*, 12(10):3133–3155, 1999.

91. B. Schölkopf, A. Smola, and K. Müller. Nonlinear component analysis as a kernel eigenvalue problem. *Neural Comput.*, 10:1299–1319, 1998.

92. B. W. Silverman. *Density Estimation for Statistics and Data Analysis*, volume 26 of *Monographs on Statistics and Applied Probability*. Chapman & Hall/CRC, Boca Raton, 1986.

93. A. Singer. From graph to manifold Laplacian: The convergence rate. *J. Appl. Comput. Harmon. Anal.*, 21:128–134, 2006.

94. K. Sobczyk. Information dynamics: Premises, challenges and results. *Mech. Syst. Signal Pr.*, 15(3):475–498, 2001.

95. A. Solomon et al. Distinguishing the roles of natural and anthropogenically forced decadal climate variability: Implications for prediction. *Bull. Am. Meteor. Soc.*, 92(2):141–156, 2011.

96. K. R. Sreenivasan and R. A. Antonia. The phenomenology of small-scale turbulence. *Annu. Rev. Fluid Mech.*, 29(1):435–472, 1997.

97. F. Takens. Detecting strange attractors in turbulence. In D. Rand and L. S. Young, editors, *Dynamical Systems and Turbulence, Warwick 1980*, volume 898 of *Lecture Notes in Mathematics*, pages 366–381. Springer, Berlin, 1981.

98. R. Talmon and R. R. Coifman. Empirical intrinsic geometry for nonlinear modeling and time series filtering. *Proc. Natl. Acad. Sci.*, 110(31):12535–12540, 2013.

99. H. Teng and G. Branstator. Initial-value predictability of prominent modes of North Pacific subsurface temperature in a CGCM. *Clim. Dyn.*, 36:1813–1834, 2010.

100. K. E. Trenberth. The definition of El Niño. *Bull. Amer. Meteor. Soc.*, 78(12):2771–2777, 1997.

101. R. S. Tsay. *Analysis of Financial Time Series*. John Wiley & Sons, Inc., Hoboken, 2010.

102. W. W. Tung, J. Gao, J. Hu, and L. Yang. Detecting chaos in heavy noise environments. *Phys. Rev. E*, 83:046210, 2011.

103. W.-w. Tung, D. Giannakis, and A. J. Majda. Symmetric and antisymmetric convection signals in the Madden-Julian oscillation. Part II: Kinematics and thermodynamics. *J. Atmos. Sci.*, 2013. Submitted.

104. W.-w. Tung, D. Giannakis, and A. J. Majda. Symmetric and antisymmetric convection signals in the Madden-Julian oscillation. Part I: Basic modes in infrared brightness temperature. *J. Atmos. Sci.*, 71: 3302–3326, 2014.

105. R. Vautard and M. Ghil. Singular spectrum analysis in nonlinear dynamics, with applications to paleoclimatic time series. *Phys. D*, 35:395–424, 1989.

106. D. Waliser. Predictability and forecasting. In W. K. M. Lau and D. E. Waliser, editors, *Intraseasonal Variability in the Atmosphere-Ocean Climate System*, pages 389–423. Springer, Berlin, 2005.

107. D. Waliser et al. MJO simulation diagnostics. *J. Clim.*, 22:3006–3030, 2009.

108. B. C. Weare and J. S. Nasstrom. Examples of extended empirical orthogonal function analyses. *Mon. Weather Rev.*, 110:481–485, 1982.

109. P. J. Webster and R. Lukas. TOGA COARE: The coupled ocean-atmosphere response experiment. *Bull. Am. Meteor. Soc.*, 73(9):1377–1416, 1992.

110. M. C. Wheeler and H. H. Hendon. An all-season real-time multivariate MJO index: Development of an index for monitoring and prediction. *Mon. Weather Rev.*, 132(8):1917–1932, 2004.

111. M. Yanai, B. Chen, and W.-w. Tung. The Madden-Julian oscillation observed during the TOGA COARE IOP: Global view. *J. Atmos. Sci.*, 57(15):2374–2396, 2000.

112. D. Zhou and C. Burges. High-order regularization on graphs. In *6th International Workshop on Mining and Learning with Graphs*, Helsinki, 2008.

8

ON SMOOTHNESS CONCEPTS IN REGULARIZATION FOR NONLINEAR INVERSE PROBLEMS IN BANACH SPACES

BERND HOFMANN

Faculty of Mathematics, Technische Universität Chemnitz, Chemnitz, Germany

8.1 INTRODUCTION

During the last 10 years, there has been a significant progress with respect to the development of stable approximate solutions to nonlinear inverse problems. This rapid expansion has been caused by requirements of applications that arise in natural sciences, engineering, imaging, and finance (see, e.g., Refs. [100, chapter 1] and [18, 23, 48, 62, 67, 71, 79, 83, 95–97, 102, 111]). The inverse problems can be written, in general, as *abstract operator equations*

$$F(x) = y, \tag{8.1}$$

where $x \in X$ is the nonobservable element to be determined from observations of the element $y \in Y$, and $F : \mathcal{D}(F) \subseteq X \to Y$, with domain $\mathcal{D}(F)$, is the nonlinear forward operator mapping between the normed spaces X and Y with norms $\| \cdot \|_X$ and $\| \cdot \|_Y$, respectively. In this chapter, we denote by X^* and Y^* the corresponding dual spaces and by $\langle \cdot, \cdot \rangle_{X^* \times X}$ the dual pairing between X and X^*. Instead of the exact right-hand side $y \in F(\mathcal{D}(F))$, only noisy data $y^\delta \in Y$ can be brought by measuring tools, and we suppose the *deterministic noise model*

$$\|y^\delta - y\|_Y \leq \delta. \tag{8.2}$$

Formulated for infinite dimensional abstract spaces X and Y, inverse problems are mostly *ill-posed*, which means that the solutions are not necessarily unique, and in any case small perturbations of the data y may lead to arbitrarily large errors in the solution. This is, in general, a consequence of the *smoothing character of the forward operator F* in the sense that the mapping $x \mapsto y$ destroys or suppresses information, and only additional subjective and objective a priori information about the expected solution and the data collection process can compensate the loss of information. In order to overcome or suppress this phenomenon of ill-posedness, for constructing approximate solutions to ill-posed equations (8.1), regularized solutions are the means of choice. In the literature, iterative regularization methods (cf., e.g., [30, chapter 11] and [4,5,49,66,69,70,72,106]) as well as variational regularization methods (cf., e.g., [30, chapter 10] and [6,9,27,28,47,65,73,75–77,86,103–105]) can be distinguished, and our focus is on the different types of *Tikhonov regularization methods* as the most prominent representative of variational regularization. Precisely, we use with some penalty functional $\mathcal{R}\colon X \to [0,\infty]$ regularized solutions $x_\alpha^\delta \in \mathcal{D}(F)$ which solve the extremal problem

$$\mathcal{T}_\alpha^\delta(x) := \frac{1}{p}\|F(x) - y^\delta\|_Y^p + \alpha\,\mathcal{R}(x) \to \min, \quad \text{subject to} \quad x \in \mathcal{D}(F) \subseteq X, \tag{8.3}$$

that is, minimizers $x_\alpha^\delta = \arg\min_{x \in \mathcal{D}(F)} \mathcal{T}_\alpha^\delta(x)$ of the Tikhonov functional $\mathcal{T}_\alpha^\delta(x)$. The nonnegative functional $\mathcal{T}_\alpha^\delta(x)$ weights by means of the regularization parameter $\alpha > 0$ the misfit term $\frac{1}{p}\|F(x) - y^\delta\|_Y^p$ with norm exponent $1 \leq p < \infty$ relative to the term $\mathcal{R}(x)$. The former term expresses the data fidelity, whereas the latter is responsible for stabilizing the regularized solutions. For adapting specific noise models like Poisson, uniform or impulsive noise, some authors generalize the structure of the misfit term (cf. [8,22,32,36,61,89,109]). We do not discuss in this chapter the case of perturbed forward operators F. Instead, we refer in this context to Refs. [78,88] and in particular for detailed results to the monograph [105], where also a very general structure of the misfit term can be found.

Modern regularization theory includes results on the *error analysis* and on *convergence rates*, and it is the aim of this chapter to give an overview of the latest developments in this field that are closely connected with the description of solution smoothness. For the error analysis it is important to evaluate a prescribed nonnegative error functional $E(x_\alpha^\delta, x^\dagger)$ that measures the deviation of regularized solutions x_α^δ from *$\mathcal{R}$-minimizing solutions*, that is, from solutions $x^\dagger$ to (8.1) with noise-free data y, for which we have

$$\mathcal{R}(x^\dagger) = \min\{\mathcal{R}(x) : x \in \mathcal{D}(F),\ F(x) = y\}.$$

As a typical example of an error measure, we will consider on the one hand the error norm

$$E(x, x^\dagger) = \|x - x^\dagger\|_X \tag{8.4}$$

or its power $E(x,x^\dagger) = \|x - x^\dagger\|_X^q$ with exponents $1 \le q < \infty$. On the other hand, preferably for uniformly convex functionals $\mathcal{R}$, the *Bregman distance*

$$E(x,x^\dagger) = D_{\xi^\dagger}^{\mathcal{R}}(x,x^\dagger) := \mathcal{R}(x) - \mathcal{R}(x^\dagger) - \langle \xi^\dagger, x - x^\dagger \rangle_{X^* \times X} \qquad (8.5)$$

is often used in Banach space regularization, where we denote by $\xi^\dagger \in \partial\mathcal{R}(x^\dagger) \subseteq X^*$ an element of the subdifferential of the convex functional $\mathcal{R}$ at the point $x^\dagger$. Since 2004 the Bregman distance has become an indispensable tool in regularization theory (cf. [15] and further refinements in Refs. [50,60,74,93,94]). If E from (8.5) is taken into account, one always assumes that $x \in \mathcal{D}(\mathcal{R}) := \{\tilde{x} \in X : \mathcal{R}(\tilde{x}) < \infty\}$ and that $x^\dagger$ belongs to the Bregman domain

$$\mathcal{D}_B(\mathcal{R}) := \{\tilde{x} \in \mathcal{D}(\mathcal{R}) : \partial\mathcal{R}(\tilde{x}) \ne \emptyset\}.$$

If, for appropriate a priori choices $\alpha = \alpha(\delta)$ or a posteriori choices $\alpha = \alpha(\delta, y^\delta)$ of the regularization parameter $\alpha > 0$, the errors $E(x_\alpha^\delta, x^\dagger)$ can be majorized by a positive function φ that only depends on the noise level δ and tends to zero as $\delta \to 0$, then the way is clear for obtaining convergence rates

$$E(x_\alpha^\delta, x^\dagger) = \mathcal{O}(\varphi(\delta)) \qquad \text{as} \qquad \delta \to 0, \qquad (8.6)$$

where we restrict in the sequel on concave index functions φ. In this context, the function $\varphi : (0, \infty) \to (0, \infty)$ is called an *index function* if it is increasing and satisfies the limit condition $\lim_{t \to +0} \varphi(t) = 0$, see, for example, Refs. [52,82]. One of the main aims of the following study is to obtain convergence rates of $E(x_\alpha^\delta, x^\dagger)$ as $\delta \to 0$ under different smoothness conditions and for several choices of the regularization parameter. The errors of Tikhonov-regularized solutions will depend on the interplay of the following four ingredients:

1. nonlinearity structure of the forward operator F,
2. smoothness of the solution $x^\dagger$,
3. properties of the functional $\mathcal{R}$, and
4. character of the error measure $E(\cdot, \cdot)$.

To derive convergence rates, *link conditions* are necessary that combine the ingredients, where solution smoothness is of particular interest with respect to F, and hence smoothness classes yielding a common type of convergence rates have to be verified. We are going to discuss the four factors in relation to associated parameter choice rules. This chapter is organized as follows: In Section 8.2, we formulate our standing assumptions and briefly mention results on existence and stability. Section 8.3 is devoted to the concept of convergence of regularized solutions and corresponding conditions concerning the choice of the regularization parameter. The recently developed concept of variational inequalities for obtaining convergence rates is presented with some new details in Section 8.4. Section 8.5 concludes the chapter with a collection of sufficient conditions for deriving the required variational inequalities.

8.2   MODEL ASSUMPTIONS, EXISTENCE, AND STABILITY

Throughout this chapter, let us assume that the common requirements on $F, \mathcal{D}(F)$ and $\mathcal{R}$, made for the Tikhonov regularization in Ref. [60] and in the recent monographs [97, 100] are fulfilled. Note that these requirements, in principle, also can be found in Refs. [105, pp. 39–40, 146], and we refer to Sections 8.2.11 and 8.3.11 ibidem for special cases of Banach spaces and penalties $\mathcal{R}$ in applications. Precisely, we make the following assumptions:

Assumption 8.1

(a) *X and Y are infinite dimensional Banach spaces, where in addition to the norm topologies $\|\cdot\|_X$ and $\|\cdot\|_Y$ weaker topologies τ_X and τ_Y are under consideration for X and Y, respectively, such that the norm $\|\cdot\|_Y$ is τ_Y-sequentially lower semicontinuous.*

(b) *The domain $\mathcal{D}(F)$ is a convex and τ_X-sequentially closed subset of X.*

(c) *The operator $F : \mathcal{D}(F) \subseteq X \to Y$ is τ_X-τ_Y-continuous, that is, for a sequence $\{x_n\}_{n=1}^{\infty} \subset \mathcal{D}(F)$ the τ_X-convergence of x_n to $x_0 \in \mathcal{D}(F)$ as $n \to \infty$ implies that $F(x_n)$ is τ_Y-convergent to $F(x_0)$.*

(d) *$\mathcal{R} : X \to [0, \infty]$ is a proper, convex, and τ_X-sequentially lower semicontinuous functional, where proper denotes that the domain of $\mathcal{R}$*

$$\mathcal{D}(\mathcal{R}) := \{x \in X : \mathcal{R}(x) < \infty\}$$

is nonempty. Moreover, we assume that $\mathcal{D} := \mathcal{D}(F) \cap \mathcal{D}(\mathcal{R}) \neq \emptyset$.

(e) *The penalty functional $\mathcal{R}$ is assumed to be stabilizing functional in the sense that the sublevel sets*

$$\mathcal{M}^{\mathcal{R}}(c) := \{x \in X : \mathcal{R}(x) \leq c\}$$

are sequentially precompact with respect to the topology τ_X in X for all $c \geq 0$.

(f) *Given $y \in F(\mathcal{D}(F))$ there is some element $x^{\dagger} \in \mathcal{D}$ solving the operator equation (8.1), consequently we have $y \in F(\mathcal{D})$.*

(g) *$F : \mathcal{D} \to Y$ attains a one-sided directional derivative with Gâteau derivative-like properties, that is, for every $x_0 \in \mathcal{D}$ there is a bounded linear operator $F'(x_0) : X \to Y$ such that*

$$\lim_{t \to +0} \frac{1}{t} \left(F(x_0 + t(x - x_0)) - F(x_0) \right) = F'(x_0)(x - x_0)$$

is valid for all $x \in \mathcal{D}(F)$.

(h) *$E : x_{\alpha}^{\delta} \times x^{\dagger} \mapsto [0, \infty)$ is a nonnegative error measure with $E(x^{\dagger}, x^{\dagger}) = 0$.*

Remark 8.1   The most important detail in Assumption 8.1 for obtaining stable approximate solutions is the *stabilizing character of the penalty functional* $\mathcal{R}$

formulated in Item (e). For the prominent case of convex norm power penalties $\mathcal{R}(x) := \|x\|_X^r$ with some exponent $r \geq 1$, the precompactness of the sublevel sets $\mathcal{M}^{\mathcal{R}}(c)$ for all $c \geq 0$ becomes evident for *reflexive Banach spaces X* if the weak topology (weak convergence $\rightharpoonup$) in X plays the role of τ_X. Namely, we know that the closed unit ball in a Banach space X is weakly sequentially compact if and only if X is reflexive. For *non-reflexive Banach spaces* $X = Z^*$, which are duals of a separable Banach space Z, based on the sequential Banach-Alaoglu theorem the weak*-topology (weak*-convergence $\rightharpoonup^*$) provides us with stabilizing norm power penalties (cf., e.g., [100, p. 89]). Non-differentiable stabilizing functionals were discussed, for example, in Refs. [107, 108].

Remark 8.2 In the literature, there are tendencies to weaken parts of Assumption 8.1. So instead of the convexity of $\mathcal{D}(F)$ from Item (b), one can sometimes only require that the domain is *star-like* with respect to the solution $x^\dagger$ (cf., e.g., [60] and [100, sect. 4.2]). In any case, no interior points of $\mathcal{D}(F)$ are required such that, for $\Omega \subset \mathbb{R}^k$ and arbitrary constants $c \in \mathbb{R}$, "half-spaces"

$$\mathcal{D}(F) = \{x \in X := L^q(\Omega) : x \geq c \text{ a.e. on } \Omega\}$$

can be included in our studies. We take in the following from Item (d) that $\mathcal{R}$ is convex, but extensions of error estimates and convergence rates results to non-convex penalties $\mathcal{R}$ for the Tikhonov regularization can be found in Refs. [12, 41–43, 74, 92, 112] and also in a recent paper [3]. The discussions below will show that additional specific conditions must be imposed on the derivative $F'(x)$ from Item (g). Such conditions characterize the structure of nonlinearity of the forward operator F in a neighborhood of the solution $x^\dagger$ to equation (8.1). Here we denote by A^* the adjoint of a bounded linear operator $A : X \to Y$, which is also a bounded linear operator defined by the equation

$$\langle \eta, Ax \rangle_{Y^* \times Y} = \langle A^* \eta, x \rangle_{X^* \times X} \qquad \text{for all} \qquad x \in X, \, \eta \in Y^*,$$

but mapping between the dual spaces Y^* and X^*. In this sense, we have for $F'(x_0)$ from Item (g) a bounded linear adjoint operator $F'(x_0)^* : Y^* \to X^*$.

   The following proposition summarizes some important assertions on existence and stability of regularized solutions, which are essential based on Item (e) of Assumption 8.1, and we refer to Refs. [60, section 3] and [100, section 4.1] for proofs. The existence of $\mathcal{R}$-minimizing solutions and regularized solutions ensures the well-posedness of the Tikhonov regularization approach in our Banach space setting as access to approximations of specially shaped solutions to the operator equation (8.1).

Proposition 8.1 *Under Assumption 8.1 and for given $y \in F(\mathcal{D})$, there is some $\mathcal{R}$-minimizing solution $x^\dagger \in \mathcal{D}$ of Equation 8.1. Furthermore, regularized solutions $x_\alpha^\delta \in \mathcal{D}$ solving the extremal problem (8.3) exist for all $\alpha > 0$ and all $y^\delta \in Y$, and they are τ_X-stable with respect to perturbations in the data.*

The kind of stability mentioned in Proposition 8.1 means that for a sequence $y_n \in Y$ with $\lim\limits_{n\to\infty} \|y_n - y^\delta\|_Y = 0$, any sequence $x_n = \arg\min\limits_{x\in\mathcal{D}}\{1/p\|F(x) - y_n\|_Y^p + \alpha\mathcal{R}(x)\}$ possesses a subsequence x_{n_k} that converges with respect to the topology τ_X to some regularized solution $x_\alpha^\delta = \arg\min\limits_{x\in\mathcal{D}}\{1/p\|F(x) - y^\delta\|_Y^p + \alpha\mathcal{R}(x)\}$ as $k \to \infty$. We will call this in the sequel *convergence in the sense of subsequences*. Note that $\mathcal{R}$-minimizing solutions $x^\dagger$ as well as regularized solutions x_α^δ are not necessarily unique. Therefore, convergence in the sense of subsequences is a typical behavior occurring and is discussed in Section 8.3, which is devoted to the convergence of regularized solutions x_α^δ to $\mathcal{R}$-minimizing solutions $x^\dagger$ when the noise level $\delta > 0$ tends to zero.

8.3   CONVERGENCE OF REGULARIZED SOLUTIONS

From mathematical point of view, it is required that approximate solutions are convergent in some sense. The following proposition formulates sufficient conditions that ensure that Tikhonov-regularized solutions possess such property. For proofs of the proposition, we refer the reader to Refs. [105] and [100, section 4.1.2]. In this proposition and in the results below, we suppose that all items of Assumption 8.1 are valid.

Proposition 8.2   *For a sequence of noisy data $y_n \in Y$ to $y \in F(\mathcal{D})$ with noise levels $\delta_n \to 0$ as $n \to \infty$, we take into account a sequence of regularized solutions $x_n = \arg\min\limits_{x\in\mathcal{D}}\{1/p\|F(x) - y_n\|_Y^p + \alpha_n\mathcal{R}(x)\}$, where the corresponding regularization parameters $\alpha_n = \alpha_n(\delta_n, y_n)$ are chosen such that*

$$\lim_{n\to\infty}\|F(x_n) - y_n\|_Y = 0 \quad \text{and} \quad \limsup_{n\to\infty}\mathcal{R}(x_n) \leq \mathcal{R}(x^\dagger) \tag{8.7}$$

for an $\mathcal{R}$-minimizing solution $x^\dagger \in \mathcal{D}$ of Equation 8.1. Then x_n converges in the sense of subsequences with respect to the topology τ_X to $\mathcal{R}$-minimizing solutions and we have

$$\lim_{n\to\infty}\mathcal{R}(x_n) = \mathcal{R}(x^\dagger). \tag{8.8}$$

Remark 8.3   Under some additional conditions on $\mathcal{R}$, one can improve the convergence result from Proposition 8.2 and even obtain norm convergence $\lim\limits_{n\to\infty}\|x_n - x^\dagger\|_X = 0$ (cf., e.g. [97, proposition 3.32]). In particular, if a generalized Kadec-Klee or Radon-Riesz property with respect to $\mathcal{R}$ is valid, then we have norm convergence due to (8.8), and we refer to Refs. [11, lemma 2.2], [43, prop. 3.6], [97, example 3.3], or [100, remark 4.4] for examples and more details.

To concretize the conditions in (8.7), we formulate the subsequent proposition. Then we are able to select appropriate a priori and a posteriori parameter choice rules for the regularization parameter α that satisfies these conditions, and we will verify this in Proposition 8.4.

Proposition 8.3 *Under the setting of Proposition 8.2 let the regularization parameters α_n satisfy the limit conditions*

$$\alpha_n \to 0 \qquad \text{and} \qquad \frac{(\delta_n)^p}{\alpha_n} \to 0 \qquad \text{as} \qquad n \to \infty. \tag{8.9}$$

Then the conditions (8.7) are fulfilled and the convergence assertions of Proposition 8.2 are valid.

Obviously, under (8.9) the validity of the conditions (8.7) is an immediate consequence of the formulae

$$\|F(x_\alpha^\delta) - y^\delta\|^p \le \delta^p + p\,\alpha\,\mathcal{R}(x^\dagger), \qquad \alpha > 0, \tag{8.10}$$

and

$$\mathcal{R}(x_\alpha^\delta) \le \frac{\delta^p}{p\,\alpha} + \mathcal{R}(x^\dagger), \qquad \alpha > 0, \tag{8.11}$$

which follow for $\mathcal{R}$-minimizing solutions $x^\dagger$ to Equation 8.1 from the minimizing property $T_\alpha^\delta(x_\alpha^\delta) \le T_\alpha^\delta(x^\dagger)$ of Tikhonov-regularized solutions.

As is easily seen, the limit conditions (8.9) give the explicit instructions

$$\alpha(\delta) \to 0 \qquad \text{and} \qquad \frac{\delta^p}{\alpha(\delta)} \to 0 \qquad \text{as} \qquad \delta \to 0 \tag{8.12}$$

for appropriate a priori parameter choices $\alpha = \alpha(\delta)$ in order to ensure convergent regularized solutions.

However, in regard to the merits of a posteriori parameter choices, it is of interest to select such parameter choice rules $\alpha = \alpha(\delta, y^\delta)$ which obey the conditions

$$\alpha(\delta, y^\delta) \to 0 \qquad \text{and} \qquad \frac{\delta^p}{\alpha(\delta, y^\delta)} \to 0 \qquad \text{as} \qquad \delta \to 0. \tag{8.13}$$

Recently, such study was performed for the *sequential discrepancy principle* in Ref. [3], and we refer also to Refs. [105, p. 137–139] and [53].

The basis of this variety of discrepancy principle is, for prescribed $0 < q < 1$ and $\alpha_0 > 0$, a sequence

$$\Delta_q := \{\alpha_j > 0 \colon \alpha_j = q^j \alpha_0, \ j \in \mathbb{Z}\} \tag{8.14}$$

of regularization parameters and the specification of some constant $\tau > 1$.

Definition 8.1 (sequential discrepancy principle) *For fixed $\delta > 0$ and $y^\delta \in Y$, we say that $\alpha = \alpha(\delta, y^\delta) \in \Delta_q$ is chosen according to the sequential discrepancy principle if*

$$\|F(x_\alpha^\delta) - y^\delta\|_Y \leq \tau\delta < \|F(x_{\alpha/q}^\delta) - y^\delta\|_Y. \tag{8.15}$$

Due to Assumption 8.1, the set $X_{\min} := \{x \in \mathcal{D} : \mathcal{R}(x) = \mathcal{R}_{\min}\}$ is nonempty for the value $\mathcal{R}_{\min} := \inf_{x\in\mathcal{D}} \mathcal{R}(x) \geq 0$ and so is $Y_{\min} := F(X_{\min})$. Moreover, we have an element $x_{\min} \in X_{\min}$ such that $\text{dist}(y^\delta, Y_{\min}) := \inf_{x\in X_{\min}} \|F(x) - y^\delta\|_Y = \|F(x_{\min}) - y^\delta\|_Y$. Using this notation, the Assumption 8.2 below still needs the following two definitions concerning the *exact penalization veto* and requirements on *data compatibility*, respectively.

Definition 8.2 (exact penalization veto) *We say that the exact penalization veto is satisfied for $y \in F(\mathcal{D})$ if, for all $\alpha > 0$, any minimizer*

$$x^\dagger = \arg\min_{x\in\mathcal{D}} \left\{ \frac{1}{p}\|F(x) - y\|_Y^p + \alpha\mathcal{R}(x) \right\},$$

which is simultaneously an $\mathcal{R}$-minimizing solution to (8.1), belongs to $X_{\min}$.

Definition 8.3 (compatible data) *For $y \in F(\mathcal{D})$ and prescribed $\tau > 1$, we say that we have data compatibility if there is some $\overline{\delta} > 0$ such that for all data $y^\delta \in Y$ fulfilling (8.2) the condition*

$$\tau\delta < \text{dist}(y^\delta, Y_{\min}) \qquad \text{for all} \qquad 0 < \delta \leq \overline{\delta}$$

is satisfied.

In Ref. [3], sufficient conditions for fulfilling the exact penalization veto have been formulated, which is mostly the case if we have $p > 1$ for the norm exponent in the misfit term of the Tikhonov functional T_α^δ. Sufficient conditions for obtaining data compatibility can also be found ibidem. For supplementary discussions in the context of the following assumption, we also refer to Ref. [105, p. 162].

Assumption 8.2 *Let for the exact right-hand side $y \in F(\mathcal{D})$ of Equation 8.1, the exact penalization veto (see Definition 8.2) be satisfied and assume for prescribed $\tau > 1$ data compatibility in the sense of Definition 8.3.*

The following proposition is a direct consequence of Proposition 8.3 for the a priori parameter choice (*I*) and of theorem 1 from Ref. [3] for the a posteriori parameter choice (*II*), and we refer to Ref. [3] for proof details in the latter case.

Proposition 8.4

(I) *Let the a priori parameter choice $\alpha = \alpha(\delta)$ satisfy condition (8.12). Then, under the setting of Proposition 8.2, that parameter choice $\alpha_n = \alpha(\delta_n)$ satisfies condition (8.9).*

(II) *If Assumption 8.2 holds, then there is some $\bar{\delta} > 0$ such that $\alpha = \alpha(\delta, y^\delta)$ can be chosen according to the sequential discrepancy principle for all $0 < \delta \leq \bar{\delta}$. Moreover this parameter choice satisfies condition (8.13), and consequently under the setting of Proposition 8.2, the corresponding sequence $\alpha_n = \alpha(\delta_n, y_n)$ satisfies condition (8.9).*

Hence, in both cases (I) and (II) the convergence assertions of Proposition 8.2 are valid.

8.4 A POWERFUL TOOL FOR OBTAINING CONVERGENCE RATES

The convergence with respect to the topology τ_X of regularized solution $x^\delta_{\alpha(\delta, y^\delta)}$ in the sense of subsequences to $\mathcal{R}$-minimizing solutions $x^\dagger$ as $\delta \to 0$ (cf. Proposition 8.2) is not worth too much. Even if the Kadec-Klee property of $\mathcal{R}$ allows us to amplify this to norm convergence (cf. Remark 8.3), the convergence can be *arbitrarily slow* for awkward solutions $x^\dagger$ as we know from Ref. [99]. To get more, namely a uniform error estimate for classes of solutions $x^\dagger$ in the sense of convergence rates (8.6) with an index function φ determining the rate, additional requirements must be imposed on all elements $x^\dagger$ from the class under consideration. In practice, one can restrict the considerations to rate functions φ which are *concave*. The above-mentioned requirements are always *smoothness conditions*. Precisely, the $\mathcal{R}$-minimizing solutions $x^\dagger$ of (8.1) to be approximated by regularized solutions must have a certain level of *smoothness with respect to the forward operator F*.

If F is nonlinear, then moreover the specific *structure of nonlinearity*, for example, expressed by inequalities of the form

$$\|F'(x^\dagger)(x - x^\dagger)\|_Y \leq \sigma(\|F(x) - F(x^\dagger)\|_Y) \qquad \text{for all} \qquad x \in \mathcal{M} \qquad (8.16)$$

for some *concave* index function σ and some set $\mathcal{M} \subseteq \mathcal{D}(F)$ around $x^\dagger$, influences the rate function φ. Smoothness and nonlinearity conditions together must be able to link in an appropriate manner the ingredients (i)–(iv) mentioned at the end of Section 8.1. Such conditions must include the error measure E. If E refers to the Bregman distance (8.5), then sometimes the nonlinearity condition

$$\|F(x) - F(x^\dagger) - F'(x^\dagger)(x - x^\dagger)\|_Y \leq \eta D^\mathcal{R}_{\xi^\dagger}(x, x^\dagger) \qquad \text{for all} \qquad x \in \mathcal{M} \quad (8.17)$$

with some constant $\eta > 0$ is used (cf., e.g., [60,94]), which often proves to be weaker than (8.16). For future progress and the acceptance of research results in this field, it is necessary to assign the occurring nonlinearity conditions, for example, (8.16) and (8.17), still more relevant classes of operators F from applied inverse problems with corresponding specific Banach spaces X, Y and penalties $\mathcal{R}$. In Example 8.1, we will mention some related difficulties for a class of quadratic nonlinear inverse problems.

A powerful tool combining the smoothness of a solution $x^\dagger$ to Equation 8.1 with respect to F and the nonlinearity structure of F around $x^\dagger$ for obtaining convergence rates has been developed originally for non-smooth operators in connection with Ref. [60]. In mathematical terms, it is written in the form of *variational inequalities* (also called variational source conditions or source inequalities by some authors, see the overview of Ref. [34]), and we discuss throughout this chapter the broadly applicable variant

$$\beta E(x, x^\dagger) \le \mathcal{R}(x) - \mathcal{R}(x^\dagger) + \varphi(\|F(x) - F(x^\dagger)\|_Y) \quad \text{for all } x \in \mathcal{M} \tag{8.18}$$

with a *concave index function* φ and some constant $0 < \beta \le 1$ which was also employed, partly with modifications, in Refs. [1, 17, 32, 37, 42, 44]. For further, partially classical, studies and alternative approaches to obtain convergence rates in Tikhonov regularization for nonlinear operator equations, we refer to the seminal paper by Engl et al. [29] and also Refs. [19, 31, 54, 55, 59, 68, 80, 81, 84, 101].

Remark 8.4   For deriving convergence rates (8.6) from the variational inequality (8.18) with domain of validity $\mathcal{M}$, the set $\mathcal{M}$ has to contain $x^\dagger$ as well as all regularized solutions x_α^δ under consideration for sufficiently small $\delta > 0$. For sublevel sets

$$\mathcal{M} = \mathcal{M}^\mathcal{R}(\mathcal{R}(x^\dagger) + \varepsilon) \cap \mathcal{D}(F) \quad \text{with some} \quad \varepsilon > 0, \tag{8.19}$$

because of the limit condition (8.8), this is the case if Proposition 8.2 applies, that is, the regularized solutions are τ_X-convergent to $\mathcal{R}$-minimizing solutions.

Moreover, note that there are good reasons to focus on concave φ. For index functions φ with $\lim\limits_{t \to +0} \frac{\varphi(t)}{t} = 0$ as for example convex functions, with the exception of the case of extremely non-smooth operators F (cf. [44]), the variational inequality (8.18) with E from (8.5) can only hold in the singular case that $\mathcal{R}(x^\dagger) \le \mathcal{R}_{\min}$ (see Ref. [32, proposition 12.10] and [57, proposition 4.3]).

Remark 8.5   If the set $\mathcal{M}$ in (8.18) contains at least one $\mathcal{R}$-minimizing solution x_0 to (8.1), then the solution $x^\dagger$ to Equation 8.1 occurring in (8.18), is always an $\mathcal{R}$-minimizing solution. With $F(x_0) = F(x^\dagger) = y$ and $\mathcal{R}(x_0) \le \mathcal{R}(x^\dagger)$, this becomes evident if one considers $0 \le E(x_0, x^\dagger) \le \mathcal{R}(x_0) - \mathcal{R}(x^\dagger)$ as a consequence of (8.18). Note that every $\mathcal{R}$-minimizing solution to (8.1) belongs to $\mathcal{M}$ if $\mathcal{M}$ is of the form (8.19).

Provided that the regularization parameter $\alpha > 0$ is chosen in an adapted manner, then the variational inequalities (8.18) lead directly to the convergence rates (8.6) where the concave index function φ carries over from (8.18) to (8.6). We are beginning with a priori parameter choices in accordance with Ref. [53], and after some discussion of associated error estimates, we will summarize the results in Theorem 8.1. In contrast to Ref. [53], we also present results for the norm exponent $p = 1$ in the misfit term of the Tikhonov functional.

As an appropriate choice of the regularization parameter under the variational inequality (8.18), we use

$$\alpha(\delta) := \alpha_0 \frac{\delta^p}{\varphi(\delta)} \tag{8.20}$$

for some prescribed constant $\alpha_0 > 0$, where we assume that $x_{\alpha(\delta)}^\delta \in \mathcal{M}$ holds for all $0 < \delta \leq \bar{\delta}$. Then we have from (8.11) the inequality $\mathcal{R}(x_{\alpha(\delta)}^\delta) - \mathcal{R}(x^\dagger) \leq \varphi(\delta)/p\,\alpha_0$ and hence the error estimate

$$E(x_{\alpha(\delta)}^\delta, x^\dagger) \leq \frac{1}{\beta} \left[\frac{\varphi(\delta)}{p\,\alpha_0} + \varphi(\|F(x_{\alpha(\delta)}^\delta) - F(x^\dagger)\|_Y) \right] \quad \text{for all } x \in \mathcal{M}. \tag{8.21}$$

To handle the second term $\varphi(\|F(x_{\alpha(\delta)}^\delta) - F(x^\dagger)\|_Y)$, we need to distinguish the cases $p > 1$ and $p = 1$ for the norm exponent in the Tikhonov functional T_α^δ.

In the case $p > 1$, we can estimate for $\|F(x_{\alpha(\delta)}^\delta) - F(x^\dagger)\|_Y > \delta$ analog to the proof of corollary 1 from Ref. [53] taking into account that the index function φ is concave, which implies that the quotient $\varphi(\delta)/\delta$ is nonincreasing for all $\delta > 0$. Precisely, we have for $p > 1$ and $0 < \delta \leq \bar{\delta}$

$$\begin{aligned}
\frac{\|F(x_{\alpha(\delta)}^\delta) - F(x^\dagger)\|_Y^p}{p} &\leq 2^p \frac{\delta^p}{p} + \alpha(\delta)\, 2^{p-1} \varphi(\|F(x_{\alpha(\delta)}^\delta) - F(x^\dagger)\|_Y) \\
&= 2^p \frac{\delta^p}{p} + \alpha_0\, \delta^p\, 2^{p-1} \frac{\varphi(\|F(x_{\alpha(\delta)}^\delta) - F(x^\dagger)\|_Y)}{\varphi(\delta)} \\
&= \frac{\delta^p}{p} \left(2^p + \alpha_0\, p\, 2^{p-1} \frac{\varphi(\|F(x_{\alpha(\delta)}^\delta) - F(x^\dagger)\|_Y)}{\varphi(\delta)} \right) \\
&\leq 2^{p-1}(2 + \alpha_0 p) \frac{\delta^p\, \varphi(\|F(x_{\alpha(\delta)}^\delta) - F(x^\dagger)\|_Y)}{p\,\varphi(\delta)} \\
&\leq 2^{p-1}(2 + \alpha_0 p) \frac{\delta^p\, \|F(x_{\alpha(\delta)}^\delta) - F(x^\dagger)\|_Y \varphi(\delta)}{p\,\delta\,\varphi(\delta)} \\
&= 2^{p-1}(2 + \alpha_0 p) \frac{\delta^{p-1}}{p} \|F(x_{\alpha(\delta)}^\delta) - F(x^\dagger)\|_Y.
\end{aligned}$$

This yields for arbitrary constants $\alpha_0 > 0$

$$\|F(x_{\alpha(\delta)}^\delta) - F(x^\dagger)\|_Y \leq 2\,(2 + \alpha_0 p)^{1/(p-1)}\,\delta, \tag{8.22}$$

and because of the inequality $2\,(2 + \alpha_0 p)^{1/(p-1)} > 1$ the estimate (8.22) is also valid for $\|F(x_{\alpha(\delta)}^\delta) - F(x^\dagger)\|_Y \leq \delta$. Consequently, we have from (8.21) together with (8.22) the error bound

$$E(x_{\alpha(\delta)}^\delta, x^\dagger) \leq \frac{1}{\beta} \left[\frac{\varphi(\delta)}{p\,\alpha_0} + 2\,(2 + \alpha_0 p)^{1/(p-1)}\,\varphi(\delta) \right] \quad \text{for all } x \in \mathcal{M}, \tag{8.23}$$

valid for $p > 1$ and all $\alpha_0 > 0$.

In the case $p = 1$, however, we can estimate for $\|F(x_{\alpha(\delta)}^\delta) - F(x^\dagger)\|_Y > \delta$ and $0 < \delta \le \bar{\delta}$ as

$$
\begin{aligned}
\|F(x_{\alpha(\delta)}^\delta) - F(x^\dagger)\|_Y &\le 2\delta + \alpha(\delta)\,\varphi(\|F(x_{\alpha(\delta)}^\delta) - F(x^\dagger)\|_Y) \\
&= 2\delta + \alpha_0\,\delta\,\frac{\varphi(\|F(x_{\alpha(\delta)}^\delta) - F(x^\dagger)\|_Y)}{\varphi(\delta)} \\
&\le 2\delta + \alpha_0\,\delta\,\frac{\|F(x_{\alpha(\delta)}^\delta) - F(x^\dagger)\|_Y\,\varphi(\delta)}{\delta\,\varphi(\delta)} \\
&= 2\delta + \alpha_0\,\|F(x_{\alpha(\delta)}^\delta) - F(x^\dagger)\|_Y.
\end{aligned}
$$

This yields

$$
\|F(x_{\alpha(\delta)}^\delta) - F(x^\dagger)\|_Y \le \frac{2}{1-\alpha_0}\,\delta, \tag{8.24}
$$

but only for $0 < \alpha_0 < 1$, that is if α_0 is sufficiently small. The inequality (8.24) is also valid for $\|F(x_{\alpha(\delta)}^\delta) - F(x^\dagger)\|_Y \le \delta$. Consequently, we have from (8.21) together with (8.24) the error bound

$$
E(x_{\alpha(\delta)}^\delta, x^\dagger) \le \frac{1}{\beta}\left[\frac{\varphi(\delta)}{\alpha_0} + \frac{2}{1-\alpha_0}\,\varphi(\delta)\right] \quad \text{for all } x \in \mathcal{M}, \tag{8.25}
$$

valid for $p = 1$ and all $0 < \alpha_0 \le 1$.

With the derivation of the formulas (8.23) and (8.25), we have proven the following theorem on convergence rates.

Theorem 8.1 *Let, for $y \in F(\mathcal{D})$ and for a nonnegative error measure E, the $\mathcal{R}$-minimizing solution $x^\dagger \in \mathcal{D}$ of Equation 8.1 obeys for the variational inequality (8.18) with some $\mathcal{M} \subseteq X$, some constant $0 < \beta \le 1$, and a concave index function φ. Then we have, for the a priori parameter choice (8.20) and provided that the Tikhonov-regularized solutions satisfy the condition $x_{\alpha(\delta)}^\delta \in \mathcal{M}$ for $0 < \delta \le \bar{\delta}$, the convergence rate*

$$
E(x_{\alpha(\delta)}^\delta, x^\dagger) = \mathcal{O}(\varphi(\delta)) \quad \text{as} \quad \delta \to 0 \tag{8.26}
$$

for arbitrarily chosen $\alpha_0 > 0$ whenever the norm exponent p in (8.3) is taken from the interval $1 < p < \infty$ and for $0 < \alpha_0 < 1$ if $p = 1$.

Remark 8.6 For $p > 1$ and concave index functions φ, the a priori parameter choice (8.20) always satisfies the limit conditions (8.12). In particular, for sublevel sets $\mathcal{M}$ of the form (8.19), we see that $x_{\alpha(\delta)}^\delta \in \mathcal{M}$ for sufficiently small $\delta > 0$. Then Theorem 8.1 provides us with convergence rates (8.26). In the case $p = 1$, however, we have to distinguish the situation $\lim_{\delta \to 0} \frac{\delta}{\varphi(\delta)} = 0$, where $\alpha(\delta)$ tends to zero as $\delta \to 0$ and (8.12) is also satisfied, and $\lim_{\delta \to +0} \frac{\delta}{\varphi(\delta)} = c_0 > 0$. In the latter case, we have $\lim_{\delta \to +0} \alpha(\delta) = \alpha_0 c_0 > 0$ and the regularization parameter does not tend to zero as $\delta \to 0$. Then the exact penalization veto (see Definition 8.2) is in general violated (cf. Proposition 8.5 below) and the condition $x^\delta_{\alpha(\delta)} \in \mathcal{M}$ for sufficiently small $\delta > 0$ is more severe and requires attention in that case. The convergence of regularized solutions in Section 8.3 is only in the sense of subsequences, optionally to different $\mathcal{R}$-minimizing solutions, whereas for error measures E such that $E(x_1, x_2) = 0$ implies $x_1 = x_2$, the formulas (8.23) and (8.25) show the convergence $\lim_{\delta \to 0} E(x^\delta_{\alpha(\delta)}, x^\dagger) = 0$ to a well-defined $\mathcal{R}$-minimizing solution $x^\dagger$ even if there exist different $\mathcal{R}$-minimizing solutions of (8.1). An explanation for this phenomenon provides the fact that then only one $\mathcal{R}$-minimizing solution can satisfy a variational inequality of type (8.18). If, for example, E from (8.4) and $x_1, x_2 \in \mathcal{M}$ with $F(x_1) = F(x_2) = y$ and $\mathcal{R}(x_1) = \mathcal{R}(x_2)$ satisfy both the variational inequality $\beta \|x - x_i\|_X \leq \mathcal{R}(x) - \mathcal{R}(x_i) + \varphi(\|F(x) - F(x_i)\|_Y)$ for all $x \in \mathcal{M}$ and $i = 1, 2$, then we have $x_1 = x_2$.

Proposition 8.5   *Let the $\mathcal{R}$-minimizing solution $x^\dagger \in \mathcal{D}$ of (8.1) for $y \in F(\mathcal{D})$ satisfy for the nonnegative error measure E the variational inequality (8.18) with some $\mathcal{M} \subseteq X$, some constant $0 < \beta \leq 1$, and a concave index function φ satisfying the limit condition*

$$\lim_{\delta \to +0} \frac{\delta}{\varphi(\delta)} = c_0 > 0. \tag{8.27}$$

Provided that we use the norm exponent $p = 1$ and have data compatibility in the sense of Definition 8.3, then the exact penalization veto from Definition 8.2 is violated when $x_\alpha = \arg\min_{x \in \mathcal{D}} \{\|F(x) - y\|_Y + \alpha \mathcal{R}(x)\} \in \mathcal{M}$ for all $0 < \alpha \leq \overline{\alpha}$ and some $\overline{\alpha} > 0$.

Proof: The minimizer x_α satisfies the inequalities

$$\|F(x_\alpha) - y\|_Y + \alpha \mathcal{R}(x_\alpha) \leq \alpha \mathcal{R}(x^\dagger)$$

and

$$\frac{1}{\alpha} \|F(x_\alpha) - y\|_Y \leq \mathcal{R}(x^\dagger) - \mathcal{R}(x_\alpha).$$

Moreover, for all $0 < \alpha \leq \overline{\alpha}$, we have from (8.18) and $E(x_\alpha, x^\dagger) \geq 0$ that

$$\mathcal{R}(x^\dagger) - \mathcal{R}(x_\alpha) \leq \varphi(\|F(x_\alpha) - y\|_Y).$$

By combining the last two formulae we find the inequality

$$\frac{1}{\alpha} \|F(x_\alpha) - y\|_Y \leq \varphi(\|F(x_\alpha) - y\|_Y)$$

and hence either $\|F(x_\alpha) - y\|_Y = 0$ or

$$0 \le \frac{\|F(x_\alpha) - y\|_Y}{\varphi(\|F(x_\alpha) - y\|_Y)} \le \alpha. \tag{8.28}$$

Then there exists some $\underline{\alpha} > 0$ such that $\|F(x_\alpha) - y\|_Y = 0$ for all $0 < \alpha \le \underline{\alpha}$. Otherwise there would exist a sequence of positive regularization parameters $\alpha_n \to 0$ as $n \to \infty$ with $\lim_{n\to\infty} \|F(x_{\alpha_n}) - y\|_Y = 0$ because of $\|F(x_{\alpha_n}) - y\|_Y \le \alpha_n \mathcal{R}(x^\dagger)$. This, however, contradicts (8.27), since (8.28) would imply that $\lim_{n\to\infty} \frac{\|F(x_{\alpha_n}) - y\|_Y}{\varphi(\|F(x_{\alpha_n}) - y\|_Y)} = 0$. We notice that in the case of data compatibility, $x^\dagger \in X_{\min}$ cannot hold (cf. [53, proposition 5]) for an $\mathcal{R}$-minimizing solution $x^\dagger$ and hence the exact penalization veto is violated.☐

In order to obtain the convergence rate (8.26) from Theorem 8.1, the function φ must be known for applying the a priori parameter choice (8.20). Therefore, it makes sense to search for a posteriori choices of the regularization parameter $\alpha > 0$ which yield the same rate without to know φ. As we will see in Theorem 8.2, this is the case for the sequential discrepancy principle introduced above in Definition 8.1. It is a consequence of Proposition 8.5 that, for the specific case (8.27) of $p = 1$, Item (II) of Proposition 8.4 does not apply, which means that the limit condition (8.13) cannot be shown for the sequential discrepancy principle. However, Theorem 8.2 yields convergence rates with respect to the error measure E also in that case.

Theorem 8.2 *Suppose that for $y \in F(\mathcal{D})$ the $\mathcal{R}$-minimizing solution $x^\dagger \in \mathcal{D}$ of Equation 8.1 obeys for the nonnegative error measure E the variational inequality (8.18) with some $\mathcal{M} \subseteq X$, some constant $0 < \beta \le 1$, and a concave index function φ. Moreover assume for prescribed $\tau > 1$ data compatibility in the sense of Definition 8.3. Then there is some $\bar{\delta} > 0$ such that $\alpha = \alpha(\delta, y^\delta)$ can be chosen according to the sequential discrepancy principle for all $0 < \delta \le \bar{\delta}$. Furthermore we have, for that a posteriori parameter choice and provided that the Tikhonov-regularized solutions satisfy the condition $x^\delta_{\alpha(\delta,y^\delta)} \in \mathcal{M}$ for $0 < \delta \le \bar{\delta}$, the convergence rate*

$$E(x^\delta_{\alpha(\delta,y^\delta)}, x^\dagger) = \mathcal{O}(\varphi(\delta)) \qquad \text{as} \qquad \delta \to 0 \tag{8.29}$$

for norm exponents p in (8.3) taken from the interval $1 \le p < \infty$.

Sketch of a proof: From Ref. [3, lemma 2], it follows that under data compatibility in the sense of Definition 8.3 for prescribed $\tau > 1$, there is some $\bar{\delta} > 0$ such that $\alpha = \alpha(\delta, y^\delta)$ can be chosen from the set Δ_q (cf. (8.14)) according to the sequential discrepancy principle for all $0 < \delta \le \bar{\delta}$. Then theorem 2 from Ref. [53] yields the convergence rate (8.29) whenever $p > 1$, and the proof is essentially based on Ref. [53, corollary 2], where a lower bound of the regularization parameter $\alpha(\delta, y^\delta)$ according to the sequential discrepancy principle was derived as

$$\alpha(\delta, y^\delta) \ge \frac{q}{p\,2^{p-1}} \left(\frac{\tau^p - 1}{\tau^p + 1}\right) \frac{[(\tau - 1)\delta]^p}{\varphi((\tau - 1)\delta)}, \tag{8.30}$$

which implies for concave index functions φ the estimate

$$\mathcal{R}(x_{\alpha(\delta,y^\delta)}) - \mathcal{R}(x^\dagger) \le \frac{\delta^p}{p\,\alpha(\delta,y^\delta)} \le \frac{1}{q}\left(\frac{2}{\tau-1}\right)^{p-1}\left(\frac{\tau^p+1}{\tau^p-1}\right)\varphi(\delta). \qquad (8.31)$$

Formula (8.31) together with the consequence $\|F(x_{\alpha(\delta,y^\delta)}) - y\|_Y \le \|F(x_{\alpha(\delta,y^\delta)}) - y^\delta\|_Y + \delta \le (\tau+1)\delta$ of (8.15), yields the rate (8.29) based on (8.18). The case $p=1$ was not under consideration in Ref. [53], and so we have to discuss only this case in the following. In contrast to the case $p > 1$, where always $\alpha(\delta,y^\delta) \to 0$ as $\delta \to 0$ is true, we have for $p=1$ also the situation (8.27) where the lower bound in (8.30) and hence the regularization parameters $\alpha(\delta,y^\delta)$ are uniformly bounded below by a positive constant. Nevertheless, (8.31) remains true in the form

$$\mathcal{R}(x_{\alpha(\delta,y^\delta)}) - \mathcal{R}(x^\dagger) \le \frac{\delta}{\alpha(\delta,y^\delta)} \le \frac{1}{q}\left(\frac{\tau+1}{\tau-1}\right)\varphi(\delta),$$

which ensures the convergence rate (8.29) also in that case. $\square$

Remark 8.7   The variational inequality approach described above is limited to rates $E(x_\alpha^\delta, x^\dagger) = \mathcal{O}(\delta)$ as $\delta \to 0$ for the Tikhonov regularization in Banach spaces. Higher rates for the Bregman distance up to $D_{\xi^\dagger}^{\mathcal{R}}(x_\alpha^\delta, x^\dagger) = \mathcal{O}(\delta^{4/3})$ can be constructed by alternative approaches. Mostly improved rates require additional conditions, in particular source conditions which are stronger than the benchmark condition (8.32) below, and we refer for details to Ref. [87] (see also Refs. [100, section 4.2.4] and references therein). Moreover, new ideas for that purpose were published recently in the paper by Grasmair [45].

8.5   HOW TO OBTAIN VARIATIONAL INEQUALITIES?

In this section we outline, partly along the lines of the recent papers ([53, appendix], [10,11] and [17]), under certain *conditions on the structure of nonlinearity* some cross connections between *source conditions*, which are usually perceived as standard conditions of solution smoothness, and *variational inequalities* presented in the previous section for obtaining convergence rates. In contrast to Hilbert space regularization, for Banach space regularization, the range of available source conditions is very poor and we extend it by employing *approximate source conditions*.

8.5.1   Bregman distance as error measure: the benchmark case

If one considers in Banach space regularization   source conditions at the $\mathcal{R}$-minimizing solution $x^\dagger \in \mathcal{D}_B(\mathcal{R})$ with subgradient $\xi^\dagger \in \partial\mathcal{R}(x^\dagger) \subset X^*$, then the focus is on the *benchmark source condition*

$$\xi^\dagger = F'(x^\dagger)^* v, \quad \text{for some source element } v \in Y^*. \qquad (8.32)$$

This allows us to bound for all $x \in X$ as

$$\langle \xi^{\dagger}, x^{\dagger} - x \rangle_{X^* \times X} = \langle F'(x^{\dagger})^* v, x^{\dagger} - x \rangle_{X^* \times X} = \langle v, F'(x^{\dagger})(x^{\dagger} - x) \rangle_{Y^* \times Y}$$
$$\leq \|v\|_{Y^*} \|F'(x^{\dagger})(x - x^{\dagger})\|_Y.$$

Adding the term $\mathcal{R}(x) - \mathcal{R}(x^{\dagger})$ on both sides provides us with the Bregman distance estimate

$$D_{\xi^{\dagger}}^{\mathcal{R}}(x, x^{\dagger}) \leq \mathcal{R}(x) - \mathcal{R}(x^{\dagger}) + \|v\|_{Y^*} \|F'(x^{\dagger})(x - x^{\dagger})\|_Y \qquad \text{for all} \qquad x \in \mathcal{D}(\mathcal{R}). \tag{8.33}$$

In the special case of a Hilbert space X with $\mathcal{R}(x) = \|x\|_X^2$ and $D_{\xi^{\dagger}}^{\mathcal{R}}(x, x^{\dagger}) = \|x - x^{\dagger}\|_X^2$ this attains the form

$$\|x - x^{\dagger}\|_X^2 \leq \|x\|_X^2 - \|x^{\dagger}\|_X^2 + \|v\|_{Y^*} \|F'(x^{\dagger})(x - x^{\dagger})\|_Y \qquad \text{for all} \qquad x \in X. \tag{8.34}$$

It was emphasized in Ref. [33] that, for bounded linear operators $F = A$ mapping between Hilbert spaces X and Y, solution smoothness can always be expressed by variational inequalities (8.18) with index functions φ and a domain of validity $\mathcal{M} = X$. In particular, the variation inequality (8.34) gives for bounded linear $F = A : X \to Y$ with $F'(x^{\dagger}) = A$ based on the Theorems 8.1 and 8.2 a convergence rate $E(x_{\alpha}^{\delta}, x^{\dagger}) = \mathcal{O}(\delta)$ as $\delta \to 0$ with $E(x, x^{\dagger}) = \|x - x^{\dagger}\|_X^2$. For the general Banach space setting, inequality (8.33) also results in that convergence rate in the case of bounded linear operators $F = A : X \to Y$, but for $E(x, x^{\dagger}) = D_{\xi^{\dagger}}^{\mathcal{R}}(x, x^{\dagger})$.

If the mapping F is nonlinear, then we have to use the structure of nonlinearity to bound $\|F'(x^{\dagger})(x - x^{\dagger})\|_Y$ in terms of $\|F(x^{\dagger}) - F(x)\|_Y$, and the validity of such structural conditions requires additional assumptions. A simple but hard-to-verify nonlinear condition is (8.16) for some concave index function σ and some set $\mathcal{M} \subset \mathcal{D}(F)$ (cf. [10]). In this case, (8.33) provides us with a variational inequality (8.18) on $\mathcal{M}$ with $\beta = 1$, $E(x, x^{\dagger}) = D_{\xi^{\dagger}}^{\mathcal{R}}(x, x^{\dagger})$ and $\varphi(t) = \|v\|_{Y^*} \sigma(t)$, $t > 0$. Hence, the index function σ carries over to φ in the variational inequality if the rather strong benchmark smoothness (8.32) is present. Based on the Theorems 8.1 and 8.2, we then can have for appropriate choices of the regularization parameter $\alpha > 0$ a convergence rate

$$D_{\xi^{\dagger}}^{\mathcal{R}}(x_{\alpha}^{\delta}, x^{\dagger}) = \mathcal{O}(\sigma(\delta)) \qquad \text{as} \qquad \delta \to 0. \tag{8.35}$$

An alternative structural condition is given in the form (8.17) with some constant $\eta > 0$ and an appropriate set $\mathcal{M} \subset \mathcal{D}(F)$. This weaker structural condition allows for the bound

$$\|F'(x^{\dagger})(x - x^{\dagger})\|_Y \leq \eta D_{\xi^{\dagger}}^{\mathcal{R}}(x, x^{\dagger}) + \|F(x) - F(x^{\dagger})\|_Y \qquad \text{for all} \qquad x \in \mathcal{M}.$$

Then (8.33) implies a variational inequality (8.18) valid on $\mathcal{M}$ with $0 < \beta = 1 - \eta \|v\|_{Y^*} \leq 1$, $E(x, x^{\dagger}) = D_{\xi^{\dagger}}^{\mathcal{R}}(x, x^{\dagger})$ and $\varphi(t) = \|v\|_{Y^*} t$, $t > 0$, and hence a convergence rate

$$D_{\xi^{\dagger}}^{\mathcal{R}}(x_{\alpha}^{\delta}, x^{\dagger}) = \mathcal{O}(\delta) \qquad \text{as} \qquad \delta \to 0. \tag{8.36}$$

under

$$\eta\|v\|_{Y^*} < 1. \tag{8.37}$$

Such smallness conditions as is (8.37) are in general not necessary when the stronger nonlinear condition (8.16) can be employed.

Example 8.1 (deautoconvolution)   For the real Hilbert space $X = Y = L^2(0,1)$ and the nonlinear forward operator

$$[F(x)](s) := \int_0^s x(s-t)x(t)\,dt \qquad 0 \le s \le 1, \tag{8.38}$$

with domain $\mathcal{D}(F) = X$, the operator equation (8.1) describes the *inverse autoconvolution problem* occurring in stochastics. We refer also to applications in physics, for example, to Refs. [7,98] for the deconvolution of appearance spectra and to Ref. [39] for an application in laser optics, where complex-valued functions x and in addition a measuring-tool-based kernel function occur. The nature of ill-posedness of the deautoconvolution problem and properties of its stable approximate solution by the Tikhonov regularization were firstly investigated comprehensively in Ref. [40]. For more recent studies concerning deautoconvolution, also exploiting alternative methods, we refer to Refs. [21, 24, 63, 64, 90, 110]. The autoconvolution operator (8.38) possesses the Fréchet derivative

$$[F'(x^\dagger)h](s) = 2\int_0^s x^\dagger(s-t)h(t)\,dt \qquad 0 \le s \le 1, \qquad h \in L^2(0,1),$$

such that evidently the nonlinearity condition (8.17) is fulfilled in the form

$$\|F(x) - F(x^\dagger) - F'(x^\dagger)(x-x^\dagger)\|_{L^2(0,1)} \le \|x-x^\dagger\|^2_{L^2(0,1)} \qquad \text{for all} \qquad x \in L^2(0,1).$$

Moreover, for $\mathcal{R}(x) := \|x - \bar{x}\|^2_X$ with reference element $\bar{x} \in X$, we have $\mathcal{D} = X$ and with $E(x,x^\dagger) := \|x - x^\dagger\|_X$ Assumption 8.1 applies in all its facets. The autoconvolution equation is very specific, because F is "smoothing", that is, the equation is ill-posed, but F is *not compact*. On the other hand, the derivative $F'(x)$ is compact for all $x \in X$. Unfortunately, stronger nonlinearity conditions

$$\|F'(x^\dagger)(x-x^\dagger)\|_{L^2(0,1)} \le \sigma(\|F(x) - F(x^\dagger)\|_{L^2(0,1)})$$

for F from (8.38) with index functions σ are completely unknown in the literature (see the discussion in Ref. [14]) and hence by now convergence rates can only be proven if $x^\dagger \in L^2(0,1)$ satisfies the benchmark source condition

$$x^\dagger(t) = \bar{x}(t) + \int_t^1 x^\dagger(s-t)v(s)\,ds, \qquad 0 \le t \le 1, \qquad v \in L^2(0,1), \tag{8.39}$$

together with the smallness condition $\|v\|_{L^2(0,1)} < 1$ on the source element. Then we have from (8.36)

$$\|x_\alpha^\delta - x^\dagger\|_{L^2(0,1)} = \mathcal{O}(\sqrt{\delta}) \qquad \text{as} \qquad \delta \to 0$$

for appropriate choices of the regularization parameter $\alpha > 0$. Proposition 2.6 in Ref. [14] formulates necessary conditions on the interplay of $x^\dagger$ and $\bar{x}$ for that case and shows that for $\bar{x} = 0$ the benchmark source condition cannot hold if $x^\dagger \neq 0$. For the handling of a wider class of quadratic nonlinear problems including the deautoconvolution problem, we should refer to Ref. [35]. ∎

Of course, it is of interest to search for *converse results* in the sense that a variational inequality (8.18) with $E(x, x^\dagger) = D_{\xi^\dagger}^{\mathcal{R}}(x, x^\dagger)$ and some index function φ directly leads to the benchmark source condition (8.32). Indeed, for Banach spaces X and Y, one can take such a result for $\varphi(t) = ct$, $c > 0$, that is, for the limiting case of a concave function φ, from proposition 3.38 in Ref. [97] when the nonlinearity condition (8.17) is present. For Hilbert spaces X and Y, the researchers are more familiar with converse results and we refer, for example, to Refs. [85] and [38].

Remark 8.8 In the literature on inverse problems for partial differential equations, *conditional stability estimates*

$$\|x - x^\dagger\|_X \leq \psi(\|F(x) - F(x^\dagger)\|_Y) \quad \text{for all } x \in \mathcal{M} \tag{8.40}$$

with index functions ψ are very popular. Obviously, such conditions are not so far from variational inequalities (8.18) of the form

$$\beta\|x - x^\dagger\|_X^2 \leq \mathcal{R}(x) - \mathcal{R}(x^\dagger) + \varphi(\|F(x) - F(x^\dagger)\|_Y) \quad \text{for all } x \in \mathcal{M} \tag{8.41}$$

occurring frequently when X is a Hilbert space. At first glance, it seems that the condition (8.40) is stronger than the condition (8.41) when φ and ψ are comparable, because the difference $\mathcal{R}(x) - \mathcal{R}(x^\dagger)$ of penalty terms in (8.41) can compensate for some, but just the fact that this difference may attain positive as well as negative values is the reason that both conditions can be stronger depending on the specific situation. It is an interesting phenomenon that for both types of conditions the index functions immediately occur as rate functions in corresponding convergence rates of the Tikhonov regularization, but there are two main differences. First, the optimal a priori parameter choice under condition (8.40) is $\alpha(\delta) = \alpha_0 \delta^p$ independent of ψ (cf. [19]), whereas (8.41) requires $\alpha(\delta) = \alpha_0(\delta^p/\varphi(\delta))$, which needs the knowledge of φ, and the regularization parameter tends slower to zero as $\delta \to 0$. Secondly, higher rates up to $\|x_\alpha^\delta - x^\dagger\|_X = \mathcal{O}(\delta)$ are possible as a consequence of (8.40), whereas $\|x_\alpha^\delta - x^\dagger\|_X = \mathcal{O}(\sqrt{\delta})$ is the common saturation rate under (8.41). For further details, we also refer to Ref. [100, Section 4.2.5] and references therein. An extension of the theory for conditional stability estimates to the general case

$$E(x, x^\dagger) \leq \psi(\|F(x) - F(x^\dagger)\|_Y) \quad \text{for all } x \in \mathcal{M}$$

is presented in the recent paper by Cheng et al. [20].

8.5.2 Bregman distance as error measure: violating the benchmark

If the benchmark source condition (8.32) is violated, which means that the $\mathcal{R}$-minimizing solution $x^\dagger$ is not smooth enough with respect to F, then we may use the *method of approximate source conditions* to derive variational inequalities and convergence rates. This method was developed originally for Hilbert space regularization (cf. [51] as well as Refs. [26, 38, 56, 58, 59, 88]) and later extended to the Banach space situation as in Ref. [50] (see also Ref. [10]). Precisely, we need a nonnegative *distance function*

$$d_{\xi^\dagger}(R) := \inf\{\|\xi^\dagger - \xi\|_{X^*} : \xi = F'(x^\dagger)^*v, v \in Y^*, \|v\|_{Y^*} \leq R\}, \quad R > 0,$$

which measures, for $x^\dagger$ and its subgradient $\xi^\dagger$ under consideration, the degree of violation with respect to the benchmark source condition (8.32). If $x^\dagger$ would satisfy (8.32) with source element $v \in Y^*$ and $\|v\|_{Y^*} = R_0 > 0$, we would have $d_{\xi^\dagger}(R) = 0$ for all $R \geq R_0$. However, we have $d_{\xi^\dagger}(R) > 0$ for all $R > 0$ if (8.32) is violated. If the distance function obeys the limit condition

$$\lim_{R \to \infty} d_{\xi^\dagger}(R) \to 0, \tag{8.42}$$

then as in the benchmark case of Section 8.5.1, a convergence rate (8.6) can be derived via the detour of a variational inequality (8.18) along the lines of Theorems 8.1 and 8.2. However, in contrast to the benchmark case, here the rate function φ makes use of the distance function $d_{\xi^\dagger}$ as we will see below. It was mentioned in Ref. [10] that the limit condition (8.42), which is equivalent to $\xi^\dagger \in \overline{\text{range}(F'(x^\dagger))^*}^{\|\cdot\|_{X^*}}$, is valid when the biadjoint operator $F'(x^\dagger)^{**} : X^{**} \to Y^{**}$ is injective. For reflexive Banach spaces X and Y, this simplifies to the condition that $F'(x^\dagger)$ is injective.

On the way to a variational inequality, the approach of approximate source conditions for Banach spaces presumes that the Bregmann distance is *q-coercive*, that is,

$$D^{\mathcal{R}}_{\xi^\dagger}(x, x^\dagger) \geq c_q \|x - x^\dagger\|_X^q \qquad \text{for all} \quad x \in \mathcal{M} \tag{8.43}$$

is satisfied for some exponent $2 \leq q < \infty$ and a corresponding constant $c_q > 0$. Such assumption is for example fulfilled if $\mathcal{R}(x) := \|x\|_X^q$ and X is a q-convex Banach space.

Then, for every $R > 0$, one can find elements $v_R \in Y^*$ and $u_R \in X^*$ such that

$$\xi^\dagger = F'(x^\dagger)^*v_R + u_R \qquad \text{with} \qquad \|v_R\|_{Y^*} = R, \ \|u_R\|_{X^*} \leq d_{\xi^\dagger}(R),$$

and we can estimate for all $R > 0$ and $x \in \mathcal{M}$ as

$$\begin{aligned}
-\langle \xi^\dagger, x - x^\dagger \rangle_{X^* \times X} &= -\langle F'(x^\dagger)^*v_R + u_R, x - x^\dagger \rangle_{X^* \times X} \\
&= -\langle v_R, F'(x^\dagger)(x - x^\dagger) \rangle_{Y^* \times Y} + \langle u_R, x^\dagger - x \rangle_{X^* \times X} \\
&\leq R\|F'(x^\dagger)(x - x^\dagger)\|_Y + d_{\xi^\dagger}(R)\|x - x^\dagger\|_X.
\end{aligned}$$

Again, adding the difference $\mathcal{R}(x) - \mathcal{R}(x^\dagger)$ on both sides is helpful and leads to the variational inequality

$$D_{\xi^\dagger}^{\mathcal{R}}(x,x^\dagger) \leq \mathcal{R}(x) - \mathcal{R}(x^\dagger) + R\|F'(x^\dagger)(x-x^\dagger)\|_Y + d_{\xi^\dagger}(R)\|x-x^\dagger\|_X, \ x \in \mathcal{M}.$$
$$(8.44)$$

By using the q-coercivity (8.43), we can further estimate the last term in (8.44) as

$$d_{\xi^\dagger}(R)\|x-x^\dagger\|_X \leq c_q^{-1/q} d_{\xi^\dagger}(R) \left(D_{\xi^\dagger}^{\mathcal{R}}(x,x^\dagger)\right)^{1/q}.$$

An application of *Young's inequality* yields

$$d_{\xi^\dagger}(R)\|x-x^\dagger\|_X \leq \frac{1}{q} D_{\xi^\dagger}^{\mathcal{R}}(x,x^\dagger) + \frac{c_q^{-q^*/q}}{q^*} \left(d_{\xi^\dagger}(R)\right)^{q^*},$$

where q^* with $1/q + 1/q^* = 1$ is the exponent conjugate to q. Substituting this into (8.44), we obtain for all $x \in \mathcal{M}$ and with $0 < \beta := 1 - 1/q < 1$ the variational inequality

$$\beta D_{\xi^\dagger}^{\mathcal{R}}(x,x^\dagger) \leq \mathcal{R}(x) - \mathcal{R}(x^\dagger) + R\|F'(x^\dagger)(x-x^\dagger)\|_Y + \frac{c_q^{-q^*/q}}{q^*} \left(d_{\xi^\dagger}(R)\right)^{q^*}.$$

The term $\|F'(x^\dagger)(x-x^\dagger)\|_Y$ may be treated under structural conditions, used before in the benchmark case. For example, the nonlinearity condition (8.16) yields

$$\beta D_{\xi^\dagger}^{\mathcal{R}}(x,x^\dagger) \leq \mathcal{R}(x) - \mathcal{R}(x^\dagger) + R\sigma(\|F(x) - F(x^\dagger)\|_Y) + \frac{c_q^{-q^*/q}}{q^*} \left(d_{\xi^\dagger}(R)\right)^{q^*}. \ (8.45)$$

We equilibrate the second and the third term, depending on R and $d_{\xi^\dagger}(R)$, respectively, by means of the auxiliary continuous and strictly decreasing function

$$\Phi(R) := \frac{\left(d_{\xi^\dagger}(R)\right)^{q^*}}{R}, \quad R > 0, \tag{8.46}$$

which fulfills the limit conditions $\lim_{R \to 0} \Phi(R) = \infty$ and $\lim_{R \to \infty} \Phi(R) = 0$; thus it has a continuous decreasing inverse $\Phi^{-1} : (0,\infty) \to (0,\infty)$. By setting $R := \Phi^{-1}\left(\sigma(\|F(x) - F(x^\dagger)\|_Y)\right)$ and introducing the index function $\zeta(t) := \left[d_{\xi^\dagger}(\Phi^{-1}(\sigma(t)))\right]^{q^*}$, $t > 0$, we get from (8.45), with some constant $K > 0$, a variational inequality of the form

$$\beta D_{\xi^\dagger}^{\mathcal{R}}(x,x^\dagger) \leq \mathcal{R}(x) - \mathcal{R}(x^\dagger) + K\zeta(\|F(x) - F(x^\dagger)\|_Y), \qquad \text{for all} \qquad x \in \mathcal{M}.$$

We observe (cf. [100, remark 3.34]) by taking into account the one-to-one correspondence $\Phi(R) = \sigma(t)$ between $t > 0$ and $R > 0$ that the limit condition

$$\frac{\sigma(t)}{\zeta(t)} = \frac{\Phi(R)}{[d_{\xi^\dagger}(R)]^{q^*}} = \frac{1}{R} \to 0 \qquad \text{as} \quad R \to \infty \text{ resp. } t \to 0 \qquad (8.47)$$

is true. This means that the index function $\zeta(t)$ tends to zero as $t \to +0$ *slower* than the concave index function σ arising from the nonlinearity structure and there exists a concave index function $\tilde{\zeta}$ which is a majorant function to ζ such that one can find along the lines of Theorems 8.1 and 8.2 a convergence rate

$$D_{\xi^\dagger}^{\mathcal{R}}(x_\alpha^\delta, x^\dagger) = \mathcal{O}(\tilde{\zeta}(\delta)) \qquad \text{as} \qquad \delta \to 0, \qquad (8.48)$$

which is *lower* than the rate (8.35) because of (8.47). This is a consequence of the fact that, for the same nonlinearity situation, the smoothness of $x^\dagger$ with respect to the forward operator is *higher* when the benchmark source condition (8.32) is satisfied. The rate $\tilde{\zeta}$ is the worse, the more (8.32) is violated, that is, the slower the distance function $d_{\xi^\dagger}(R)$ tends to zero as $R \to \infty$.

We conclude this subsection with two examples that highlight the most important classes of convergence rates, *Hölder rates* and *logarithmic rates*. In this context, we get some insight into the interplay of the index function σ from the nonlinearity condition (8.16) and the decay rate of the distance function $d_{\xi^\dagger}(R)$ as $R \to \infty$.

Example 8.2 (Hölder convergence rates) In this example, we assume that the concave index function σ from (8.16) is of power-type, that is,

$$\sigma(t) = t^\kappa, \quad t > 0, \qquad \text{with some exponent} \qquad 0 < \kappa \le 1. \qquad (8.49)$$

If the solution $x^\dagger$ under consideration satisfies the benchmark source condition (8.32), then we have as a special case of (8.35) the Hölder convergence rate

$$D_{\xi^\dagger}^{\mathcal{R}}(x_\alpha^\delta, x^\dagger) = \mathcal{O}(\delta^\kappa) \qquad \text{as} \qquad \delta \to 0, \qquad (8.50)$$

where the rate is improved if κ grows and the highest attainable rate occurs for $\kappa = 1$.

Provided that (8.32) is violated and we have a distance function of power type

$$d_{\xi^\dagger}(R) \le \frac{\hat{K}}{R^{\frac{\mu}{1-\mu}}}, \quad R \ge R_0, \qquad \text{with } \hat{K} > 0, \text{ and some } 0 < \mu < 1, \qquad (8.51)$$

satisfying the limit condition (8.42), then it follows from (8.43) for $2 \le q < \infty$ and (8.48) with $\Phi(R) \sim R^{\frac{\mu(1-q^*)-1}{1-\mu}}$, $\tilde{\zeta}(\delta) = \zeta(\delta) \sim \delta^{\frac{\mu q^* \kappa}{1+(q^*-1)\mu}} = \delta^{\frac{\mu q \kappa}{\mu+q-1}}$ that the associated convergence rate is also a Hölder rate, but because of lower smoothness of $x^\dagger$ reduced in comparison with formula (8.50), and attains the form

$$D_{\xi^\dagger}^{\mathcal{R}}(x_\alpha^\delta, x^\dagger) = \mathcal{O}\left(\delta^{\frac{\mu q \kappa}{\mu+q-1}}\right) \qquad \text{as} \qquad \delta \to 0 \qquad (8.52)$$

(see also Ref. [50, formula (4.15)]). Evidently, (8.52) represents the optimal rate

$$D^{\mathcal{R}}_{\xi^{\dagger}}(x^{\delta}_{\alpha}, x^{\dagger}) = \mathcal{O}\left(\delta^{\frac{2\mu\kappa}{\mu+1}}\right) \qquad \text{as} \qquad \delta \to 0 \tag{8.53}$$

whenever $q = q^* = 2$, for example, in the case of Hilbert spaces.   ■

Example 8.3 (logarithmic convergence rates)   Now we are going to discuss logarithmic convergence rates,

$$D^{\Omega}_{\xi^{\dagger}}(x^{\delta}_{\alpha}, x^{\dagger}) = \mathcal{O}\left(\frac{1}{\left[\log\left(\frac{1}{\delta}\right)\right]^{\nu}}\right) \qquad \text{as} \quad \delta \to 0, \tag{8.54}$$

where for all exponents $\nu > 0$ the occurring rates are always lower than any Hölder
rate (8.50) introduced in Example 8.2 with arbitrarily small exponent $\kappa > 0$. We shall
see that a rate (8.54) may occur for two completely different extreme situations with
respect to the assumptions on the smoothness of $x^{\dagger}$ and the nonlinearity of F.

On the one hand, we suppose a rather smooth solution $x^{\dagger}$, where the benchmark
source condition (8.32) is satisfied, but the nonlinearity structure is poor in the sense
that the index function σ in the nonlinearity condition (8.16),

$$\sigma(t) = \frac{1}{\left[\log\left(\frac{1}{\delta}\right)\right]^{\nu}}, \quad t > 0,$$

is of logarithmic type. Then formula (8.35) yields immediately the rate (8.54) in that
case.

On the other hand, the other extreme situation occurs if $\sigma(t) = t$, that is, the condition (8.16) represents the most advantageous structure of nonlinearity, but the solution
smoothness is poor in the sense that a logarithmic decay rate

$$d_{\xi^{\dagger}}(R) \leq \widetilde{K}(\log R)^{-\lambda} \qquad \text{with some} \quad \widetilde{K} > 0, \ \lambda > 0, \quad \text{and for all} \quad R \geq R_0,$$

expresses a situation that $\xi^{\dagger}$ strongly violates the benchmark source condition, independent of the choice of $\lambda > 0$. However, since we have for sufficiently large $R > 0$
and for $\varepsilon > 0$ a constant $\overline{K} > 0$ with

$$\Phi(R) \sim \frac{1}{R(\log R)^{\lambda q^*}} \geq \frac{\overline{K}}{R^{1+\varepsilon}},$$

this implies $\Phi^{-1}(t) \geq \hat{K} t^{-1/(1+\varepsilon)}$ for some constant $\hat{K} > 0$ and sufficiently
small $t > 0$. Hence, also for that combination of smoothness and nonlinearity struc-
ture, we have a logarithmic rate (8.54), here with exponent $\nu = \lambda q^*$.   ■

8.5.3   Norm distance as error measure: ℓ^1-regularization

Recent developments in different fields of application, for example, in imaging, compressed sensing, and finance (cf., e.g., [13, 16, 25, 91, 96, 97, 100]), are connected with

the a priori information of *sparsity* with respect to the solution $x^\dagger$ to an ill-posed operator equation (8.1). For a normalized Schauder basis $\{u_k\}_{k\in\mathbb{N}}$ in the Banach space X, for example an orthonormal basis if X is a Hilbert space, such that every element $x = \sum_{k=1}^{\infty} x_k u_k$ is uniquely determined by the infinite sequence $\underline{x} := (x_1, x_2, \dots)$ of its coefficients, sparsity means that there is a *finite* index set $J \subset \mathbb{N}$ such that $x^\dagger = \sum_{k\in J} x_k^\dagger u_k$. We know that $x^\dagger$ is a finite linear combination of basis elements, but the present index set J is completely unknown. If we have that $\underline{x}^\dagger := (x_1^\dagger, x_2^\dagger, \dots) \in \ell^1$, then a popular approach for the stable approximate solution of (8.1) under *sparsity constraints* is the ℓ^1-regularization, where instead of $F : \mathcal{D}(F) \subseteq X \to Y$ the mapping $G : \underline{x} \in \ell^1 \mapsto y = F\left(\sum_{k=1}^{\infty} x_k u_k\right) \in Y$ is under consideration and a variant of the Tikhonov regularization with the optimization problem

$$T_\alpha^\delta(\underline{x}) := \frac{1}{p}\|G(\underline{x}) - y^\delta\|_Y^p + \alpha\|\underline{x}\|_{\ell^1} \to \min, \quad \text{subject to} \quad \underline{x} \in \mathcal{D}(G) \subseteq \ell^1, \quad (8.55)$$

and penalty functional

$$\mathcal{R}(\underline{x}) = \|\underline{x}\|_{\ell^1} = \sum_{k=1}^{\infty} |x_k|. \quad (8.56)$$

is used. For example, Refs. [17] and [11] show that for the Banach spaces $X = \ell^1$ and Y, the penalty (8.56) satisfies the requirements in Assumption 8.1, and τ_X-convergence discussed in Section 8.3 arises as weak*-convergence since ℓ^1 is nonreflexive, but the space of sequences tending to zero $Z = c_0$ is a predual space of ℓ^1.

In the past years, Grasmair et al. [46] and many others have shown convergence and convergence rates results on ℓ^1-regularization under sparsity constraints. However, in practice often the problem is not really sparse, but almost sparse. We mean this in the sense that the number of nonzero components $x_k^\dagger$ in the solution to (8.1) is infinite but its decay $|x_k^\dagger| \to 0$ as $k \to \infty$ is relevant such that $\underline{x}^\dagger \in \ell^1$. Then ℓ^1-regularization makes sense, too. Unfortunately, as shown in Ref. [17, section 4], the corresponding benchmark source condition cannot hold if the sparsity assumption on the solution fails and moreover distance functions constructed for using the method of approximate source conditions do not tend to zero as $R \to \infty$. Nevertheless, the approach of variational inequalities is also successful here. For example, with theorem 3.2 in Ref. [11] it was proven that a variational inequality

$$\|\underline{x} - \underline{x}^\dagger\|_{\ell^1} \le \|\underline{x}\|_{\ell^1} - \|\underline{x}^\dagger\|_{\ell^1} + \varphi(\|F(\underline{x}) - F(\underline{x}^\dagger)\|_Y) \qquad \text{for all} \qquad \underline{x} \in \mathcal{M} \quad (8.57)$$

can be verified when the additional assumption of source condition type

$$\underline{e}_k = G'(\underline{x}^\dagger)^* f_k \qquad \text{for some} \quad f_k \in Y^* \quad \text{and all} \quad k \in \mathbb{N}$$

is fulfilled, where $\underline{e}_k = (0, 0, \dots, 0, 1, 0, \dots)$ is the unit sequence having just 1 in the k-th component. For applications meeting this assumption in the case of linear forward operators, we refer to Ref. [2]. Moreover under (8.16) $\mathcal{M}$ in (8.57) contains $\underline{x}^\dagger$

and all regularized solutions x_α^δ under consideration for sufficiently small $\delta > 0$. The concave index function φ in (8.57) attains the form

$$\varphi(t) = 2 \inf_{n \in \mathbb{N}} \left(\sum_{k=n+1}^{\infty} |\underline{x}_k^\dagger| + \left(\sum_{k=1}^{n} \|f_k\|_{Y^*} \right) \sigma(t) \right),$$

which implies a corresponding convergence rate when the sequential discrepancy principle is applied. The rate function depends on three ingredients: the decay rate of $|\underline{x}_k^\dagger|$ as $k \to \infty$, the growth of the sum $\sum_{k=1}^{n} \|f_k\|_{Y^*}$ as $n \to \infty$, and the function σ characterizing the nonlinearity profile. In Ref. [11, Examples 3.4 and 3.5], the functions φ are made explicit for the case of Hölder rates and for exponentially decaying components $|\underline{x}_k^\dagger|$ as $k \to \infty$.

8.6   SUMMARY

We have given an overview of some new aspects and recent developments in Tikhonov regularization for nonlinear inverse problems. These problems, formulated as operator equations in Banach spaces, are in general ill-posed. Therefore, stabilizing penalty functionals are necessary for constructing stable and convergent approximate solutions. The interplay of convergence and appropriate choices of the regularization parameters have been discussed intensively as well as cross connections between solution smoothness, nonlinearity structure, and convergence rates. As crucial point of this chapter, we have presented specific variational inequalities as a sophisticated tool for expressing both solution smoothness and the structure of nonlinearity. In detail, we have shown the consequences of this tool for obtaining convergence rates and have given a series of examples how to construct such inequalities. Moreover, we outlined some cross connections to classical source conditions and to approximate source conditions.

Acknowledgments

This research was partially supported by the German Research Foundation (DFG) under grants HO 1454/8-1 and HO 1454/9-1. The author expresses his thanks to Jens Flemming and Stephan W. Anzengruber (TU Chemnitz), Peter Mathé (WIAS Berlin), as well as Radu I. Boţ (University of Vienna) for the fruitful cooperation and numerous hints being incorporated into the present work. Valuable comments of an anonymous referee were also incorporated with thanks.

REFERENCES

1. S. W. Anzengruber and R. Ramlau, "Convergence rates for Morozov's discrepancy principle using variational inequalities," *Inverse Problems*, **27**(10), 105007 (18pp) (2011).

2. S. W. Anzengruber, B. Hofmann, and R. Ramlau, "On the interplay of basis smoothness and specific range conditions occurring in sparsity regularization," *Inverse Problems*, **29**(12), 125002 (21pp) (2013).

3. S. W. Anzengruber, B. Hofmann, and P. Mathé, "Regularization properties of the discrepancy principle for Tikhonov regularization in Banach spaces," *Applicable Analysis*, **93**(7), 1382–1400 (2014).

4. A. B. Bakushinsky and M. Y. Kokurin, *Iterative Methods for Approximate Solution of Inverse Problems*, Springer, Dordrecht, 2004.

5. A. B. Bakushinsky, M. Yu. Kokurin, and A. Smirnova, *Iterative Methods for Ill-Posed Problems – An Introduction*, volume 54 of *Inverse and Ill-Posed Problems Series*, Walter de Gruyter, Berlin, 2011.

6. J. Baumeister, *Stable Solution of Inverse Problems*, Vieweg, Braunschweig, 1987.

7. J. Baumeister, "Deconvolution of appearance potential spectra," in: *Direct and Inverse Boundary Value Problems* (ed. R. Kleinman et al.), Oberwolfach, 1989, Lang, Frankfurt am Main, 1991, pp. 1–13.

8. M. Benning and M. Burger, "Error estimates for general fidelities," *Electronic Transactions on Numerical Analysis,* **38**, 44–68 (2011).

9. M. Benning and M. Burger, "Ground states and singular vectors of convex variational regularization methods," *Methods and Applications of Analysis*, **20**(4), 295–334 (2013).

10. R. I. Boţ and B. Hofmann, "An extension of the variational inequality approach for obtaining convergence rates in regularization of nonlinear ill-posed problems," *Journal of Integral Equations and Applications*, **22**(3), 369–392 (2010).

11. R. I. Boţ and B. Hofmann, "The impact of a curious type of smoothness conditions on convergence rates in ℓ^1-regularization," *Eurasian Journal of Mathematical and Computer Applications*, **1**(1), 29–40 (2013).

12. K. Bredies and D. A. Lorenz, "Regularization with non-convex separable constraints," *Inverse Problems*, **25**(8), 085011 (14pp) (2009).

13. J. Brodie, I. Daubechies, C. De Mol, D. Giannone, and I. Loris, "Sparse and stable Markowitz portfolios," *PNAS*, **106**, 12267–12272 (2009).

14. S. Bürger and B. Hofmann, "About a deficit in low order convergence rates on the example of autoconvolution," *Applicable Analysis* (to appear), published electronically 2014: http://dx.doi.org/10.1080/00036811.2014.886107 (accessed October 30, 2014).

15. M. Burger and S. Osher, "Convergence rates of convex variational regularization," *Inverse Problems*, **20**(5), 1411–1421 (2004).

16. M. Burger, E. Resmerita, and L. He, "Error estimation for Bregman iterations and inverse scale space methods in image restoration," *Computing*, **81**(2–3), 109–135 (2007).

17. M. Burger, J. Flemming, and B. Hofmann, "Convergence rates in ℓ^1-regularization if the sparsity assumption fails," *Inverse Problems*, **29**(2), 025013 (16pp) (2013).

18. G. Chavent, *Nonlinear Least Squares for Inverse Problems*, Springer, New York, 2009.

19. J. Cheng and M. Yamamoto, "One new strategy for a priori choice of regularizing parameters in Tikhonov's regularization," *Inverse Problems*, **16**(4), L31–L38 (2000).

20. J. Cheng, B. Hofmann, and S. Lu, "The index function and Tikhonov regularization for ill-posed problems," *Journal of Computational and Applied Mathematics*, **265**, 110–119 (2014).

21. K. Choi and A. D. Lanterman, "An iterative deautoconvolution algorithm for nonnegative functions," *Inverse Problems* **21**(3), 981–995 (2005).

22. C. Clason, "L^∞ fitting for inverse problems with uniform noise," *Inverse Problems*, **28**(10), 104007 (18pp) (2012).

23. D. Colton and R. Kress, *Inverse Acoustic and Electromagnetic Scattering Theory*, 3rd Edition, Springer, New York, 2013.

24. Z. Dai and P. K. Lamm, "Local regularization for the nonlinear inverse autoconvolution problem," *SIAM Jornal on Numerical Analysis*, **46**(2), 832–868 (2008).

25. I. Daubechies, M. Defrise, and C. De Mol, "An iterative thresholding algorithm for linear inverse problems with a sparsity constraint," *Communications in Pure and Applied Mathematics*, **57**(11), 1413–1457 (2004).

26. D. Düvelmeyer, B. Hofmann, and M. Yamamoto, "Range inclusions and approximate source conditions with general benchmark functions," *Numerical Functional Analysis and Optimization*, **28**(11–12), 1245–1261 (2007).

27. H. Egger and A. Leitão, "Nonlinear regularization methods for ill-posed problems with piecewise constant or strongly varying solutions," *Inverse Problems*, **25**(11), 115014 (19pp) (2009).

28. H. W. Engl and J. Zou, "A new approach to convergence rate analysis of Tikhonov regularization for parameter identification in heat conduction," *Inverse Problems*, **16**(6), 1907–1923 (2000).

29. H. W. Engl, K. Kunisch, and A. Neubauer, "Convergence rates for Tikhonov regularisation of nonlinear ill-posed problems," *Inverse Problems*, **5**(4), 523–540 (1989).

30. H. W. Engl, M. Hanke, and A. Neubauer, *Regularization of Inverse Problems*, Kluwer Academic Publishers, Dordrecht, 1996, 2nd Edition 2000.

31. G. Fleischer and B. Hofmann, "On inversion rates for the autoconvolution equation," *Inverse Problems* **12**(4), 419–435 (1996).

32. J. Flemming, *Generalized Tikhonov Regularization and Modern Convergence Rate Theory in Banach Spaces*, Shaker Verlag, Aachen, 2012.

33. J. Flemming, "Solution smoothness of ill-posed equations in Hilbert spaces: four concepts and their cross connections," *Applied Analysis*, **91**(5), 1029–1044 (2012).

34. J. Flemming, "Variational smoothness assumptions in convergence rate theory—an overview," *Journal of Inverse and Ill-Posed Problems*, **21**(3), 395–409 (2013).

35. J. Flemming, "Regularization of autoconvolution and other ill-posed quadratic equations by decomposition," *Journal of Inverse and Ill-Posed Problems*, **22**(4), 551–567 (2014).

36. J. Flemming and B. Hofmann, "A new approach to source conditions in regularization with general residual term," *Numerical Functional Analysis and Optimization*, **31**(3), 254–284 (2010).

37. J. Flemming and B. Hofmann, "Convergence rates in constrained Tikhonov regularization: equivalence of projected source conditions and variational inequalities," *Inverse Problems*, **27**(8), 085001 (11pp) (2011).

38. J. Flemming, B. Hofmann, and P. Mathé, "Sharp converse results for the regularization error using distance functions," *Inverse Problems*, **27**(2), 025006 (18pp) (2011).

39. D. Gerth, B. Hofmann, S. Birkholz, S. Koke, and G. Steinmeyer, "Regularization of an autoconvolution problem in ultrashort laser pulse characterization," *Inverse Problems in Science and Engineering*, **22**(2), 245–266 (2014).

40. R. Gorenflo and B. Hofmann, "On autoconvolution and regularization," *Inverse Problems*, **10**(2), 353–373 (1994).

41. M. Grasmair, "Well-posedness and convergence rates for sparse regularization with sublinear l^q penalty term," *Inverse Problems and Imaging*, **3**(3), 383–387 (2009).

42. M. Grasmair, " Generalized Bregman distances and convergence rates for non-convex regularization methods," *Inverse Problems*, **26**(11), 115014 (16pp) (2010).

43. M. Grasmair, "Non-convex sparse regularisation," *Journal of Mathematical Analysis and Applications*, **365**(1), 19–28 (2010).

44. M. Grasmair, "An application of source inequalities for convergence rates of Tikhonov regularization with a non-differentiable operator," 2012, http://arxiv.org/pdf/1209. 2246v1.pdf (accessed October 30, 2014).

45. M. Grasmair, "Variational inequalities and improved convergence rates for Tikhonov regularization on Banach spaces," *Journal of Inverse and Ill-Posed Problems*, **21**(3), 379–394 (2013).

46. M. Grasmair, M. Haltmeier, and O. Scherzer, "Sparse regularization with l^q penalty term," *Inverse Problems*, **24**(5), 055020 (13pp) (2008).

47. C. W. Groetsch, *The Theory of Tikhonov Regularization for Fredholm Integral Equations of the First Kind*, Pitman, Boston, 1984.

48. C. W. Groetsch, *Inverse Problems in the Mathematical Sciences*, Vieweg, Braunschweig, 1993.

49. M. Hanke, "The regularizing Levenberg-Marquardt scheme is of optimal order," *Journal of Integral Equations and Applications*, **22**(2), 259–283 (2010).

50. T. Hein and B. Hofmann, "Approximate source conditions for nonlinear ill-posed problems—chances and limitations," *Inverse Problems*, **25**(3), 035003 (16pp) (2009).

51. B. Hofmann, "Approximate source conditions in Tikhonov-Phillips regularization and consequences for inverse problems with multiplication operators," *Mathematical Methods in the Applied Science*, **29**(3), 351–371 (2006).

52. B. Hofmann and P. Mathé, "Analysis of profile functions for general linear regularization methods," *SIAM Jorunal on Numerical Analysis*, **45**(3), 1122–1141 (2007).

53. B. Hofmann and P. Mathé, "Parameter choice in Banach space regularization under variational inequalities," *Inverse Problems*, **28**(10), 104006 (17pp) (2012).

54. B. Hofmann and O. Scherzer, "Factors influencing the ill-posedness of nonlinear problems," *Inverse Problems*, **10**(6), 1277–1297 (1994).

55. B. Hofmann and O. Scherzer, "Local ill-posedness and source conditions of operator equations in Hilbert spaces," *Inverse Problems*, **14**(5), 1189–1206 (1998).

56. B. Hofmann and M. Yamamoto, "Convergence rates for Tikhonov regularization based on range inclusions," *Inverse Problems*, **21**(3), 805–820 (2005).

57. B. Hofmann and M. Yamamoto, "On the interplay of source conditions and variational inequalities for nonlinear ill-posed problems," *Application Analysis*, **89**(11), 1705–1727 (2010).

58. B. Hofmann, D. Düvelmeyer, and K. Krumbiegel, "Approximate source conditions in Tikhonov regularization—new analytical results and some numerical studies," *Mathematical Modelling and Analysis*, **11**(1), 41–56 (2006).

59. B. Hofmann, P. Mathé, and S. V. Pereverzev, "Regularization by projection: approximation theoretic aspects and distance functions," *Journal of Inverse and Ill-Posed Problems*, **15**(5), 527–545 (2007).

60. B. Hofmann, B. Kaltenbacher, C. Pöschl, and O. Scherzer, "A convergence rates result for Tikhonov regularization in Banach spaces with non-smooth operators," *Inverse Problems*, **23**(3), 987–1010 (2007).

61. T. Hohage and F. Werner, "Convergence rates for inverse problems with impulsive noise," *SIAM Journal on Numerical Analysis*, **52**(3), 1203–1221 (2014).

62. V. Isakov, *Inverse Problems for Partial Differential Equations*, 2nd Edition, Springer, New York, 2006.

63. J. Janno, "On a regularization method for the autoconvolution equation," *Zeitschrift für Angewandte Mathematik and Mechanik*, **77**(5), 393–394 (1997).

64. J. Janno, "Lavrent'ev regularization of ill-posed problems containing nonlinear near-to-monotone operators with application to autoconvolution equation," *Inverse Problems*, **16**(2), 333–348 (2000).

65. D. Jiang, H. Feng, and J. Zou, "Convergence rates of Tikhonov regularizations for parameter identification in a parabolic-elliptic system," *Inverse Problems*, **28**(10), 104002 (20pp) (2012).

66. Q. Jin, "A general convergence analysis of some Newton-type methods for nonlinear inverse problems," *SIAM Journal on Numerical Analysis*, **49**(2), 549–573 (2011).

67. S. I. Kabanikhin, *Inverse and ill-posed problems—Theory and Applications*, volume 55 of *Inverse and Ill-posed Problems Series*, Walter de Gruyter, Berlin, 2012.

68. B. Kaltenbacher, "A note on logarithmic convergence rates for nonlinear Tikhonov regularization," *Journal of Inverse and Ill-Posed Problems*, **16**(1), 79–88 (2008).

69. B. Kaltenbacher and B. Hofmann, "Convergence rates for the iteratively regularized Gauss-Newton method in Banach spaces," *Inverse Problems*, **26**(3), 035007 (21pp) (2010).

70. B. Kaltenbacher, A. Neubauer, and O. Scherzer, *Iterative Regularization Methods for Nonlinear Ill-posed Problems*, Walter de Gruyter, Berlin, 2008.

71. A. Kirsch, *An Introduction to the Mathematical Theory of Inverse Problems*, 2nd Edition, Springer, New York, 2011.

72. A. Lechleiter and A. Rieder, "Towards a general convergence theory for inexact Newton regularizations," *Numerische Mathematik*, **114**(3), 521–548 (2010).

73. F. Liu and M. Z. Nashed, "Regularization of nonlinear ill-posed variational inequalities and convergence rates," *Set-Valued Analysis*, **6**(4), 313–344 (1998).

74. D. A. Lorenz, "Convergence rates and source conditions for Tikhonov regularization with sparsity constraints," *Journal of Inverse and Ill-Posed Problems*, **16**(5), 463–478 (2008).

75. D. A. Lorenz and N. Worliczek, "Necessary conditions for variational regularization schemes," *Inverse Problems*, **29**(7), 075016 (19pp) (2013).

76. A. K. Louis, *Inverse und schlecht gestellte Probleme[Inverse and Ill-Posed Problems]* (in German), Teubner, Stuttgart, 1989.

77. A. K. Louis, "Approximate inverse for linear and some nonlinear problems," *Inverse Problems*, **11**(6), 1211–1223 (1995).

78. S. Lu and J. Flemming, "Convergence rate analysis of Tikhonov regularization for nonlinear ill-posed problems with noisy operators," *Inverse Problems*, **28**(10), 104003 (19pp) (2012).

79. S. Lu and S. V. Pereverzev, *Regularization Theory for Ill-posed Problems,* Walter de Gruyter, Berlin, 2013.

80. S. Lu, S. V. Pereverzev and R. Ramlau, "An analysis of Tikhonov regularization for non-linear ill-posed problems under a general smoothness assumption," *Inverse Problems*, **23**(1), 217–230 (2007).

81. P. Mahale and M. T. Nair, "Tikhonov regularization of nonlinear ill-posed equations under general source conditions," *Journal of Inverse and Ill-Posed Problems*, **15**(8), 813–829 (2007).

82. P. Mathé and S. V. Pereverzev, "Geometry of linear ill-posed problems in variable Hilbert scales," *Inverse Problems*, **19**(3), 789–803 (2003).

83. J. Mueller and S. Siltanen, *Linear and Nonlinear Inverse Problems with Practical Applications*, SIAM, Philadelphia, 2012.

84. A. Neubauer, "Tikhonov regularisation for nonlinear ill-posed problems: optimal convergence rates and finite-dimensional approximation," *Inverse Problems*, **5**(4), 541–557 (1989).

85. A. Neubauer, "On converse and saturation results for Tikhonov regularization of linear ill-posed problems," *SIAM Journal on Numerical Analysis*, **34**(2), 517–527 (1997).

86. A. Neubauer, "Modified Tikhonov regularization for nonlinear ill-posed problems in Banach spaces," *Journal of Integral Equations and Applications*, **22**(2), 341–351 (2010).

87. A. Neubauer, T. Hein, B. Hofmann, S. Kindermann, and U. Tautenhahn, "Improved and extended results for enhanced convergence rates of Tikhonov regularization in Banach spaces," *Applicable Analysis*, **89**(11), 1729–1743 (2010).

88. S. V. Pereverzev and B. Hofmann, "Estimation of linear functionals from indirect noisy data without knowledge of the noise level," *GEM International Journal on Geomathematics*, **1**(1), 121–131 (2010).

89. C. Pöschl. *Tikhonov Regularization with General Residual Term*, PhD thesis, University of Innsbruck, Austria, 2008.

90. R. Ramlau, "Morozov's discrepancy principle for Tikhonov-regularization of nonlinear operators," *Numerical Functional Analysis and Optimization*, **23**(1–2), 147–172 (2002).

91. R. Ramlau and G. Teschke, "Sparse recovery in inverse problems," in *Theoretical Foundations and Numerical Methods for Sparse Recovery*, volume 9 of *Radon Ser. Comput. Appl. Math.*, pp. 201–262, Walter de Gruyter, Berlin, 2010.

92. R. Ramlau and C. A. Zarzer, "On the minimization of a Tikhonov functional with a non-convex sparsity constraint," *Electronic Transactions on Numerical Analysis*, **39**, 476–507 (2012).

93. E. Resmerita, "Regularization of ill-posed problems in Banach spaces: convergence rates," *Inverse Problems*, **21**(4), 1303–1314 (2005).

94. E. Resmerita and O. Scherzer, "Error estimates for non-quadratic regularization and the relation to enhancement," *Inverse Problems*, **22**(3), 801–814 (2006).

95. V. G. Romanov, *Investigation Methods for Inverse Problems,* in *Inverse and Ill-posed Problems Series*, VSP, Utrecht, 2002.

96. O. Scherzer (ed.), *Handbook of Mathematical Methods in Imaging*, 3 volumes, Springer, New York, 2011.

97. O. Scherzer, M. Grasmair, H. Grossauer, M. Haltmeier, and F. Lenzen, *Variational Methods in Imaging*, volume 167 of *Applied Mathematical Sciences*, Springer, New York, 2009.

98. K.-Th. Schleicher, S. W. Schulz, R. Gmeiner, and H.-U. Chun, "A computational method for the evaluation of highly resolved DOS functions from APS measurements," *Journal of Electron Spectroscopy and Related Phenomena*, **31**, 33–56 (1983).

99. E. Schock, "Approximate solution of ill-posed equations: arbitrarily slow convergence vs. superconvergence," in *Constructive Methods for the Practical Treatment of Integral Equations* (eds. G. Hämmerlin and K. H. Hoffmann), Birkhäuser, Basel, 1985, pp. 234–243.

100. T. Schuster, B. Kaltenbacher, B. Hofmann, and K. S. Kazimierski, *Regularization Methods in Banach Spaces*, volume 10 of *Radon Ser. Comput. Appl. Math.*, Walter de Gruyter, Berlin/Boston, 2012.

101. T. I. Seidman and C. R. Vogel, "Well posedness and convergence of some regularization methods for nonlinear ill posed problems," *Inverse Problems*, **5**(2), 227–238 (1989).

102. J. K. Seo and E. J. Woo, *Nonlinear Inverse Problems in Imaging*, John Wiley & Sons, Inc., Hoboken, 2012.

103. U. Tautenhahn, "On a general regularization scheme for nonlinear ill-posed problems," *Inverse Problems*, **13**(5), 1427–1437 (1997).

104. A. N. Tikhonov and V. Y. Arsenin, *Solutions of Ill-posed Problems*, John Wiley & Sons, Inc. New York, 1977.

105. A. N. Tikhonov, A. S. Leonov, and A. G. Yagola, *Nonlinear Ill-posed Problems, Volumes. 1 and 2*, volume 14 of *Applied Mathematics and Mathematical Computation*, Chapman & Hall, London, 1998.

106. G. M. Vaĭnikko and A. Yu. Veretennikov, *Iteratsionnye protsedury v nekorrektnykh zadachakh [Iteration Procedures in Ill-Posed Problems]* (in Russian), "Nauka", Moscow, 1986.

107. V. V. Vasin, "Some tendencies in the Tikhonov regularization of ill-posed problems," *Jounal of Inverse and Ill-Posed and Problems*, **14**(8), 813–840 (2006).

108. V. V. Vasin and M. A. Korotkii, "Tikhonov regularization with nondifferentiable stabilizing functionals," *Journal of Inverse and Ill-Posed Problems*, **15**(8), 853–865 (2007).

109. F. Werner and T. Hohage, "Convergence rates in expectation for Tikhonov-type regularization of inverse problems with Poisson data," *Inverse Problems*, **28**(10), 104004 (15pp) (2012).

110. L. von Wolfersdorf, "Autoconvolution equations and special functions," *Integral Transforms and Special Functions*, **19**(9), 677–686 (2008).

111. A. G. Yagola, I. V. Kochikov, G. M. Kuramshina, and Yu. A. Pentin, *Inverse Problems of Vibrational Spectroscopy*, in *Inverse and Ill-posed Problems Series*, VSP, Utrecht, 1999.

112. C. A. Zarzer, "On Tikhonov regularization with non-convex sparsity constraints", *Inverse Problems*, **25**(2), 025006 (13pp) (2009).

9

INITIAL AND INITIAL-BOUNDARY VALUE PROBLEMS FOR FIRST-ORDER SYMMETRIC HYPERBOLIC SYSTEMS WITH CONSTRAINTS

Nicolae Tarfulea

Department of Mathematics, Computer Science, and Statistics, Purdue University Calumet, Hammond, IN, USA

9.1 INTRODUCTION

Various real-life applications lead to first-order symmetric hyperbolic (FOSH) systems of differential equations whose solutions must satisfy certain constraint equations. Frequently, for the initial value problem, the constraints are preserved by the evolution (e.g., for Maxwell's equations or Einstein's field equations in various FOSH formulations), but for an initial-boundary value problem, this is not true in general. A strong motivation for considering such initial-boundary value problems comes from numerical simulations. Often, the numerical solutions to such evolution problems are computed on artificial space cut offs because of the necessary boundedness of computational domains. Therefore, boundary conditions have to be posed in such a way that the numerical solution of the cut off system approximates as best as possible the solution of the original problem on infinite space, and this includes the preservation of constraints. It has become increasingly clear that in order for constraints to be preserved during evolution, the boundary conditions have to be chosen in an appropriate way. With this motivation, several

Mathematical and Computational Modeling: With Applications in Natural and Social Sciences, Engineering, and the Arts, First Edition. Roderick Melnik.

authors [6,7,20,21,32,35,36,38,40,41,51,61,66,74,75], e.g., have sought constraint-preserving boundary conditions (especially in the context of Maxwell's and Einstein's equations), that is, boundary conditions that guarantee that the solution satisfies the given constraints for all time. See also Refs. [44, 53] for methods to control constraint violation. An informative review of results concerning continuum and discrete initial-boundary value problems arising from hyperbolic partial differential equations, including constrained FOSH systems, can be found in Ref. [67]. There are also important results on initial value problems in the case of nonlinear first-order systems of conservation laws because it was proven that all systems are symmetric if they satisfy an entropy principle with convex entropy density. In fact, in these cases there exists a privileged "main field" such that the original system becomes symmetric (see Ref. [64]). In this context, nonlinear symmetric hyperbolic systems with involutive constraints have been studied by many authors (see Refs. [13, 14, 28, 29] and references therein). The problem of conditions at the boundary for an initial-boundary value problem for systems of conservation laws has also been considered. For example, in Ref. [46] the boundary conditions are obtained in the form of so-called boundary entropy inequalities. These inequalities contain constraints for the boundary layer which give the relations between the unknown function and its normal derivative at the boundary.

This chapter is organized as follows: In Section 9.2, we discuss the well-posedness for initial value problems involving FOSH systems of equations, with special emphasis on FOSH systems with constraints. In particular, we consider a second (extended) FOSH system which we show is equivalent to the original one under certain assumptions. Next, in Section 9.3, we direct our attention toward a characterization of maximal nonnegative boundary conditions for FOSH systems. We also continue the analysis of the extended FOSH system and its equivalence with the original one in presence of boundary conditions. In Section 9.4, we apply our results and the afferent techniques for finding constraint-preserving boundary conditions for a system of wave equations in a first-order formulation subject to divergence constraints and to Einstein's equations in both Einstein–Christoffel and Alekseenko–Arnold formulations linearized about Minkowski's spacetime.

9.2 FOSH INITIAL VALUE PROBLEMS WITH CONSTRAINTS

In this section, some basic results on FOSH systems of partial differential equations are briefly reviewed, with special attention being given to systems with constraints. Section 9.2.1 is intended to enlist the basic definition of FOSH systems and some relevant existence and uniqueness results. Sections 9.2.2 and 9.2.3 are dedicated to the analysis of constrained initial value problems in a more abstract framework and in the case of FOSH systems of partial differential equations, respectively. The emphasis is on the equivalence between a given system subject to constraints and a corresponding extended unconstrained system. This approach originates in a presentation given by Arnold [6] on Einstein's field equations. It was further analyzed and developed by Arnold and Tarfulea [8], and Tarfulea [76]. All these results represent background material relevant to the discussions that follow in the next sections of the chapter. Much more information on hyperbolic systems can be found in the books by John [45], Kreiss and Lorenz [49], Gustafsson et al. [42], and Evans [30], among many others. In particular, see the recent article by Sarbach and Tiglio [67] for an ample review of hyperbolic evolution problems.

9.2.1   FOSH initial value problems

In this section, we will be concerned with a linear first-order system of equations for a column vector $u = u(x,t) = u(x_1,\ldots,x_n,t)$ with m components $u_1,\ldots,u_m$. Such a system can be written as

$$\partial_t u = \sum_{i=1}^{n} A_i(x,t)\partial_i u + f(x,t) \text{ in } \mathbb{R}^n \times (0,T], \tag{9.1}$$

where $T > 0$. Here $A_1,\ldots,A_n$ are given $m \times m$ matrix functions, and f is a given m-vector field. We will further assume that A_i are of class C^2, with bounded derivatives over $\mathbb{R}^n \times [0,T]$, and $f \in H^1(\mathbb{R}^n \times (0,T);\mathbb{R}^m)$.

As initial data we prescribe the values of u on the hyperplane $t = 0$

$$u = u_0 \text{ on } \mathbb{R}^n \times \{t = 0\}, \tag{9.2}$$

with $u_0 \in H^1(\mathbb{R}^n;\mathbb{R}^m)$. For each $y \in \mathbb{R}^n$, define

$$A(y)(x,t) = \sum_{i=1}^{n} y_i A_i(x,t) \quad (x \in \mathbb{R}^n, t \geq 0).$$

The system (9.1) is called symmetric hyperbolic if $A_i(x,t)$ is a symmetric $m \times m$ matrix for each $x \in \mathbb{R}^n$, $t \geq 0$ $(i = 1,\ldots,n)$. Thus, the $m \times m$ matrix $A(y)(x,t)$ has only real eigenvalues and the corresponding eigenvectors form a basis of $\mathbb{R}^N$ for each $y \in \mathbb{R}^n$, $x \in \mathbb{R}^n$, and $t \geq 0$.

Remark 9.1   More general systems having the form

$$A_0 \partial_t u = \sum_{i=1}^{n} A_i(x,t)\partial_i u + B(x,t)u + f(x,t)$$

are also called symmetric, provided the matrix functions A_i are symmetric for $i = 0,\ldots,n$, and A_0 is positive definite. The results set forth below can be easily extended to such systems.

As in Ref. [30, section 7.3.2.], let us define the bilinear form

$$A[u,v;t] := \int_{\mathbb{R}^n} \sum_{i=1}^{n} (A_i(\cdot,t)\partial_i u) \cdot v \, dx$$

for $0 \leq t \leq T$, $u, v \in H^1(\mathbb{R}^n;\mathbb{R}^m)$.

Definition 9.1 *We say*

$$u \in L^2((0,T); H^1(\mathbb{R}^n; \mathbb{R}^m)), \text{ with } u' \in L^2((0,T); L^2(\mathbb{R}^n; \mathbb{R}^m)),$$

is a weak solution of the initial value problem (9.1), (9.2) provided

(i) $(u', v) = A[u, v; t] + (f, v)$ *for each* $v \in H^1(\mathbb{R}^n; \mathbb{R}^m)$ *and a.e.* $0 \le t \le T$, *and*
(ii) $u(0) = u_0$.

Here $(\cdot, \cdot)$ *denotes the inner product in* $L^2(\mathbb{R}^n; \mathbb{R}^m)$.

By using energy methods and the vanishing viscosity technique (see Ref. [30, section 7.3.2]), the following existence and uniqueness result can be proven:

Theorem 9.1 *The initial value problem (9.1), (9.2) has a unique weak solution.*

In what follows, we will be more interested in FOSH systems with constant coefficients. For such systems, a more general result (including regularity) is valid. It is a slight generalization of Ref. [30, section 7.3.3, theorem 5].

Theorem 9.2 *Assume* $u_0, f(\cdot, t) \in H^s(\mathbb{R}^n; \mathbb{R}^m)$, *for* $t \ge 0$, *with* $s > n/2 + m$. *Then there is a unique solution* $u \in C^1(\mathbb{R}^n \times [0, \infty); \mathbb{R}^m)$ *of the initial value problem (9.1) and (9.2).*

The main tool used for proving this theorem is the Fourier transform. The unique C^1 solution is given by (see Ref. [30, section 7.3.3, theorem 5] for details):

$$u(x,t) = \frac{1}{(2\pi)^{n/2}} \int_{\mathbb{R}^n} e^{ix \cdot y} [e^{itA(y)} \hat{u}_0(y) + \int_0^t e^{i(t-s)A(y)} \hat{f}(y,s) ds] \, dy. \qquad (9.3)$$

9.2.2 Abstract formulation

In order to highlight the structure, we consider a more general and abstract problem than a FOSH system, namely a linear ordinary differential equation in Hilbert space. In this setting, we define what it means for the evolution to be constraint preserving and establish sufficient conditions for constraint preservation. We then introduce the extended evolution problem which will be an important concept throughout this paper and show that the extended problem is, in a certain sense, equivalent to the original one.

Let Y be a Hilbert space with inner product $\langle \cdot, \cdot \rangle_Y$ (the subscript Y will be dropped when clear from context), and let A be a closed, linear operator from Y to itself with domain $D(A)$ which is dense in Y. We assume that A is skew-adjoint ($A^* = -A$). Given $u_0 \in D(A)$ and $f : [0, T] \to Y$ continuous, we consider the initial value problem

$$\dot{u}(t) = Au(t) + f(t), \quad t \in [0, T], \tag{9.4}$$
$$u(0) = u_0. \tag{9.5}$$

Definition 9.2 *A linear operator A on the Hilbert space Y with domain $D(A)$ is m-dissipative if*

1. $\|x - \lambda Ax\| \geq \|x\|$, *for all $x \in D(A)$ and all $\lambda > 0$,*
2. *for all $\lambda > 0$ and all $f \in H$, there exists $x \in D(A)$ such that $x - \lambda Ax = f$.*

Lemma 9.1 *[23, corollaire 2.4.9] The operator A is m-dissipative.*

The following existence and uniqueness result shows that under appropriate conditions the initial-value problem (9.4) and (9.5) is solvable. It is a direct consequence of the fact that A is *m*-dissipative (see [31, pp. 51–53] or [19, pp. 116–118]).

Theorem 9.3 *Let $u_0 \in D(A)$ and $f \in C([0,T], Y)$. Assume that either $f \in L^1([0,T], D(A))$ or $f \in W^{1,1}([0,T], Y)$. Then there exists a unique solution $u \in C([0,T], D(A)) \cap C^1([0,T], Y)$ to the initial-value problem (9.4) and (9.5). Moreover, the solution u is given by $u(t) = T(t)u_0 + \int_0^t T(t-s)f(s)\, ds$, where $T(t)$ is the semigroup of contractions generated by A.*

In many applications of interest, the solution u of (9.4)–(9.5) must also satisfy a constraint

$$Bu(t) = 0, \quad t \in [0, T], \tag{9.6}$$

where $B : D(B) \subseteq Y \to X$ is a densely defined, closed, linear operator from Y into a second Hilbert space X. Since the problem (9.4) and (9.5) determines the solution uniquely, the constraint (9.6) must be a consequence of the problem, that is, of the initial data u_0, the forcing function f, and the operator A. We now consider when this is the case.

Throughout this section, we shall assume that the operator A is *constraint-preserving*, that is,

$$A(D(A) \cap \ker B) \subseteq \ker B, \tag{9.7}$$

and its restriction to $D(A) \cap \ker B$, denoted by $\overline{A} : D(A) \cap \ker B \to \ker B$, is *m*-dissipative. According to Ref. [23, proposition 2.4.2 and theorem 2.4.5], the last condition is satisfied if

$$\langle \overline{A}^* u, u \rangle \leq 0, \quad \forall u \in D(\overline{A}^*). \tag{9.8}$$

Here $\overline{A}^*$ is the adjoint of $\overline{A}$, and so $D(\overline{A}^*) \subseteq \ker B$. Observe that, since the linear operator B is closed, its kernel $\ker B$ is a closed linear subspace of Y. Moreover, $D(A) \cap \ker B$ is dense in $\ker B$ and $D(A) \cap \ker B \subseteq D(\overline{A}^*)$.

An obvious necessary condition for (9.6) is compatibility of the initial data,

$$Bu_0 = 0. \tag{9.9}$$

The forcing function must be compatible as well, namely, we must have

$$Bf(t) = 0, \quad \text{for all } t \in [0,T]. \tag{9.10}$$

This is because, for any fixed $\bar{t} \in [0,T]$, we have $B([u(t) - u(\bar{t})]/(t - \bar{t})) = 0$, for all $t \in [0,T]$, $t \neq \bar{t}$. Letting $t \to \bar{t}$, this implies that $B(\dot{u}(\bar{t})) = 0$ (since B is a closed operator). By operating on (9.4) with B, it follows that $0 = B(\dot{u}(\bar{t})) = BAu(\bar{t}) + Bf(\bar{t}) = Bf(\bar{t})$, where the last equality comes from (9.7).

The following theorem gives simple but useful conditions for the constraints to be satisfied by the solution of the evolution problem.

Theorem 9.4 *Let $u_0 \in D(A) \cap \ker B$ and $f \in C([0,T], \ker B)$. Assume that either $f \in L^1([0,T], D(A) \cap \ker B)$ or $f \in W^{1,1}([0,T], \ker B)$. We also suppose that operator A is constraint-preserving in the sense of (9.7) and $\bar{A}^*$ satisfies (9.8). Then the constraint (9.6) is satisfied by the solution of (9.4)–(9.5).*

Proof: Consider the problem:

$$\dot{u}(t) = \bar{A}u(t) + f(t), \quad t \in [0,T], \tag{9.11}$$
$$u(0) = u_0. \tag{9.12}$$

From Ref. [23, proposition 4.1.6], there is a unique solution $u \in C^1([0,T], \ker B) \cap C([0,T], D(A) \cap \ker B)$ to the initial value problem (9.11) and (9.12). In particular, $u(t) \in D(A) \cap \ker B$, for all $t \in [0,T]$. Since $\bar{A}$ is a restriction of A, it follows that u is the solution of (9.4) and (9.5).

Now we introduce the extended evolution system which we shall study:

$$\begin{pmatrix} \dot{u} \\ \dot{z} \end{pmatrix} = \begin{pmatrix} A & -B^* \\ B & 0 \end{pmatrix} \begin{pmatrix} u \\ z \end{pmatrix} + \begin{pmatrix} f \\ 0 \end{pmatrix}, \quad t \in [0,T], \tag{9.13}$$

$$\begin{pmatrix} u \\ z \end{pmatrix} (0) = \begin{pmatrix} u_0 \\ 0 \end{pmatrix}. \tag{9.14}$$

Here $B^* : D(B^*) \subseteq X \to Y$ is the adjoint of B, which is a closed, densely defined linear operator.

The following theorem shows that under certain conditions the extended system is equivalent to the original one.

Theorem 9.5 *Let $u_0 \in \ker B \subseteq D(A)$ and $f \in C([0,T], \ker B)$. We also assume that operator A is constraint-preserving in the sense of (9.7). Then the pair $(u,z) : [0,T] \to$*

$Y \times X$ *is a solution of the extended problem* (9.13) *and* (9.14) *if and only if* $z \equiv 0$ *in* $[0,T]$ *and* u *satisfies the initial value problem* (9.4) *and* (9.5).

Proof: If $z \equiv 0$ and u satisfies (9.4) and (9.5), then $Bu \equiv 0$, and so (u,z) satisfies (9.13) and (9.14). Now suppose (u,z) satisfies (9.13) and (9.14). Since the linear operator B is closed, its kernel $\ker B$ is a closed linear subspace of Y. This implies $Y = \ker B \oplus (\ker B)^{\perp} = \ker B \oplus \overline{\operatorname{Im} B^*}$. According to this decomposition of Y, $u(t) = u_1(t) + u_2(t)$, where $u_1(t) \in \ker B$ and $u_2(t) \in \overline{\operatorname{Im} B^*}$. Since both $\ker B$ and $\overline{\operatorname{Im} B^*}$ are closed and the corresponding projections are continuous, we have $\dot{u}(t) = \dot{u}_1(t) + \dot{u}_2(t)$, with $\dot{u}_1(t) \in \ker B$ and $\dot{u}_2(t) \in \overline{\operatorname{Im} B^*}$. From (9.7) and (9.13), we obtain that u_1 is the solution of the Cauchy problem $\dot{u}_1(t) = Au_1(t) + f(t)$, $u_1(0) = u_0$, while (u_2, z) satisfies $\dot{u}_2(t) = Au_2(t) - B^*z(t)$, $\dot{z}(t) = Bu_2(t)$, $u_2(0) = 0$, $z(0) = 0$. Observe that

$$\frac{1}{2}\frac{d}{dt}(\|z\|_X^2) = \langle \dot{z}, z \rangle_X = \langle Bu_2, z \rangle_X = \langle u_2, B^*z \rangle_Y$$
$$= \langle u_2, Au_2 - \dot{u}_2 \rangle_Y$$
$$= -\langle u_2, \dot{u}_2 \rangle_Y \quad \text{(since } A \text{ is skew-adjoint)}$$
$$= -\frac{1}{2}\frac{d}{dt}(\|u_2\|_Y^2).$$

Therefore, $\|z(t)\|_X^2 + \|u_2(t)\|_Y^2 = \|z(0)\|_X^2 + \|u_2(0)\|_Y^2 = 0$, which implies $z \equiv 0$ and $u_2 \equiv 0$. Thus, $u = u_1$ and (9.4)–(9.6) are satisfied.

9.2.3   FOSH initial value problems with constraints

In this section, we will prove a result similar to Theorem 9.5 for the initial value problem

$$\partial_t u = Au + f, \tag{9.15}$$
$$Bu = 0, \tag{9.16}$$
$$u(x,0) = u_0, \tag{9.17}$$

where $A = \sum_{i=1}^{n} A_i \partial_i$, with $A_i \in \mathbb{R}^{m \times m}$ constant symmetric matrices, and $B = \sum_{i=1}^{n} B_i \partial_i$, with $B_i \in \mathbb{R}^{p \times m}$ constant matrices. Of course, we assume that (9.7) and the compatibility conditions $Bu_0 = 0$, $Bf(\cdot, t) = 0$, $\forall t \geq 0$, hold.

Theorem 9.6   *If u_0 and $f(\cdot, t)$, $\forall t \geq 0$, are smooth functions in $H^s(\mathbb{R}^n; \mathbb{R}^m)$, for $s > n/2 + m$, then the pair $(u,z)^T \in C^1(\mathbb{R}^n \times [0,\infty); \mathbb{R}^{m+p})$ is a solution of the associated unconstrained problem*

$$\frac{\partial}{\partial t}\begin{pmatrix} u \\ z \end{pmatrix} = \begin{pmatrix} A & -B^* \\ B & 0 \end{pmatrix}\begin{pmatrix} u \\ z \end{pmatrix} + \begin{pmatrix} f \\ 0 \end{pmatrix}, \tag{9.18}$$
$$u(x,0) = u_0, \ z(x,0) = 0, \tag{9.19}$$

if and only if $z \equiv 0$, and $u \in C^1(\mathbb{R}^n \times [0,\infty); \mathbb{R}^m)$ satisfies the constrained problem (9.15)–(9.17).

Proof: Obviously, if $z \equiv 0$ and u satisfies (9.15)–(9.17), then $(u,0)^T$ satisfies (9.18) and (9.19). Now, let us prove the converse. Denote by

$$\overline{A} = \begin{pmatrix} A & -B^* \\ B & 0 \end{pmatrix} = \sum_{i=1}^{n} \overline{A}_i \frac{\partial}{\partial x_i}.$$

From (9.3), we know that the solution of (9.18) and (9.19) has the following expression

$$
\begin{pmatrix} u \\ z \end{pmatrix}(x,t) = \frac{1}{(2\pi)^{n/2}} \int_{\mathbb{R}^n} e^{ix\cdot y} \left[e^{it\overline{A}(y)} \begin{pmatrix} \hat{u}_0 \\ 0 \end{pmatrix}(y) \right.
$$
$$
\left. + \int_0^t e^{i(t-s)\overline{A}(y)} \begin{pmatrix} \hat{f} \\ 0 \end{pmatrix}(y,s)ds \right] dy,
$$

(9.20)

where $\overline{A}(y) = \sum_{i=1}^n y_i \overline{A}_i$. The next step in the proof is to show that

$$\overline{A}^k(y) \begin{pmatrix} \hat{u}_0 \\ 0 \end{pmatrix} = \begin{pmatrix} A^k(y)\hat{u}_0 \\ 0 \end{pmatrix}$$

(9.21)

for all positive integer k. We are going to prove (9.21) by induction. For $k = 1$, we have

$$\overline{A}(y) \begin{pmatrix} \hat{u}_0 \\ 0 \end{pmatrix} = \begin{pmatrix} A(y)\hat{u}_0 \\ B(y)\hat{u}_0 \end{pmatrix}.$$

Because $Bu_0 = 0$, it follows that $B(y)\hat{u}_0 = 0$ by taking the Fourier transform. Hence

$$\overline{A}(y) \begin{pmatrix} \hat{u}_0 \\ 0 \end{pmatrix} = \begin{pmatrix} A(y)\hat{u}_0 \\ 0 \end{pmatrix}.$$

Assume that (9.21) is true for $k-1$ and let us prove it for k. Observe that

$$\overline{A}^k(y) \begin{pmatrix} \hat{u}_0 \\ 0 \end{pmatrix} = \overline{A}(y) \begin{pmatrix} A^{k-1}(y)\hat{u}_0 \\ 0 \end{pmatrix} = \begin{pmatrix} A^k(y)\hat{u}_0 \\ B(y)A^{k-1}(y)\hat{u}_0 \end{pmatrix}.$$

Since $u_0 \in \ker B$, from (9.7) we can see that

$$BA^{k-1}u_0 = 0.$$

(9.22)

Applying the Fourier transform to (9.22), we get

$$B(y)A^{k-1}(y)\hat{u}_0 = 0.$$

Thus,

$$\overline{A}^k(y)\begin{pmatrix}\hat{u}_0 \\ 0\end{pmatrix} = \begin{pmatrix}A^k(y)\hat{u}_0 \\ 0\end{pmatrix}$$

and the proof of (9.21) is complete.

From (9.21), observe that

$$e^{it\overline{A}(y)}\begin{pmatrix}\hat{u}_0 \\ 0\end{pmatrix}(y) = \sum_{k=0}^{\infty}\frac{(it)^k}{k!}\overline{A}^k(y)\begin{pmatrix}\hat{u}_0 \\ 0\end{pmatrix} = \begin{pmatrix}e^{itA(y)}\hat{u}_0 \\ 0\end{pmatrix}. \qquad (9.23)$$

Similarly one can show that

$$e^{i(t-s)\overline{A}(y)}\begin{pmatrix}\hat{f} \\ 0\end{pmatrix}(y) = \sum_{k=0}^{\infty}\frac{[i(t-s)]^k}{k!}\overline{A}^k(y)\begin{pmatrix}\hat{f} \\ 0\end{pmatrix} = \begin{pmatrix}e^{i(t-s)A(y)}\hat{f} \\ 0\end{pmatrix}. \qquad (9.24)$$

From (9.23) to (9.24), it follows that

$$\begin{pmatrix}u \\ z\end{pmatrix} = \begin{pmatrix}u \\ 0\end{pmatrix},$$

with

$$u(x,t) = \frac{1}{(2\pi)^{n/2}}\int_{\mathbb{R}^n} e^{ix\cdot y}[e^{itA(y)}\hat{u}_0(y) + \int_0^t e^{i(t-s)A(y)}\hat{f}(y,s)ds]\,dy$$

the unique C^1 solution of (9.15)–(9.17).   □

9.3 FOSH INITIAL-BOUNDARY VALUE PROBLEMS WITH CONSTRAINTS

Motivated in part by the necessity of computer solutions to constrained FOSH problems, we now turn to FOSH systems defined on finite domains. For dealing with finite domains, we need suitable boundary conditions, which we will take in the well known class of maximal nonnegative boundary conditions. To be more specific, consider the FOSH initial-boundary value problem on the bounded domain Ω

$$\partial_t u = Au + f(x,t) \text{ in } \Omega \times (0,T], \qquad (9.25)$$

$$u(x,0) = g(x) \text{ in } \Omega, \qquad (9.26)$$

$$u(x,t) \in N(x,t) \text{ for } (x,t) \in \partial\Omega \times [0,T], \qquad (9.27)$$

where $u(x,t) \in \mathbb{R}^m$ is the state vector function, $A := \sum_{i=1}^n A_i(x,t)\partial_i$ with $A_i(x,t)$ real symmetric $m \times m$ matrix functions, $f(x,t) \in \mathbb{R}^m$ is the forcing term, $g(x) \in \mathbb{R}^m$ is the initial data, and N is a Lipschitz continuous map from $\partial\Omega \times [0,T]$ to the subspaces of $\mathbb{R}^m$. The boundary conditions (9.27) will also be viewed as

$$E(x,t)u(x,t) = 0, \quad \text{on } \partial\Omega \times [0,T], \qquad (9.28)$$

where $E(x,t)$ is a matrix function whose null space (i.e., $\ker E(x,t)$) is $N(x,t)$, for all $(x,t) \in \partial\Omega \times [0,T]$.

Set $\mathbf{n} = (n_1, n_2, \ldots, n_n)^T$ be the outer unit normal to $\Gamma := \partial\Omega \times (0,T)$, and denote by $A_n(x,t)$ the *boundary matrix* $A_n(x,t) := -\sum_{i=1}^n n_i(x)A_i(x,t)$ in the direction of the unit normal $\mathbf{n}$. We assume that the signature (number of positive, zero, and negative eigenvalues) of A_n is constant on each component of $\Gamma := \partial\Omega \times (0,T)$.

Definition 9.3 *We say that N is nonnegative for A_n if the quadratic form associated to A_n is nonnegative when restricted to N, that is,*

$$\langle A_n(x,t)v, v \rangle \geq 0, \ \forall (x,t) \in \bar{\Gamma}, \ \forall v \in N(x,t). \qquad (9.29)$$

We say that N is maximal nonnegative if it is nonnegative and $\dim N(x,t)$ is equal to the number of nonnegative eigenvalues of $A_n(x,t)$ counting multiplicity.

Note that the maximality condition means that the boundary subspace $N(x,t)$ cannot be enlarged while preserving (9.29). If N is maximal nonnegative, we say that the boundary conditions (9.27) are maximal nonnegative boundary conditions for the FOSH system (9.4). It is well known that FOSH initial-boundary value problems with maximal nonnegative boundary conditions are well-posed (see Refs. [22, 34, 51, 52, 55, 59, 67, 69–71], and references therein).

Now consider the constraint

$$Bu := \sum_{i=1}^n B_i(x,t)\partial_i u = 0, \qquad (9.30)$$

where $B_i(x,t) \in \mathbb{R}^{p \times m}$, $1 \leq i \leq n$, are matrix functions.

Definition 9.4 *A well-posed set of boundary conditions (9.27) is called constraint-preserving if the solution to the initial-boundary value problem (9.25)–(9.27) satisfies the constraint (9.30) for all time, whenever the compatibility conditions $Bg = 0$ and $Bf(\cdot, t) = 0$ hold.*

9.3.1 FOSH initial-boundary value problems

In this section, we indicate necessary and sufficient conditions for having maximal nonnegative boundary conditions. These conditions can be used to obtain constraint preserving boundary conditions for constrained FOSH systems (as explained at the end of this section). In order to do this, we need a more specific characterization. Let $(x,t) \in \partial\Omega \times (0,T)$ arbitrary, but fixed for the moment. Since the boundary matrix $A_n(x,t)$ is symmetric, its eigenvalues are real and the corresponding eigenvectors can be chosen orthonormal. Suppose that the boundary matrix $A_n(x,t)$ has l_0 null eigenvalues $\lambda_1,\dots,\lambda_{l_0}$, l_- negative eigenvalues $\lambda_{l_0+1},\dots,\lambda_{l_0+l_-}$, and l_+ positive eigenvalues $\lambda_{l_0+l_-+1},\dots,\lambda_m$ (counted with their multiplicities). Let $\mathbf{e}_1,\dots,\mathbf{e}_{l_0}$; $\mathbf{e}_{l_0+1},\dots,\mathbf{e}_{l_0+l_-}$; $\mathbf{e}_{l_0+l_-+1},\dots,\mathbf{e}_m$ be the corresponding orthonormal (column) eigenvectors.

The following theorem gives a more practical form of maximal nonnegative boundary conditions; it was proven by Tarfulea [78] and can be viewed as an extension of Ref. [49, Theorem 8.2.2]. For convenience, we include the proof here.

Theorem 9.7 *The boundary conditions* (9.28) *are maximal nonnegative if and only if there exists a* $l_- \times l_+$ *matrix function* $M(x,t)$ *such that*

$$E(x,t) = \begin{pmatrix} \mathbf{e}_{l_0+1}^T(x,t) \\ \vdots \\ \mathbf{e}_{l_0+l_-}^T(x,t) \end{pmatrix} - M(x,t) \begin{pmatrix} \mathbf{e}_{l_0+l_-+1}^T(x,t) \\ \vdots \\ \mathbf{e}_m^T(x,t) \end{pmatrix}, \tag{9.31}$$

with

$$\left\| \begin{pmatrix} \sqrt{|\lambda_{l_0+1}|} & \cdots & 0 \\ \vdots & & \vdots \\ 0 & \cdots & \sqrt{|\lambda_{l_0+l_-}|} \end{pmatrix} M(x,t) \begin{pmatrix} 1/\sqrt{\lambda_{l_0+l_-+1}} & \cdots & 0 \\ \vdots & & \vdots \\ 0 & \cdots & 1/\sqrt{\lambda_m} \end{pmatrix} \right\|_2 \leq 1. \tag{9.32}$$

(Here $\|\cdot\|_2$ *is the operatorial* L^2 *norm.)*

Proof: First, assume that the boundary conditions (9.28) are maximal nonnegative. Any vector $u \in \mathbb{R}^m$ can be written as $u = \sum_{i=1}^m u_i \mathbf{e}_i$. Observe that $E(x,t)u = S_0 u_0 + S_- u_- - S_+ u_+$, where $u_0 = (u_1,\dots,u_{l_0})^T$, $u_- = (u_{l_0+1},\dots,u_{l_0+l_-})^T$, $u_+ = (u_{l_0+l_-+1},\dots,u_m)^T$, $S_0 = E(x,t)(\mathbf{e}_1,\dots,\mathbf{e}_{l_0})$, $S_- = E(x,t)(\mathbf{e}_{l_0+1},\dots,\mathbf{e}_{l_0+l_-})$, and $S_+ = -E(x,t)(\mathbf{e}_{l_0+l_-+1},\dots,\mathbf{e}_m)$. Because $N(x,t)$ is maximal nonnegative, it follows that $\ker A_n(x,t) \subseteq N(x,t)$, and so $S_0 = 0_{l_- \times l_0}$ (since $E(x,t)\mathbf{e}_i = 0$, $1 \leq i \leq l_0$).

If u belongs to the boundary subspace $N(x,t)$, condition (9.29) implies that the coefficients u_i, $1 \leq i \leq m$, satisfy

$$u^T A_n(x,t)u = \sum_{i=1}^m \lambda_i u_i^2 \geq 0,$$

or

$$\sum_{i=l_0+l_-+1}^{m} \lambda_i u_i^2 \geq - \sum_{i=l_0+1}^{l_0+l_-} \lambda_i u_i^2. \tag{9.33}$$

Moreover, since $N(x,t) = \ker E(x,t)$, we have

$$S_- u_- - S_+ u_+ = 0. \tag{9.34}$$

Let us show that the $l_- \times l_-$ matrix S_- is invertible. Arguing by contradiction, assume that there exists $\alpha_- = (\alpha_{l_0+1}, \dots, \alpha_{l_0+l_-})^T \neq 0$ so that $S_- \alpha_- = 0$. Then, $v = \sum_{i=l_0+1}^{l_0+l_-} \alpha_i e_i$ belongs to $N(x,t)$ (since $E(x,t)v = 0$). But this is in contradiction with (9.33). Hence, S_- is invertible and Equation (9.34) recasts into $u_- = S_-^{-1} S_+ u_+$, which is (9.28) with $E(x,t)$ of the form (9.31) for $M(x,t) = S_-^{-1} S_+$. From this and (9.33), one obtains for all $u_+ \in \mathbb{R}^{l_+}$

$$\left\| \begin{pmatrix} \sqrt{\lambda_{l_0+l_-+1}} & \cdots & 0 \\ & \vdots & \vdots \\ 0 & \cdots & \sqrt{\lambda_m} \end{pmatrix} u_+ \right\|_2$$

$$\geq \left\| \begin{pmatrix} \sqrt{|\lambda_{l_0+1}|} & \cdots & 0 \\ & \vdots & \vdots \\ 0 & \cdots & \sqrt{|\lambda_{l_0+l_-}|} \end{pmatrix} S_-^{-1} S_+ u_+ \right\|_2.$$

Finally, through the substitution

$$y = \begin{pmatrix} \sqrt{\lambda_{l_0+l_-+1}} & \cdots & 0 \\ & \vdots & \vdots \\ 0 & \cdots & \sqrt{\lambda_m} \end{pmatrix} u_+,$$

one obtains

$$\|y\|_2 \geq \left\| \begin{pmatrix} \sqrt{|\lambda_{l_0+1}|} & \cdots & 0 \\ & \vdots & \vdots \\ 0 & \cdots & \sqrt{|\lambda_{l_0+l_-}|} \end{pmatrix} M(x,t) \begin{pmatrix} 1/\sqrt{\lambda_{l_0+l_-+1}} & \cdots & 0 \\ & \vdots & \vdots \\ 0 & \cdots & 1/\sqrt{\lambda_m} \end{pmatrix} y \right\|_2,$$

for all $y \in \mathbb{R}^{l_+}$, which proves the inequality (9.32).

The fact that (9.31) and (9.32) imply that the boundary conditions (9.28) are maximal nonnegative can be proven by reversing the above arguments.

Based on Theorem 9.7, a systematic procedure for obtaining well-posed constraint preserving boundary conditions for constant-coefficient constrained FOSH systems can be developed. This procedure was employed in Refs. [7,76,77] to find boundary conditions for certain FOSH formulations of Einstein's equations in the linearized regime. Its main ideas and general guidelines were explained in Ref. [78]. Let us review the procedure here, as it plays an important role in the next sections of this work. Assume that the constraints satisfy a constant-coefficient FOSH system as well; this is true in many cases of interest. The key point of the technique is the matching of maximal nonnegative boundary conditions given in Theorem 9.7 for the main system and the system of constraints. In order to do this, the first step is to write down the general form of maximal nonnegative boundary conditions for the system of constraints. Then, replace the constraints according to their definitions in terms of the main variables, decompose all derivatives of the main variables into their transverse and tangential parts, and trade all transverse derivatives for tangential and temporal derivatives by using the homogeneous form of the main system. After grouping the terms, equate with zero the quantities under tangential and temporal differentiation. The result is a system of linear equations in the main variables whose coefficients depend on the parameters introduced by the general form of the boundary conditions for the system of constraints. Finally, choose these coefficients so that the resulting system represents a set of maximal nonnegative boundary conditions for the main system. The boundary conditions for the applications to be presented in Section 9.4 were found by employing the technique just described.

9.3.2 FOSH initial-boundary value problems with constraints

Throughout this section, we consider that the FOSH system (9.25) is with constant coefficients, that is, $A = \sum_{i=1}^{n} A_i \partial_i$, with $A_i \in \mathbb{R}^{m \times m}$ constant symmetric matrices. Remember that, if $\mathbf{n} \in \mathbb{R}^n$ denote the outer normal to Ω at $x \in \partial\Omega$, then A_n is the boundary matrix associated to the FOSH system (9.25), given by $A_n(x) = -\sum_{i=1}^{n} n_i(x) A_i$. Note that for sufficiently smooth $u, v : \Omega \to \mathbb{R}^m$,

$$\int_\Omega Au \cdot v\, dx = -\int_\Omega Av \cdot u\, dx - \int_{\partial\Omega} A_n u \cdot v\, ds, \quad \int_\Omega Au \cdot u\, dx = -\frac{1}{2}\int_{\partial\Omega} A_n u \cdot u\, ds. \quad (9.35)$$

We also assume that the constraint operator B has constant coefficients, that is, $B = \sum_{i=1}^{n} B_i \partial_i$, with $B_i \in \mathbb{R}^{p \times m}$ constant matrices. In what follows, we wish to relate constraint-preserving boundary conditions for the original system (9.25) to maximal nonnegative boundary conditions for the extended system (9.13) in which B^* denotes the *formal* adjoint of B, given by $B^* = -\sum_{i=1}^{n} B_i^T \partial_i$.

Analogous to (9.27), we consider boundary conditions for the extended system in the form

$$(u(x,t), z(x,t)) \in \bar{N}(x,t) \text{ for } (x,t) \in \partial\Omega \times [0,T], \tag{9.36}$$

where $\bar{N}$ is a smooth map from $\partial\Omega \times [0,T]$ to the subspaces of $\mathbb{R}^{m+p}$. Let $B_n = -\sum_{i=1}^{n} n_i B_i$, so

$$\int_{\Omega} Bu \cdot z\, dx = \int_{\Omega} B^* z \cdot u\, dx - \int_{\partial\Omega} B_n u \cdot z\, ds. \tag{9.37}$$

For $x \in \partial\Omega$, the boundary matrix associated to the FOSH system in (9.13) is given by

$$\bar{A}_n(x) = \begin{pmatrix} A_n(x) & B_n^T(x) \\ B_n(x) & 0 \end{pmatrix},$$

where, as before, $A_n = -\sum_{i=1}^{n} n_i A_i$. We begin by noting that a boundary subspace N is nonnegative for A_n if and only if $\bar{N} := N \times [B_n(N)]^{\perp}$ is nonnegative to the $\bar{A}_n$. This is a consequence of the following elementary lemma in which the notation is independent of the rest of the chapter.

Lemma 9.2   *Let $A \in \mathbb{R}^{m \times m}$, $B \in \mathbb{R}^{p \times m}$, and $N \in \mathbb{R}^m$ a subspace. Then N is non-negative for A if and only if the subspace $\bar{N} := N \times [B(N)]^{\perp}$ is nonnegative for $\bar{A} := \begin{pmatrix} A & B^T \\ B & 0 \end{pmatrix}$.*

Proof: If N is nonnegative for A and $(u,z) \in \bar{N}$, then $z^T Bu = 0$, and so

$$(u,z)^T \bar{A}(u,z) = u^T A u \geq 0.$$

On the other hand, if $(u,0)^T \bar{A}(u,0) \geq 0$, then $u^T A u \geq 0$, which gives the converse.   □

We are now ready to present our equivalence result.

Theorem 9.8   *Suppose that N is maximal nonnegative for A_n, the boundary conditions (9.27) are constraint-preserving for (9.25) and (9.26), and $\dim(N \times [B_n(N)]^{\perp})$ is equal to the number of nonnegative eigenvalues of $\bar{A}_n$. Then the original initial-boundary value problem (9.25)–(9.27) with constraint (9.30) and the corresponding extended initial-boundary value problem (9.13), (9.14), and (9.36), with $\bar{N} := N \times [B_n(N)]^{\perp}$, are equivalent.*

Proof: Since N is nonnegative for A_n, by Lemma 9.2, $\bar{N}$ is nonnegative for $\bar{A}_n$, and so the hypotheses say that both N and $\bar{N}$ are maximal nonnegative, for A_n and $\bar{A}_n$, respectively. Let v be a solution of (9.25)–(9.27) and (u, z) be a solution of (9.13) and (9.14), and (9.36). A formal calculation shows that

$$
\begin{aligned}
\frac{1}{2}\frac{d}{dt}(\|u - v\|_{L^2(\Omega)}^2 + \|z\|_{L^2(\Omega)}^2) &= \langle \dot{u} - \dot{v}, u - v\rangle_{L^2(\Omega)} + \langle \dot{z}, z\rangle_{L^2(\Omega)} \\
&= \langle A(u - v) - B^* z, u - v\rangle_{L^2(\Omega)} + \langle Bu, z\rangle_{L^2(\Omega)} \\
&= \langle A(u - v), u - v\rangle_{L^2(\Omega)} - \langle B^* z, u\rangle_{L^2(\Omega)} \\
&\quad + \langle B^* z, v\rangle_{L^2(\Omega)} + \langle Bu, z\rangle_{L^2(\Omega)} \\
&= -\frac{1}{2}\int_{\partial\Omega} (u - v)^T A_n(u - v)\, ds + \langle B^* z, v\rangle_{L^2(\Omega)}.
\end{aligned}
$$

where the last identity follows from the last equation in (9.35) combined with (9.37). Since the boundary conditions (9.27) are constraint preserving, we have $Bv = 0$, and so, by (9.37),

$$
\langle B^* z, v\rangle_{L^2(\Omega)} = \langle z, Bv\rangle_{L^2(\Omega)} + \int_{\partial\Omega} B_n v \cdot z\, ds = 0.
$$

Therefore,

$$
\frac{1}{2}\frac{d}{dt}(\|u - v\|_{L^2(\Omega)}^2 + \|z\|_{L^2(\Omega)}^2) = -\frac{1}{2}\int_{\partial\Omega} (u - v)^T A_n(u - v)\, ds \leq 0,
$$

which implies $z \equiv 0$ and $u \equiv v$ in $\Omega \times [0, T]$. Hence, the constrained initial-boundary value problem (9.25)–(9.27) and (9.30) and the initial-boundary value problem (9.13), (9.14), and (9.36) (with $\bar{N} = N \times [B_n(N)]^\perp$) are equivalent. $\qquad\square$

Thus, if we choose boundary conditions (9.27) which satisfy the hypotheses of the theorem, we have two equivalent systems to choose from, the original FOSH system with these boundary conditions, or the extended FOSH system with the associated boundary conditions. As discussed earlier, we believe that the latter system may be more appropriate for maintaining constraints after numerical discretization.

9.4 APPLICATIONS

In this section, we discuss a few applications of the results presented in the previous sections. We start with a simpler example involving a constrained FOSH system originating from a system of wave equations with constraints. Then we analyze two more complex applications to Einstein's equations in the Einstein–Christoffel [5] and Alekseenko–Arnold [3] formulations, respectively.

Throughout this section, we assume that Ω is a bounded polyhedral domain in $\mathbb{R}^3$. For an arbitrary face of $\partial\Omega$, let n^i denote its exterior unit normal. Let m^i and l^i be two additional vectors which together with n^i form an orthonormal basis. The projection operator orthogonal to n^i is then given by $\tau^{ij} := m^i m^j + l^i l^j$ (and does not depend on the particular choice of these tangential vectors). Hereafter, repeated indices one in subscript and the other in superscript means summation.

9.4.1 System of wave equations with constraints

This first example of application is for a constrained system of wave equations factored in first-order form. Despite being relatively simple, this model problem involves similar ideas and techniques that could be used for more difficult problems (such as the ones presented next).

Consider the following FOSH system: given $(f,g) : \overline{\Omega} \times [0,T] \to \mathbb{R}^3 \times \mathbb{R}^{3\times3}$, we seek $(v,u) : \overline{\Omega} \times [0,T] \to \mathbb{R}^3 \times \mathbb{R}^{3\times3}$ such that

$$\dot{v} = \operatorname{div} u + f, \quad \dot{u} = \operatorname{grad} v + g, \quad \text{in } \Omega \times (0,T]. \tag{9.38}$$

Here

$$(\operatorname{div} u)_i = \sum_{j=1}^{3} \frac{\partial u_{ij}}{\partial x_j}, \quad (\operatorname{grad} v)_{ij} = \frac{\partial v_i}{\partial x_j}.$$

We impose a four-dimensional system of constraints, namely

$$C := \operatorname{div} v = 0, \ D := \operatorname{div} u^T = 0, \ \text{i.e.,} \ C := \sum_{i=1}^{3} \frac{\partial v_i}{\partial x_i} = 0, \ D_j := \sum_{i=1}^{3} \frac{\partial u_{ij}}{\partial x_i} = 0. \tag{9.39}$$

Furthermore, we assume that the forcing terms f and g are compatible with the constraints (9.39), that is, $\operatorname{div} f = 0$ and $\operatorname{div} g^T = 0$.

To motivate the system (9.38) subject to the constraints (9.39), consider a vector field $U : \Omega \times (0,T] \to \mathbb{R}^3$, each component of which solves a wave equation

$$\ddot{U}_i = \Delta U_i + F_i, \quad \text{in } \Omega \times (0,T], \tag{9.40}$$

for some forcing function $F : \Omega \times (0,T] \to \mathbb{R}^3$. Choose some convenient matrix fields $G : \Omega \times (0,T] \to \mathbb{R}^{3\times3}$ (e.g., $G \equiv 0$) and set $v = \dot{U}$, $u = \operatorname{grad} U + G$. Then (v,u) solves (9.38) with $f = F - \operatorname{div} G$, $g = \dot{G}$. If the solution U to the wave equations (9.40) satisfy the constraint $\operatorname{div} U = 0$ and G is chosen so that $\operatorname{div} G^T = 0$, then (v,u) satisfies (9.39). In short, (9.38) is a first-order formulation of a vector-valued wave equation and (9.39) is the constraint that the solution be divergence-free.

As initial data for the FOSH system (9.38), we consider

$$v(x,0) = v_0(x), \quad u(x,0) = u_0(x), \quad \text{in } \Omega, \tag{9.41}$$

which is assumed compatible with the constraints (9.39), that is, $\operatorname{div} v_0 = 0$ and $\operatorname{div} u_0^T = 0$.

Let us verify that in the case of this model problem, the operator A is constraint-preserving. If (v, u) belongs to $\ker B$, that is, satisfies (9.39), we wish to show that $(\operatorname{div} u, \operatorname{grad} v)$ also belongs to $\ker B$. This follows easily by differentiating (9.39): $\operatorname{div} \operatorname{div} u = \operatorname{div} \operatorname{div} u^T = 0$ and $\operatorname{div}(\operatorname{grad} v)^T = \operatorname{grad} \operatorname{div} v = 0$. Thus we can apply Theorem 9.4 to conclude that if the initial data satisfies the constraints and the forcing terms satisfy the compatibility conditions, then the solution (v, u) of (9.38) satisfies (9.39) for all time. Of course, the application of Theorem 9.4 requires the verification of the other hypotheses, which needs rigorous definitions of the differential operators involved. Alternately, one can consider the energy functional associated to the FOSH system satisfied by the constraints (see (9.42)) and prove that it is zero for all time if it is zero initially; hence the evolution preserves the constraints for the pure Cauchy problem.

We now show that the constraints (9.39) are satisfied for all time, if the boundary conditions for the system (9.38) are chosen appropriately. Our next result, Theorem 9.9, introduces a set of well-posed constraint-preserving boundary conditions for the above model problem. These boundary conditions were obtained by using Theorem 9.7 and following step-by-step the technique described immediately below its statement. Let us briefly reexplain the procedure in this concrete case. First of all, observe that the constraints C and D satisfy a FOSH system. Indeed, if one applies the constraint operator C to the first evolution equation in (9.38) and applies the constraint operator D to the second evolution equation in (9.38), one obtains the FOSH system

$$\dot{C} = \operatorname{div} D, \quad \dot{D} = \operatorname{grad} C. \tag{9.42}$$

The next step is to match the general forms of maximal nonnegative boundary conditions for the main system (9.38) and the system of constraints (9.42). For this, one first writes down the general form of maximal nonnegative boundary conditions for the system of constraints (9.42), which is a linear system in C and D_j on the boundary $\partial\Omega$ and involves coefficients depending on certain parameters (see Theorem 9.7). Then, one replaces C and D_j in this system according to their definitions (9.39), decomposes all derivatives of the main variables v_i and u_{ij} in their transverse and tangential parts, and trades all transverse derivatives for tangential and temporal derivatives by using the homogeneous form of the main system (9.38). After grouping the terms, one equates with zero the quantities under tangential and temporal differentiation. The result is a system of linear equations in the main variables v_i and u_{ij} whose coefficients depend on the parameters introduced by the general form of the boundary conditions for the system of constraints (9.42). Finally, one chooses these coefficients so that the resulting system represents a set of maximal nonnegative boundary conditions for the main system (9.38). The outcome is stated in the following theorem, whose proof can be found in Ref. [78]. Recall that, for an arbitrary face of $\partial\Omega$, n^i denotes its exterior unit normal and τ^{ij} is the corresponding projection operator orthogonal to n^i. Again, repeated subscript–superscript indices means summation.

Theorem 9.9 *([78, theorem 3.1]) The boundary conditions*

$$n^i n^j u_{ij} = 0, \quad \tau^{ij} v_j = 0, \tag{9.43}$$

are maximal nonnegative for (9.38) *and preserve the constraints* (9.39).

The extended system corresponding to (9.38) is

$$\dot{v} = \operatorname{div} u + \operatorname{grad} p + f, \quad \dot{u} = \operatorname{grad} v + (\operatorname{grad} q)^T + g, \quad \dot{p} = \operatorname{div} v, \quad \dot{q} = \operatorname{div} u^T. \tag{9.44}$$

Here $(p, q) : \mathbb{R}^3 \times [0, T] \to \mathbb{R} \times \mathbb{R}^3$ is a pair of new dynamic variables (corresponding to z earlier in the chapter). Theorem 9.5 tells us that if we initialize them to zero and solve (9.44), then they remain zero for all time and (v, u) satisfy (9.38). Furthermore, since the set of boundary conditions (9.43) is maximal nonnegative and constraint preserving for the original system (9.38), we can follow the conclusion of Theorem 9.8 in order to prescribe maximal nonnegative boundary conditions for which the original and extended problems are equivalent. More precisely, we can prove the following result.

Theorem 9.10 *The boundary conditions*

$$n^i n^j u_{ij} = 0, \quad \tau^{ij} v_j = 0, \quad p = 0, \quad \tau^{ij} q_j = 0. \tag{9.45}$$

are maximal nonnegative for the system (9.44), *and so there exists a unique solution* (v, u, p, q) *to the corresponding initial-boundary value problem. Moreover,* $p \equiv 0$, $q \equiv 0$, *and* (v, u) *is the unique solution to* (9.38), (9.41), (9.43), *which satisfies the constraints* (9.39) *for all time.*

Proof: Let N denote the subspace of $\mathbb{R}^3 \times \mathbb{R}^{3 \times 3}$ given by the boundary conditions (9.43). Observe that the system of boundary conditions (9.45) is nothing but $N \times [B_n(N)]^\perp$. Thus, it suffices, by Theorem 9.8, to prove that $\dim N + \dim[B_n(N)]^\perp$ is equal to the number of nonnegative eigenvalues of $\bar{A}_n$, the boundary matrix for the extended system (9.13) which, for this model problem, is given by (9.44). The operator B_n acts on $\mathbb{R}^3 \times \mathbb{R}^{3 \times 3}$ by $(v, u) \mapsto (v^T n, u^T n)$, so

$$B_n(N) = \{ (a, w) \in \mathbb{R} \times \mathbb{R}^3 \mid w^T n = 0 \}, \quad \dim B_n(N) = 3, \quad \dim[B_n(N)]^\perp = 1.$$

and $\dim N + \dim[B_n(N)]^\perp = 10$.

Now we examine the eigenvalues of the extended boundary matrix $\bar{A}_n$. This can be viewed as the operator from $\mathbb{R}^3 \times \mathbb{R}^{3 \times 3} \times \mathbb{R} \times \mathbb{R}^3$ to itself given by

$$(v, u, p, q) \mapsto (-un - pn, -vn^T - nq^T, -v^T n, -u^T n).$$

Its null space has dimension four and is given by

$$\{(0,u,0,0) \in \mathbb{R}^3 \times \mathbb{R}^{3\times3} \times \mathbb{R} \times \mathbb{R}^3 \,|\, un = u^T n = 0\}.$$

Furthermore, if λ is an eigenvalue with corresponding eigenvector $(v,u,p,q) \in \mathbb{R}^3 \times \mathbb{R}^{3\times3} \times \mathbb{R} \times \mathbb{R}^3$, then $-\lambda$ is also an eigenvalue with eigenvector $(v,-u,-p,q) \in \mathbb{R}^3 \times \mathbb{R}^{3\times3} \times \mathbb{R} \times \mathbb{R}^3$. Therefore, this operator has six positive, six negative, and four null eigenvalues (counted with their multiplicities). Thus there are 10 nonnegative eigenvalues, as claimed.

9.4.2 Applications to Einstein's equations

In general relativity, spacetime is a four-dimensional manifold of events endowed with a pseudo-Riemannian metric $g_{\alpha\beta}$. Einstein's equations $G_{\alpha\beta} = 8\pi T_{\alpha\beta}$ connect the spacetime curvature represented by the Einstein tensor $G_{\alpha\beta}$ with the stress-energy tensor $T_{\alpha\beta}$. In essence, according to the ultra-condensed definition of general relativity due to the famous American physicist John Wheeler, Einstein's equations say that matter and energy tell spacetime how to curve, while the curvature of spacetime dictates how matter and energy flow through it. The simple tensorial form is deceiving; Einstein's equations in local coordinates represent one of the most difficult and richest systems of partial differential equations describing a viable physical theory. In fact, they are equations for geometries, that is, their solutions are equivalent classes under spacetime diffeomorphisms of metric tensors. To break this diffeomorphisms invariance, Einstein's equations must be first transformed into a system having a well-posed Cauchy problem. That is, the spacetime is foliated and each slice Σ_t is characterized by its intrinsic geometry γ_{ij} and extrinsic curvature K_{ij}, which is essentially the "velocity" of γ_{ij} in the unit normal direction to the slice. Subsequent slices are connected via the lapse function N and shift vector β^i corresponding to the Arnowitt–Deser–Misner (ADM) decomposition [9] of the line element $ds^2 = -N^2 dt^2 + \gamma_{ij}(dx^i + \beta^i dt)(dx^j + \beta^j dt)$. This decomposition allows one to express six of the 10 components of Einstein's equations in vacuum, that is, for $T_{\alpha\beta} = 0$, as a constrained system of evolution equations for the metric γ_{ij} and the extrinsic curvature K_{ij}:

$$\dot{\gamma}_{ij} = -2NK_{ij} + 2\nabla_{(i}\beta_{j)},$$
$$\dot{K}_{ij} = N[R_{ij} + (K_l^l)K_{ij} - 2K_{il}K_j^l] + \beta^l \nabla_l K_{ij} + K_{il}\nabla_j\beta^l + K_{lj}\nabla_i\beta^l - \nabla_i\nabla_j N,$$
$$R_i^i + (K_i^i)^2 - K_{ij}K^{ij} = 0, \qquad (9.46)$$
$$\nabla^j K_{ij} - \nabla_i K_j^j = 0.$$

where we use a dot to denote time differentiation and ∇_j for the covariant derivative associated to γ_{ij}. The spatial Ricci tensor R_{ij} has components given by second-order spatial differential operators applied to the spatial metric components γ_{ij}. Indices are raised and traces taken with respect to the spatial metric γ_{ij}, and parenthesized indices are used to denote the symmetric part of a tensor (e.g., $\nabla_{(i}\beta_{j)} = (\nabla_i\beta_j + \nabla_j\beta_i)/2$,

$\lambda_{k(ij)} := (\lambda_{kij} + \lambda_{kji})/2$, etc.). We will also use bracketed indices to denote the antisymmetric part of a tensor such as $\partial_{[k}\gamma_{i]j} := (\partial_k\gamma_{ij} - \partial_i\gamma_{kj})/2$, $(M\gamma)_{[k}\delta_{i]j} := [(M\gamma)_k\delta_{ij} - (M\gamma)_i\delta_{kj}]/2$, $\partial_{[k}\beta_{i]} := (\partial_k\beta_i - \partial_i\beta_k)/2$, and so on.

The evolution equations of the ADM system (9.4.2) form a subsystem that is only weakly hyperbolic. However, there are numerous strongly or symmetric hyperbolic formulations of Einstein's equations derived from the ADM system (9.4.2) typically by adding the constraints in specific combinations to the evolution equations and introducing new unknowns involving the first partial derivatives of the spatial metric components (see Refs. [1,3,5,10,12,15,17,26,27,31,33,37,39,47,60,72,79], among many others).

A trivial solution to the ADM system (9.4.2) is the Minkowski spacetime in Cartesian coordinates, given by $\gamma_{ij} = \delta_{ij}$, $K_{ij} = 0$, $\beta^i = 0$, $N = 1$. Both the linearized Einstein–Christoffel (EC) [5] and Alekseenko–Arnold (AA) [3] formulations can be derived from the ADM system (9.4.2) linearized about this solution. To derive the linearized ADM formulation, we write $\gamma_{ij} = \delta_{ij} + \bar{\gamma}_{ij}$, $K_{ij} = \bar{\kappa}_{ij}$, $\beta^i = \bar{\beta}^i$, $N = 1 + \bar{\alpha}$, where the bars indicate perturbations, assumed to be small. If we substitute these expressions into (9.4.2), and ignore terms which are at least quadratic in the perturbations and their derivatives, then we obtain a linear system for the perturbations. Dropping the bars, the system is

$$\dot{\gamma}_{ij} = -2\kappa_{ij} + 2\partial_{(i}\beta_{j)}, \tag{9.47}$$

$$\dot{\kappa}_{ij} = \partial^l\partial_{(i}\gamma_{j)l} - \frac{1}{2}\partial^l\partial_l\gamma_{ij} - \frac{1}{2}\partial_i\partial_j\gamma^l_l - \partial_i\partial_j\alpha, \tag{9.48}$$

$$C := \partial^i(\partial^l\gamma_{lj} - \partial_j\gamma^l_l) = 0, \tag{9.49}$$

$$C_j := \partial^l\kappa_{lj} - \partial_j\kappa^l_l = 0, \tag{9.50}$$

where we use a dot to denote time differentiation. The usual approach to solving the system (9.47)–(9.50) is to begin with initial data $\gamma_{ij}(0)$ and $\kappa_{ij}(0)$ defined on $\mathbb{R}^3$ and satisfying the constraint equations (9.49), (9.50), and to define γ_{ij} and κ_{ij} for $t > 0$ via the Cauchy problem for the evolution equations (9.47), (9.48). It can be easily shown that the constraints are then satisfied for all times. Indeed, if we apply the Hamiltonian constraint operator defined in (9.49) to the evolution Equation (9.47) and apply the momentum constraint operator defined in (9.50) to the evolution equation (9.48), we obtain the first-order symmetric hyperbolic system

$$\dot{C} = -2\partial^j C_j, \quad \dot{C}_j = -\frac{1}{2}\partial_j C.$$

Thus if C and C_j vanish at $t = 0$, they vanish for all time.

Here, our main interest is in providing viable boundary conditions at the artificial boundary for the linearized EC and AA formulations that can be used for numerical simulations. This is motivated by the fact that most numerical schemes for Einstein's equations try to approximate solutions on a generically infinite domain by computations on a truncated finite domain. The question that arises is what boundary conditions to provide at the artificial boundary corresponding to the truncated finite domain. In general, most numerical approaches have been made using carefully chosen initial data that satisfies the constraints. On the other hand, finding appropriate boundary conditions that lead to well-posedness and consistent with constraints is a difficult problem and has been subject to intense investigations in the recent years. Stewart [73] has addressed this subject within Frittelli–Reula formulation [39] linearized around flat space with unit lapse and zero shift in the quarter plane. Both main system and constraints evolve in time as first-order strongly hyperbolic systems. Stewart deduces boundary conditions for the main system in terms of Fourier–Laplace transforms that preserve the constraints by imposing the incoming modes for the system of constraints to vanish and translating these conditions in terms of Laplace–Fourier transforms of the main system variables. One of the more recent works where these ideas are further pursued and refined is [66], where Sarback and Tiglio use the Fourier–Laplace transformation techniques to find necessary conditions for well-posedness of the initial-boundary value problem corresponding to the formulation presented in Ref. [65], which is a generalization of the EC. type formulations [5, 39, 43, 47] with a Bona–Masso-type gauge condition [16, 17] for the lapse. In 1999, a well-posed initial-boundary value formulation was given by Friedrich and Nagy [32] in terms of a tetrad-based Einstein–Bianchi formulation. In essence, they obtained a symmetric hyperbolic system by adding appropriate combinations of the constraints to the evolution equations so that the constraints propagate tangentially to the boundary and so the problem of preserving the constraints is solved automatically. In view of our work which is to be presented here, of particular interest are the more recent investigations regarding special boundary conditions that prevent the influx of constraint-violating modes into the computational domain for various hyperbolic formulations of Einstein's equations (see Refs. [2, 4, 11, 18, 20, 21, 35, 36, 44, 48, 50, 54, 56, 63, 66, 76, 77] among others). Most of these approaches start with the imposition of maximal nonnegative boundary conditions to the constraint propagation system (i.e., the evolution system for the constraint variables) if it can be cast into a closed first-order symmetric hyperbolic system. These boundary conditions usually translate into differential equations for the main variables on the boundary (because the constraint variables depend on derivatives of the main variables). One of the most difficult problems that arises is to prove the well-posedness of the corresponding initial-boundary value problem. For symmetric hyperbolic systems with maximal nonnegative boundary conditions there are well-known results on well-posedness. However, the differential conditions obtained for the main variables on the boundary are not of this type and the well-posedness of the corresponding initial-boundary value problem can be very difficult to prove.

Of course, specifying constraint-preserving boundary conditions for a certain formulation of Einstein's equations does not solve entirely the complicated problem of numerical relativity. There are other aspects that have to be addressed in order to obtain good numerical simulations, as for example, the existence of bulk constraint violations, in which existing violations are amplified by the evolution equations (see Refs. [24, 25, 53, 68] and references therein). See the introductory section in Ref. [48] for more information and a review of some work done in this direction. Before we end this very brief review, it is important also to mention the work done on boundary conditions for Einstein's equations in harmonic coordinates, when Einstein's equations become a system of second-order hyperbolic equations for the metric components. The question of the constraints preservation does not appear here, as it is hidden in the gauge choice, that is, the constraints have to be satisfied only at the initial surface, the harmonic gauge guarantees their preservation in time (see Refs. [57, 58, 62, 74, 75], and references therein).

9.4.2.1 Einstein–Christoffel formulation

In this section, we address the boundary conditions problem for the classical Einstein–Christoffel or EC equations derived in Ref. [5], linearized with respect to the flat Minkowski spacetime, and with arbitrary lapse density and shift perturbations. This problem has been addressed before in Ref. [20] in the case of spherically symmetric black-hole spacetimes in vacuum or with a minimally coupled scalar field, within the EC formulation of Einstein's equations. Here Stewart's idea of imposing the vanishing of the ingoing constraint modes as boundary conditions is employed once again. Then, the radial derivative is eliminated in favor of the time derivative in the expression of the ingoing constraints by using the main evolution system. The emerging set of boundary conditions depends only on the main variables and their time derivative and preserves the constraints. In Ref. [21], this technique is refined and employed for the generalized EC formulation [47] when linearized around Minkowski spacetime with vanishing lapse and shift perturbations on a cubic box. Again, the procedure consists in choosing well-posed boundary conditions to the evolution system for the constraint variables and translating them into well-posed boundary conditions for the variables of the main evolution system. The scheme proposed in Ref. [21] ends up giving two sets, called "Dirichlet and Neumann-like," of constraint-preserving boundary conditions. However, the energy method used in Ref. [21] works only for symmetric hyperbolic constraint propagation, which forces the parameter η of the generalized EC system to satisfy the condition $0 < \eta < 2$. The analysis in Ref. [21] does not cover the case $\eta = 4$ required for the standard EC formulation introduced in Ref. [5]. Arnold and Tarfulea [7] presented their results on the boundary conditions problem for the standard EC formulation ($\eta = 4$) linearized around the Minkowski spacetime with arbitrary lapse density and shift perturbations. Much of the material of this section is from Ref. [7].

The linearized EC formulation provides an alternate approach to obtaining a solution of (9.47)–(9.50) with the given initial data, based on solving a system with better hyperbolicity properties. If γ_{ij}, κ_{ij} solve (9.47)–(9.50), define

$$f_{kij} = \frac{1}{2}[\partial_k \gamma_{ij} - (\partial^l \gamma_{li} - \partial_i \gamma_l^l)\delta_{jk} - (\partial^l \gamma_{lj} - \partial_j \gamma_l^l)\delta_{ik}]. \tag{9.51}$$

Then $-\partial^k f_{kij}$ coincides with the first three terms of the right-hand side of (9.48), so

$$\dot{\kappa}_{ij} = -\partial^k f_{kij} - \partial_i \partial_j \alpha. \tag{9.52}$$

Differentiating (9.63) in time, substituting (9.47), and using the constraint equation (9.50), we obtain

$$\dot{f}_{kij} = -\partial_k \kappa_{ij} + L_{kij}, \tag{9.53}$$

where

$$L_{kij} = \partial_k \partial_{(i}\beta_{j)} - \partial^l \partial_{[l}\beta_{i]}\delta_{jk} - \partial^l \partial_{[l}\beta_{j]}\delta_{ik} \tag{9.54}$$

The evolution equations (9.52) and (9.53) for κ_{ij} and f_{kij}, together with the evolution equation (9.47) for γ_{ij}, form the linearized EC system. As initial data for this system, we use the given initial values of γ_{ij} and κ_{ij} and derive the initial values for f_{kij} from those of γ_{ij} based on (9.51):

$$f_{kij}(0) = \frac{1}{2}\{\partial_k \gamma_{ij}(0) - [\partial^l \gamma_{li}(0) - \partial_i \gamma_l^l(0)]\delta_{jk} - [\partial^l \gamma_{lj}(0) - \partial_j \gamma_l^l(0)]\delta_{ik}\}. \tag{9.55}$$

Here we study the preservation of constraints by the linearized EC system and the closely related question of the equivalence of that system and the linearized ADM system. Our main interest is in the case when the spatial domain is bounded and appropriate boundary conditions are imposed, but first we consider the result for the pure Cauchy problem in the remainder of this section.

Suppose that κ_{ij} and f_{kij} satisfy the evolution equations (9.52) and (9.53) (which decouple from (9.47)). If κ_{ij} satisfies the momentum constraint (9.50) for all time, then from (9.52) we obtain a constraint which must be satisfied by f_{kij}:

$$\partial^k(\partial^l f_{klj} - \partial_j f_{kl}{}^l) = 0. \tag{9.56}$$

Note that (9.51) is another constraint that must be satisfied for all time. The following theorem shows that the constraints (9.50), (9.51), and (9.56) are preserved by the linearized EC evolution.

Theorem 9.11   *([7, theorem 1]) Let initial data $\gamma_{ij}(0)$ and $\kappa_{ij}(0)$ be given satisfying the constraints (9.49) and (9.50), respectively, and $f_{kij}(0)$ be defined by (9.55). Then the unique solution of the evolution equations (9.47), (9.52), and (9.53) satisfies (9.50), (9.51), and (9.56) for all time.*

In view of this theorem, it is relatively easy to establish the key result that for given initial data satisfying the constraints, the unique solution of the linearized EC evolution equations satisfies the linearized ADM system, and so the linearized ADM system and the linearized EC system are equivalent.

Theorem 9.12   *([7, theorem 2]) Suppose that initial data $\gamma_{ij}(0)$ and $\kappa_{ij}(0)$ are given satisfying the Hamiltonian constraint (9.49) and momentum constraint (9.50), respectively, and that initial data $f_{kij}(0)$ is defined by (9.55). Then the unique solution* *of the linearized EC evolution equations (9.47), (9.52), (9.53) satisfies the linearized ADM system (9.47)–(9.50).*

Next, we provide maximal nonnegative boundary conditions for the linearized EC system which are constraint-preserving in the sense that the analogue of Theorem 9.11 is true for the initial-boundary value problem. This will then imply the analogue of Theorem 9.12.

Consider $\Omega \subset \mathbb{R}^3$ a bounded polyhedral domain. For an arbitrary face of $\partial\Omega$, let n^i denote its exterior unit normal. Denote by m^i and l^i two additional vectors which together with n^i form an orthonormal basis. The projection operator orthogonal to n^i is then given by $\tau_i^j := m_i m^j + l_i l^j$ (and does not depend on the particular choice of these tangential vectors).

By using the technique explained in the first part of this chapter, one can obtain two sets of maximal nonnegative boundary conditions for the hyperbolic system (9.47), (9.64), and (9.65), and so, according to the classical theory of Friedrichs [34] and Phillips [52] (also Refs. [42, 49, 55, 59, 69, 70] among others), the initial-boundary value problem is well-posed:

$$ n^i \tau^{jk} \kappa_{ij} = 0, \quad n^k n^i n^j f_{kij} = 0, \quad n^k \tau^{il} \tau^{jm} f_{kij} = 0, \tag{9.57} $$

and

$$ n^i n^j \kappa_{ij} = 0, \quad \tau^{il} \tau^{jm} \kappa_{ij} = 0, \quad n^k n^i \tau^{jl} f_{kij} = 0. \tag{9.58} $$

Observe that these two sets of boundary conditions do not depend on the choice of basis for the tangent space.

The next theorem asserts that the above sets of boundary conditions are constraint-preserving.

Theorem 9.13 *([7, theorem 4]) Let Ω be a polyhedral domain. Given $\gamma_{ij}(0)$ and $\kappa_{ij}(0)$ on Ω satisfying the constraints (9.49) and (9.50), respectively, and $f_{kij}(0)$ defined by (9.55), define γ_{ij}, κ_{ij}, and f_{kij} for positive time by the evolution equations (9.47), (9.52), and (9.53) and the boundary conditions (9.57) (or (9.58)). Then the constraints (9.50), (9.51), and (9.56) are satisfied for all time.*

The analogue of Theorem 9.12 for the initial-boundary value problem follows from the preceding theorem exactly as before.

Theorem 9.14 *([7, theorem 5]) Let Ω be a polyhedral domain. Suppose that initial data $\gamma_{ij}(0)$ and $\kappa_{ij}(0)$ are given satisfying the Hamiltonian constraint (9.49) and momentum constraint (9.50), respectively, and that initial data $f_{kij}(0)$ is defined by (9.55). Then the unique solution of the linearized EC initial-boundary value problem (9.47), (9.52), (9.53), together with the boundary conditions (9.57) (or (9.58)) satisfies the linearized ADM system (9.47)–(9.50) in Ω.*

Now we indicate an extended initial boundary value problem whose solution solves the linearized ADM system (9.47)–(9.50) in Ω. This approach could present advantages from the numerical point of view since the momentum constraint is "built-in," and so controlled for all time. The new system consists of (9.47), (9.53), and two new sets of equations corresponding to (9.52)

$$\dot{\kappa}_{ij} = -\partial^k f_{kij} + \frac{1}{2}(\partial_i p_j + \partial_j p_i) - \partial^k p_k \delta_{ij} - \partial_i \partial_j \alpha, \tag{9.59}$$

and to a new three-dimensional vector field p_i defined by

$$\dot{p}_i = \partial^l \kappa_{li} - \partial_i \kappa_l^l. \tag{9.60}$$

Observe that the additional terms that appear on the right-hand side of (9.59) compared with (9.52) are nothing but the negative components of the formal adjoint of the momentum constraint operator applied to p_i.

It was proven in Ref. [7] that the following two sets of boundary conditions are maximal nonnegative for (9.47), (9.59), (9.53), and (9.60):

$$n^i \tau^{jk} \kappa_{ij} = 0, \quad n^k n^i n^j f_{kij} = 0, \quad n^k (\tau^{il} \tau^{jm} f_{kij} + \tau^{lm} p_k) = 0, \tag{9.61}$$

and

$$n^i n^j \kappa_{ij} = 0, \quad \tau^{il} \tau^{jm} \kappa_{ij} = 0, \quad n^k n^i \tau^{jl} f_{kij} - \tau^{kl} p_k = 0. \tag{9.62}$$

Theorem 9.15 *([7, theorem 6]) Let Ω be a polyhedral domain. Suppose that the initial data $\gamma_{ij}(0)$ and $\kappa_{ij}(0)$ are given satisfying the Hamiltonian (9.49) and momentum constraints (9.50), respectively, $f_{kij}(0)$ is defined by (9.51), and $p_i(0) = 0$. Then the unique solution $(\gamma_{ij}, \kappa_{ij}, f_{kij}, p_i)$ of the initial boundary value problem (9.47), (9.59), (9.53), and (9.60), together with the boundary conditions (9.61) (or (9.62)), satisfies the properties $p_i = 0$ for all time, and $(\gamma_{ij}, \kappa_{ij})$ solves the linearized ADM system (9.47)–(9.50) in Ω.*

9.4.2.2 Alekseenko–Arnold formulation

We consider the AA formulation Ref. [3] of Einstein's equations. This formulation involves fewer unknowns (20) than other first-order formulations that have been proposed and does not require any arbitrary parameters. Our interest is in providing viable boundary conditions at the artificial boundary for the linearized AA formulation that can be used for numerical simulations. The linearized AA formulation provides an alternate way to obtaining a solution of (9.47)–(9.50) with the given initial data, based on solving a system with better hyperbolicity properties. If γ_{ij}, κ_{ij} solve (9.47)–(9.50), define

$$\lambda_{kij} = -\frac{1}{\sqrt{2}}[\partial_{[k}\gamma_{i]j} + (M\gamma)_{[k}\delta_{i]j}], \tag{9.63}$$

where $(M\gamma)_i := \partial^l\gamma_{il} - \partial_i\gamma_l^l$ and square paranthesized indices are used to denote the antisymmetric part of a tensor. Then, proceeding as in Ref. [3], we obtain an evolution system for κ_{ij} and λ_{kij}

$$\dot{\kappa}_{ij} = \sqrt{2}\partial^k\lambda_{k(ij)} - \partial_i\partial_j\alpha \tag{9.64}$$

$$\dot{\lambda}_{kij} = \sqrt{2}\partial_{[k}\kappa_{i]j} - \tau_{kij}, \tag{9.65}$$

where

$$\tau_{kij} = \frac{1}{\sqrt{2}}(\partial_j\partial_{[k}\beta_{i]} + \partial^m\partial_{[m}\beta_{k]}\delta_{ij} - \partial^m\partial_{[m}\beta_{i]}\delta_{kj}) \tag{9.66}$$

As initial data for the system (9.64), (9.65) we use the given initial values of γ_{ij} and κ_{ij} and derive the initial values for λ_{kij} from those of γ_{ij} based on (9.63):

$$\lambda_{kij}(0) = -\frac{1}{\sqrt{2}}[\partial_{[k}\gamma(0)_{i]j} + (M\gamma(0))_{[k}\delta_{i]j}]. \tag{9.67}$$

The foregoing derivation shows that if γ_{ij} and κ_{ij} satisfy the ADM system and λ_{kij} is defined by (9.63), then κ_{ij} and λ_{kij} satisfy the symmetric hyperbolic system (9.64), (9.65). Conversely, to recover the solution to the ADM system from (9.64) and (9.65), the same κ_{ij} should be taken at time 0, and λ_{kij} should be given by (9.67). Having κ_{ij} determined, the metric perturbation γ_{ij} is defined as follows from (9.47)

$$\gamma_{ij} = \gamma_{ij}(0) - 2\int_0^t (\kappa_{ij} - \partial_{(i}\beta_{j)})(s)\,ds \tag{9.68}$$

The equivalence of (9.64) and (9.65) and the linearized ADM system has been studied in the second section of Ref. [3] for the case of pure initial value problem with the result that for given initial data satisfying the constraints, the unique solution of the linearized AA evolution equations satisfies the linearized ADM system, and so the linearized ADM system and the linearized AA system are equivalent.

Theorem 9.16   *([3, theorem 1]) Let the lapse perturbation α and shift perturbation β^i be given. Suppose that initial data $\gamma_{ij}(0)$ and $\kappa_{ij}(0)$ are specified satisfying the constraint equations (9.49) and (9.50) at time $t = 0$. Define $\lambda_{kij}(0)$ by (9.67), and determine κ_{ij} and λ_{kij} from the first-order symmetric hyperbolic system by (9.64) and (9.65). Finally, define γ_{ij} by (9.68). Then the ADM system (9.47)–(9.50) is satisfied.*

However, our interest is in the case when the spatial domain is bounded and appropriate boundary conditions are imposed. We provide two sets of well-posed (more precisely, maximal nonnegative) boundary conditions for the linearized AA system (9.47), (9.64), and (9.65) that are also consistent with the constraints (9.49), (9.50), and (9.63). This will then imply the analogue of Theorem 9.16 for the case of bounded domains. Although the Equation (9.47) of the linearized AA system decouples from the other two and its solution can be recovered by (9.68) once we know k_{ij}, we prefer to solve the AA system all together as opposed to the approach offered in Ref. [3] for the pure Cauchy case. One of the main reasons is that this approach could be more useful for extending the results presented in this chapter to the more difficult non-linearized case. It is also easier to explain the equivalence between the AA and ADM systems in this framework.

By using the method explained in the first part of this chapter, one can find two sets of maximal nonnegative boundary conditions for the FOSH system (9.47), (9.64), and (9.65):

$$ n^i \tau^{jk} \kappa_{ij} = 0, \quad n^k \tau^{il} \tau^{jm} \lambda_{k(ij)} = 0, \tag{9.69} $$

and

$$ \tau^{li} \tau^{mj} \kappa_{ij} = 0, \quad \tau^{lk} n^i n^j \lambda_{kij} = 0. \tag{9.70} $$

Observe that the above sets of boundary conditions do not depend on the choice of basis for the tangent space. The next theorem asserts that the boundary conditions (9.69) (or (9.70)) are constraint-preserving should a solution to the initial-boundary value problem exist.

Theorem 9.17   *([77, theorem 3]) Let Ω be a polyhedral domain. Given $\gamma_{ij}(0)$, $\kappa_{ij}(0)$ on Ω satisfying the constraints (9.49) and (9.50), define $\lambda_{kij}(0)$ by (9.67). Having $\gamma_{ij}(0)$, $\kappa_{ij}(0)$, and $\lambda_{kij}(0)$, define γ_{ij}, κ_{ij}, and λ_{kij} for positive time by the evolution equations (9.47), (9.64), (9.65), and the boundary conditions (9.69) (or (9.70)). Then the constraints (9.49), (9.50), and (9.63) are satisfied for all time.*

The analogue of Ref. [3, theorem 1] in the case of a bounded (polyhedral) domain is given next.

Theorem 9.18   *([77, theorem 4]) Let Ω be a polyhedral domain. Suppose that initial data $\gamma_{ij}(0)$ and $\kappa_{ij}(0)$ are given satisfying the Hamiltonian constraint (9.49) and momentum constraint (9.50), respectively, and that initial data $\lambda_{kij}(0)$ is defined by (9.67). Then the unique solution of the linearized AA initial-boundary value problem (9.47), (9.64), and (9.65), together with the boundary conditions (9.69) (or (9.70)) satisfies the linearized ADM system (9.47)–(9.50) in Ω.*

By using similar ideas and techniques as in the first part of this chapter, we indicate an extended initial-boundary value problem corresponding to the AA formulation whose solution solves the linearized ADM system (9.47)–(9.50) in Ω. This approach could have a potential impact on the accuracy of numerical simulations since the momentum constraints violations are kept under control for all time.

The new system consists of (9.47) and (9.65), and two new sets of equations, one replacing (9.64)

$$\dot{\kappa}_{ij} = \sqrt{2}\partial^k\lambda_{k(ij)} + \frac{1}{2}(\partial_i p_j + \partial_j p_i) - \partial^k p_k \delta_{ij} - \partial_i \partial_j \alpha, \qquad (9.71)$$

and the other one corresponding to a new three-dimensional vector field p_i defined by

$$\dot{p}_i = \partial^l \kappa_{li} - \partial_i \kappa_l^l. \qquad (9.72)$$

Observe that the additional terms that appear on the right-hand side of (9.71) compared with (9.64) are precisely the negative components of the formal adjoint of the momentum constraint operator (9.50) applied to p_i.

By employing the technique explained in the first part of the chapter, we found the following two sets of boundary conditions (corresponding to (9.69) and (9.70), respectively) are maximal nonnegative for (9.47), (9.71), (9.65), and (9.72):

$$n^i \tau^{jk} \kappa_{ij} = 0, \quad n^k (\tau^{il} \tau^{jm} \lambda_{k(ij)} - \frac{1}{\sqrt{2}} \tau^{lm} p_k) = 0, \qquad (9.73)$$

and

$$\tau^{li} \tau^{mj} \kappa_{ij} = 0, \quad \tau^{lk} (n^i n^j \lambda_{kij} - \frac{1}{\sqrt{2}} p_k) = 0, \qquad (9.74)$$

which shows that there is no dependence on the choice of basis for the tangent space.

Theorem 9.19   ([77, theorem 5]) *Let Ω be a polyhedral domain. Suppose that the initial data $\gamma_{ij}(0)$ and $\kappa_{ij}(0)$ are given satisfying the Hamiltonian (9.49) and momentum constraints (9.50), respectively, $\lambda_{kij}(0)$ is defined by (9.67), and $p_i(0) = 0$. Then the unique solution $(\gamma_{ij}, \kappa_{ij}, \lambda_{kij}, p_i)$ of the initial boundary value problem (9.47), (9.71), (9.65), and (9.72), together with the boundary conditions (9.73) (or (9.74)), satisfies the properties $p_i = 0$ for all time, and $(\gamma_{ij}, \kappa_{ij})$ solves the linearized ADM system (9.47)–(9.50) in Ω.*

Acknowledgments

The author thanks the anonymous referee for signaling relevant work done in the nonlinear case (e.g., for quasi-linear symmetric hyperbolic systems with involutive constraints). This work was partially supported by a grant from the Simons Foundation (#209782 to N. Tarfulea).

REFERENCES

1. A. Abrahams, A. Anderson, Y. Choquet–Bruhat, and J. W. York, Jr., "Geometrical hyperbolic systems for general relativity and gauge theories," Class. Quantum Grav. **14** (1997), A9–A22.

2. A. M. Alekseenko, "Constraint-preserving boundary conditions for the linearized Baumgarte-Shapiro-Shibata-Nakamura Formulation," Abstr. Appl. Anal. **2008** (2008), 742040.

3. A. M. Alekseenko and D. N. Arnold, "New first-order formulation for the Einstein equations," Phys. Rev. D **68** (2003), no. 6, 064013, 6pp.

4. M. Anderson and R. A. Matzner, "Extended lifetime in computational evolution of isolated black holes," Found. Phys. **35** (2005), 1477–1495.

5. A. Anderson and J. W. York, Jr., "Fixing Einstein's equations," Phys. Rev. Lett. **82** (1999), 4384–4387.

6. D. N. Arnold, "Numerical Relativity," ICM 2002 Satellite Conference on Nonlinear PDE: Theory and Approximation, Hong Kong, 2002.

7. D. N. Arnold and N. Tarfulea, "Boundary conditions for the Einstein-Christoffel formulation of Einstein's equations," Electron. J. Differ. Equ. Conf. **15** (2007), 11–27.

8. D. N. Arnold and N. Tarfulea, "Constrained First Order Hyperbolic Systems and Applications," WAVES 2011 Conference Proceedings, Pacific Institute for Mathematical Sciences (PIMS), Canada, 2011.

9. A. Arnowitt, S. Deser, and C. Misner, 'The dynamics of general relativity," Gravitation: An Introduction to Current Research, ed. L. Witten, New York, John Wiley & Sons, Inc., pp. 227–265, 1962.

10. A. Ashtekar, "New Hamiltonian formulation of general relativity," Phys. Rev. D **36** (1987), 1587–1602.

11. J. M. Bardeen and L. T. Buchman, "Numerical tests of evolution systems, gauge conditions, and boundary conditions for 1D colliding gravitational plane waves," Phys. Rev. D **65** (2002), 064037.

12. T. W. Baumgarte and S. L. Shapiro, "Numerical integration of Einstein's field equations," Phys. Rev. D **59** (1999), 024007.

13. G. Boillat, "Nonlinear hyperbolic fields and waves." Recent mathematical methods in nonlinear wave propagation (Montecatini Terme, 1994), 1–47, Lecture Notes in Mathematics, 1640, Springer, Berlin, 1996.

14. G. Boillat, "Involutions des systèmes conservatifs," C. R. Acad. Sci. Paris Sér. I Mathematics **307** (1988), no. 17, 891–894.

15. C. Bona and J. Masso, "Hyperbolic evolution system for numerical relativity," Phys. Rev. Lett. **68** (1992), 1097–1099.

16. C. Bona, J. Masso, E. Seidel, and J. Stela, "A new formalism for numerical relativity," Phys. Rev. Lett. **75** (1995), 600–603.

17. C. Bona, J. Masso, E. Seidel, and J. Stela, "First order hyperbolic formalism for numerical relativity," Phys. Rev. D **56** (1997), 3405–3415.

18. C. Bona, T. Ledvinka, C. Palenzuela-Luque, and M. Žaček, "Constraint-preserving boundary conditions in the Z4 numerical relativity formalism," Class. Quantum Grav. **22** (2005), 2615–2633.

19. H. Brezis, "Analyse fonctionnelle. Théorie et applications." Collection Mathématiques Appliquées pour la Matris, Masson, Paris, 1983.

20. G. Calabrese, L. Lehner, and M. Tiglio, "Constraint-preserving boundary conditions in numerical relativity," Phys. Rev. D **65** (2002), 104031.

21. G. Calabrese, J. Pullin, O. Reula, O. Sarbach, and M. Tiglio, "Well posed constraint-preserving boundary conditions for the linearized Einstein equations," Commun. Math. Phys. **240** (2003), 377–395.

22. E. Casella, P. Secchi, and P. Trebeschi, "Non-homogeneous linear symmetric hyperbolic systems with characteristic boundary," Differ. Integral Equ. **19** (2006), 51–74.

23. T. Cazenave and A. Haraux, "Introduction aux problèmes d'évolution semi-linaires." Mathématiques & Applications, 1. Ellipses, Paris, 142 pp., 1990.

24. M. W. Choptuik, E. W. Hirschmann, L. S. Liebling, and F. Pretorius, "An axisymmetric gravitational collapse code," Class. Quant. Grav. **20** (2003), 1857–1878.

25. M. W. Choptuik, E.W. Hirschmann, L. S. Liebling, and F. Pretorius, "Critical collapse of the massless scalar field in axisymmetry," Phys. Rev. D **68** (2003), 044007.

26. Y. Choquet–Bruhat, "The cauchy problem," in Gravitation: An Introduction to Current Research, ed. L. Witten, John Wiley & Sons, Inc., New York, 1973.

27. Y. Choquet–Bruhat and T. Ruggeri, "Hyperbolicity of the 3+1 system of Einstein equations," Commun. Math. Phys. **89** (1983), 269–275.

28. C. M. Dafermos, "Quasilinear hyperbolic systems with involutions," Arch. Rational Mech. Anal. **94** (1986), no. 4, 373–389.

29. C. M. Dafermos, Hyperbolic Conservation Laws in Continuum Physics. Third edition. Springer-Verlag, Berlin, 2010. xxxvi+708pp.

30. L. C. Evans, "Partial Differential Equations," Graduate Studies in Mathematics, vol. **19**, AMS, Providence, Rhode Island, 1998.

31. H. Friedrich, "Hyperbolic reductions for Einstein's equations," Class. Quantum Grav. **13** (1996), 1451–1469.

32. H. Friedrich and G. Nagy, "The initial boundary value problem for Einstein's vacuum field equations," Comm. Math. Phys. **201** (1999), 619–655.

33. H. Friedrich and A. Rendall, "The cauchy problem for the Einstein equations," Lect. Notes Phys. **540** (2000), 127–223.

34. K. O. Friedrichs, "Symmetric positive linear differential equations," Comm. Pure Appl. Math. **11** (1958), 333–418.

35. S. Frittelli and R. Gomez, "Boundary conditions for hyperbolic formulations of the Einstein equations," Class. Quant. Grav. **20** (2003), 2379–2392.

36. S. Frittelli and R. Gomez, "Einstein boundary conditions for the $3 + 1$ Einstein equations," Phys. Rev. D **68** (2003), 044014.

37. S. Frittelli and R. Gomez, "First-order quasilinear canonical representation of the characteristic formulation of the Einstein equations," Phys. Rev. D **68** (2003), 084013.

38. S. Frittelli and R. Gomez, "Einstein boundary conditions in relation to constraint propagation for the initial-boundary value problem of the Einstein equations," Phys. Rev. D **69** (2004), 124020.

39. S. Frittelli and O. S. Reula, "First-order symmetric-hyperbolic Einstein equations with arbitrary fixed gauge," Phys. Rev. Lett. **76** (1996), 4667–4670.

40. C. Gundlach and J. M. Martín-García, "Symmetric hyperbolic form of systems of second-order evolution equations subject to constraints," Phys. Rev. D **70** (2004), 044031.

41. C. Gundlach and J. M. Martín-García, "Symmetric hyperbolicity and consistent boundary conditions for second-order Einstein equations," Phys. Rev. D **70** (2004), 044032.

42. B. Gustafsson, H. O. Kreiss, and J. Oliger, Time Dependent Problems and Difference Methods. Pure and Applied Mathematics. John Wiley & Sons, Inc., New York, 1995.

43. S. D. Hern, "Numerical relativity and inhomogeneous cosmologies," Ph.D. thesis, University of Cambridge, 1999, arXiv:gr-qc/0004036.

44. M. Holst, L. Lindblom, R. Owen, H. P. Pfeiffer, M. A. Scheel, and L. E. Kidder, "Optimal constraint projection for hyperbolic evolution systems," Phys. Rev. D **70** (2004), 084017.

45. F. John, Partial Differential Equations. (Fourth edition.) Applied Mathematical Sciences, vol. 1. Springer-Verlag, New York, x+249 pp., 1982.

46. K. T. Joseph and P. G. LeFloch, "Boundary layers in weak solutions of hyperbolic conservation laws," Arch. Ration. Mech. Anal. **147** (1999), no. 1, 47–88.

47. L. E. Kidder, M. Scheel, and S. Teukolsky, "Extending the lifetime of 3D black hole computations with a new hyperbolic system of evolution equations," Phys. Rev. D **64** (2001), 064017–064030.

48. L. E. Kidder, L. Lindblom, M. A. Scheel, L. T. Buchman, and H. P. Pfeiffer, "Boundary conditions for the Einstein evolution system," Phys. Rev. D **71** (2005), 064020.

49. H. O. Kreiss and J. Lorenz, Initial Boundary Value Problems and Navier-Stokes Equations. Academic Press, Boston, 1989.

50. H. O. Kreiss and J. Winicour, "Problems which are well-posed in a generalized sense with applications to the Einstein equations," Class. Quantum Grav. **23** (2006), S405–S420.

51. H.-O. Kreiss, O. A. Reula, O. Sarbach, and J. Winicour, "Boundary conditions for coupled quasilinear wave equations with applications to isolated systems," Commun. Math. Phys. **289** (2009), 1099–1129.

52. P. D. Lax and R. S. Phillips, "Local boundary conditions for dissipative symmetric linear differential operators,"Comm. Pure Appl. Math. **13** (1960), 427–455.

53. L. Lindblom, M. A. Scheel, L. E. Kidder, H. P. Pfeiffer, D. Shoemaker, and S. A. Teukolsky, "Controlling the growth of constraints in hyperbolic evolution systems," Phys. Rev. D **69** (2004), 124025.

54. L. Lindblom, M. A. Scheel, L. E. Kidder, R. Owen, and O. Rinne, "A new generalized harmonic evolution system," Class. Quantum Grav. **23** (2006), S447–S462.

55. A. Majda and S. Osher, "Initial-boundary value problem for hyperbolic equations with uniformly characteristic boundary," Comm. Pure Appl. Math. 28 (1975), 607–675.

56. G. Nagy and O. Sarbach, "A minimization problem for the lapse and the initial-value problem for Einstein's field equations," Class. Quantum Grav. **23** (2006), S447–S504.

57. F. Pretorius, "Numerical relativity using a generalized harmonic decomposition," Classical Quantum Gravity **22** (2005), 425–451.

58. F. Pretorius, "Simulation of binary black hole spacetimes with a Harmonic Evolution Scheme," Class. Quant. Grav. **23** (2006), S529–S552.

59. J. Rauch, "Symmetric positive systems with boundary characteristic of constant multiplicity," Trans. Am. Math. Soc. **291** (1985), 167–187.

60. O. Reula, "Hyperbolic methods for Einstein's equations," Living Rev. Relativ. **1** (1998), 37 pp. (electronic).

61. O. Reula and O. Sarbach, "A model problem for the initial-boundary value formulation of Einstein's field equations," J. Hyperbolic Differ. Equ. **2** (2005), 397–435.

62. O. Rinne, "Stable radiation-controlling boundary conditions for the generalized harmonic Einstein equations," Class. Quantum Grav. 23(2006), 6275–6300.

63. O. Rinne and J. M. Stewart, "A strongly hyperbolic and regular reduction of Einstein's equations for axisymmetric spacetimes," Class. Quantum Grav. **22** (2005), 1143–1166.

64. T. Ruggeri and A. Strumia, "Main field and convex covariant density for quasilinear hyperbolic systems: relativistic fluid dynamics," Ann. Inst. H. Poincar Sect. A (N.S.) **34** (1981), no. 1, 65–84.

65. O. Sarbach and M. Tiglio, "Exploiting gauge and constraint freedom in hyperbolic formulations of Einstein's equations," Phys. Rev. D **66** (2002), 064023.

66. O. Sarbach and M. Tiglio, "Boundary conditions for Einstein's field equations: Mathematical and numerical analysis," J. Hyperbolic Differ. Equ. **2** (2005), 839–883.

67. O. Sarbach and M. Tiglio, "Continuum and discrete initial-boundary value problems and Einstein's field equations," Living Rev. Relativ. **15** (2012), no. 9, 194 pp.

68. E. Schnetter, "Gauge fixing for the simulation of black hole spacetimes," Ph.D. thesis, University of Tuebingen, 2003.

69. P. Secchi, "The initial-boundary value problem for linear symmetric hyperbolic systems with characteristic boundary of constant multiplicity," Diff. Integral Equ. **9** (1996), 671–700.

70. P. Secchi, "Well-posedness of characteristic symmetric hyperbolic systems," Arch. Ration. Mech. Anal. **134** (1996), 155–197.

71. P. Secchi and P. Trebeschi, "Non-homogeneous quasi-linear symmetric hyperbolic systems with characteristic boundary," Int. J. Pure Appl. Math. **23** (2005), 39–59.

72. M. Shibata and T. Nakamura, "Evolution of three-dimensional gravitational waves: Harmonic slicing case," Phys. Rev. D **52** (1995), 5428–5444.

73. J. M. Stewart, "The Cauchy problem and the initial-boundary value problem in numerical relativity," Class. Quantum Grav. **15** (1998), 2865–2889.

74. B. Szilagyi and J. Winicour, "Well-posed initial-boundary evolution in general relativity," Phys. Rev. D **68** (2003), 041501.

75. B. Szilagyi, B. Schmidt, and J. Winicour, "Boundary conditions in linearized Harmonic gravity," Phys. Rev. D **65** (2002), 064015.

76. N. Tarfulea, "Constraint preserving boundary conditions for some hyperbolic formulations of Einstein's equations," Ph.D. Thesis, University of Minnesota, 2004, gr-qc/0508014.

77. N. Tarfulea, "Well-posed constraint-preserving boundary conditions for the AA formulation of Einstein's equations," J. Math. Anal. Appl. **359** (2009), 711–721.

78. N. Tarfulea, "On boundary conditions for first order symmetric hyperbolic systems with constraints," J. Hyperbolic Diff. Equ. **10** (2013), no. 4, 725–734.

79. G. Yoneda and H. Shinkai, "Symmetric hyperbolic system in the Ashtekar formulation," Phys. Rev. Lett. **82** (1999), 263–266.

10

INFORMATION INTEGRATION, ORGANIZATION, AND NUMERICAL HARMONIC ANALYSIS

RONALD R. COIFMAN[1], RONEN TALMON[1], MATAN GAVISH[2], AND ALI HADDAD[1]

[1]*Mathematics Department, Yale University, New Haven, CT, USA*
[2]*Statistics Department, Stanford University, Palo Alto, CA, USA*

10.1 INTRODUCTION

Our goal is to present an overview of recent developments in methodologies for empirical organization of data. We present a geometric/analytic mathematical framework for learning, which revolves around building a network or a graph whose nodes are observations. In our framework, connections between observations are constantly reconfigured and calibrated in order to achieve learning for specific tasks.

In particular, we will provide a synthesis of a range of ideas from mathematics and machine learning, which address the transition from a local similarity model to a global configuration. This is analogous to Newtonian Calculus, which from a local linear model of variability, calculates a global solution to a differential or partial differential equation. We first apply these fundamentals to find an intrinsic model for data generated through a "physical" unknown stochastic process, and then describe a data agnostic methodology to jointly organize the rows and columns of a general seemingly unstructured matrix, viewed either as the matrix of a linear operator or as a database. Here the rows are viewed as functions on the columns and the columns

Mathematical and Computational Modeling: With Applications in Natural and Social Sciences, Engineering, and the Arts, First Edition. Roderick Melnik.
© 2015 John Wiley & Sons, Inc. Published 2015 by John Wiley & Sons, Inc.

as functions of the rows, and a mutually consistent dual geometry is built to optimize prediction and processing.

We relate these methods to ideas from classical Harmonic Analysis and indicate tools to measure success of information extraction. In particular, we introduce methodologies that perform "signal processing" on data matrices, enabling functional regression, prediction, denoising, compression fast numerics, and so on [2, 15].

We illustrate these ideas to organize and map out in an automatic and purely data-driven fashion, text documents, psychological questionnaires, medical profiles, physical sensor data, financial data.

The aim of this chapter, which is based on an invited talk of the first author at the AMMCS-2013 Conference in Waterloo, Canada, is to discuss how a blend of Geometry and Harmonic Analysis generalizes classical tools of Mathematics, to provide a synthesis of many seemingly unrelated approaches to data analysis and processing.

Data analysis challenges can be broadly split into two classes. The first one includes problems about the geometry (or efficient tabulation, data model etc.) of the data: through low-dimensional embeddings (i.e., finding a small number of parameters that describe or tabulate the data variability effectively) of data sets in high-dimensions or on graphs, with a goal to permit efficient interaction and information extraction; clustering, denoising of data; outlier detection; and so on [1,3,4,8–10,18].

The second class includes problems about the approximation/fitting/learning of functions on the data or responding to queries, based on a few samples, with the goal of predicting the values of functions at new data points. An important role has been played by methods based on the assumption that the intrinsic geometry of the data plays an important role and that smoothness of relevant functions on the data should be defined intrinsically in a way that is adapted to such geometry, that is, we want to use proximity of points to infer proximity of functional values. Here it is clear that we should simultaneously organize the data and collections of functions of interest on the data; in particular, we will address here two classes of frequently occurring functions, the data coordinate functions, and their contextual building blocks, e.g., their natural local statistics, or histograms on related observational metrics [12, 22, 23].

Usually data sets of observations, or databases of documents, consist of high dimensional observations, whose coordinates represent features of the data, say the frequency or power of a quantity, or a qualitative description. We claim that when we are dealing with a large number of coordinates, it is necessary to view them as an initial collection of functions on the documents and organize them into their own dual geometry, which should be consistent with the geometry of the observations. As we will see, this geometry helps improve and redefine the organization of the initial data. An important issue becomes the choice of metrics comparing "functions" or densities, as we will see both spaces of smooth functions and their duals enable; efficient and flexible data similarity matches (e.g., Earth Mover's distances).

Ideas from harmonic analysis and spectral graph theory have played a fundamental role in recent advances in the field. As an example, diffusion geometries provide the ability to integrate local relations into global links starting from the premise that a similarity measure $A(x,y)$ between any pair (x,y) of nearby data points can be meaningfully defined and measured. A typical choice for data points lying in $\mathbb{R}^D$ is $A(x,y) = exp(-\|x-y\|^2/t)$, where t is a fixed scale parameter. In general, the choice of A is both data- and goal-dependent. If N is the number of data points, A is a $N \times N$ matrix, which we think as being sparse since only nearby data points are connected with an edge with weight above a threshold. One can renormalize A to obtain a Markov matrix P that we use to generate a random walk on the data points. We view $P(x,y)$ as a probability of jumping from x to y in one step. In reality, $P(x,y)$ should be a data-derived estimate of the probability that (x,y) are two samples of the same "object."

The advantage of a true empirical probabilistic kernel is that as it is independent of the actual measurements or observational device, we could observe any nonlinear transformation of the data, as long as the probability estimate is stable, leading to an intrinsic data graph invariant under changes of variables on observations. We call the corresponding diffusion geometry Empirical Intrinsic Geometry (EIG) of the data.

In diffusion geometry, one uses P and its powers to gain insight into the multiscale geometry of the data, for example, by finding coordinate systems, as well as to construct dictionaries of functions, as in Fourier or Wavelet Analysis, for learning and processing functions on the data.

Digital data organization by diffusion geometry can be roughly broken up into three methodologies:

1. A dimension-reduction approach that embeds the data into low-dimensional Euclidean space through the use of eigenvectors of the affinity martix/kernel P (or a normalized related matrix) to build an EIG model of the data.

2. Hierarchical folder building and clustering, a bottom-up agglomeration approach that propagates or diffuses affinity between documents. This can be achieved through probabilistic model building and statistical/combinatorial "book keeping" on the data.

3. A bigeometric approach in which we simultaneously organize observations (digital documents) and their features. We view the data as a questionnaire, in which we have a demographic geometry based on similarity of response profiles, and organize the questions conceptually, by the similarity of responses across the population.

The first approach is based on the eigenfunctions of P, or of the random walk P, let them be $P\varphi_i = \lambda_i\varphi_i$, where we may assume $\lambda_1 \geq \lambda_2 \geq \cdots \lambda_i \geq \ldots$. One may use the eigenfunctions φ_i to map the data to m-dimensional Euclidean space by $\Phi_m^t(x) := (\lambda_1^{t/2}\varphi_1(x), \ldots, \lambda_m^{t/2}\varphi_m)$. This is closely related to the so-called spectral graph embedding, extensively used for decades for graph layouts.

Observe that the probability of a transition in t time steps from x to y is $P^t(x,y) = \sum_i \lambda_i^t\varphi_i(x)\varphi_i(y)$. For t large, since all λs are smaller than 1, λ_i^t is very small for i larger than, say, m. But then the Euclidean distance between $\Phi_m^t(x)$ and $\Phi_m^t(y)$ is equal to the Euclidean distance, in the N-dimensional space of data points, between the probability distributions $P^t(x, \cdot)$ and $P^t(y, \cdot)$, called the diffusion distance between x and y at time t. Therefore, this embedding into Euclidean m-dimensional space reflects diffusion distances on the original data, which are intrinsic geometric properties of the data.

The eigenvectors are integrating and linking together the information, through the propagation of local similarities between nearby points (the usual iterative algorithm to compute eigenvectors is a process of diffusion on the data).

The second and third approaches lead to a generalized wavelet analysis, or tensor wavelet analysis based on so-called diffusion wavelets, which are associated with, and tuned to the random walk P described above [10, 16, 20]. It shares some similarities with classical multiscale and wavelet analysis, and it allows one to generalize classical signal processing techniques, for example, for compression, denoising, fitting, and regularization, to functions on data sets. We can show that these two seemingly different approaches are mathematically related, in a way which is very similar to the relationships between Fourier and wavelet analysis in Euclidean spaces, in which multiscale constructions lead to basis functions which are hierarchically organized according to diffusion distances at different scales.

Techniques based on the above ideas of diffusion on data sets have led to machine-learning algorithms that perform at state-of-art or better on standard benchmarks in the community. We refer the reader to Ref. [21] and references therein.

10.2   EMPIRICAL INTRINSIC GEOMETRY

In this section, we describe in detail a setup where empirical analysis reveals the underlying intrinsic model that gave rise to it. Our basic assumption is that we are observing a stochastic time series governed by a Langevin equation on a Riemannian manifold. These observations are transformed through a nonlinear transformation into high dimensions in an ambient unknown independent noisy environment. Our goal is to recover the original Riemannian manifold as well as the potential governing the dynamics of the observations.

As an example, consider a molecule (alanine dipeptide) consisting of 10 atoms and oscillating stochastically in water. It is known that the configuration at any given time is essentially described by two angle parameters. We assume that we observe five atoms of the molecule for a certain period of time, and five other atoms in the remainder time. The task is to describe the position of all atoms at all times, or more precisely, discover the angle parameters and their relation to the position of all atoms. See Figures 10.1 and 10.2.

The main point is that the observations are quite different, perhaps using completely different sensors in different environments (but measuring the same dynamic phenomenon) and that we derive an identical "natural" intrinsic manifold parameterizing the observations.

An important remark is that we observe stochastic data constrained to lie on an unknown Riemannian manifold, which we need somehow to reconstruct explicitly, without having any coordinate system on the manifold. This is achieved through the

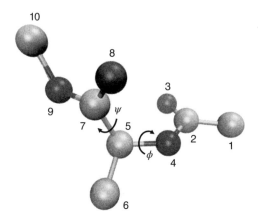

FIGURE 10.1 A representative molecular structure of alanine dipeptide, excluding the hydrogens. The atoms are numbered and the two dihedral angles ϕ and ψ are indicated.

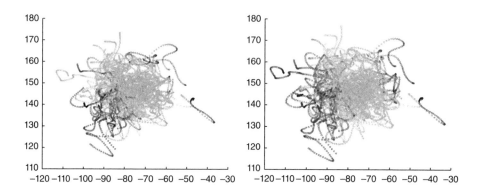

FIGURE 10.2 A two-dimensional scatter plot of random trajectories of the dihedral angles ϕ and ψ. Based on observations of the corresponding random trajectories of merely 5 out of 10 atoms of the molecule, we infer a model describing one of the angles. The points are colored according to the values of the inferred model from the five even atoms (left) and the five odd atoms (right). We observe that the gradient of the color is parallel to the x-axis, indicating an adequate representation of one of the angles. In addition, the color patterns are similar, indicating that the models are independent of the particular atoms observed, and describe the common intrinsic parameterization of the molecule dynamics.

explicit construction of the eigenvectors of the Laplace operator on the manifold (observations). These can be used to obtain a low-dimensional canonical observation invariant embedding or to obtain local charts on the manifold. This invariant description of the dynamics is similar to the formulation of Newton's law through invariant Hamiltonian equations.

10.2.1 Manifold formulation

Specifically and for simplicity of exposition, we consider a flat manifold for which we adopt the state-space formalism to provide a generic problem formulation that may be adapted to a wide variety of applications.

Let θ_t be a d-dimensional underlying coordinates of a process in time index t. The dynamics of the process are described by normalized stochastic differential equations as follows[1]

$$d\theta_t^i = a^i(\theta_t^i)dt + dw_t^i, \quad i = 1,\ldots,d, \tag{10.1}$$

where a^i are unknown drift functions and $\dot{w}_t^i$ are independent white noises. For simplicity, we consider here normalized processes with unit variance noises. Since a^i are any drift functions, we may first apply normalization without affecting the following derivation. See Ref. [19] for details. We note that the underlying process is equivalent to the system state in the classical terminology of the state-space approach.

Let $\mathbf{y}_t$ denote an n-dimensional observation process in time index t, drawn from a probability density function (pdf) $f(\mathbf{y}; \theta)$. The statistics of the observation process is time-varying and depends on the underlying process θ_t. We consider a model in which the clean observation process is accessible only via a noisy n-dimensional measurement process $\mathbf{z}_t$, given by

$$\mathbf{z}_t = g(\mathbf{y}_t, \mathbf{v}_t) \tag{10.2}$$

where g is an unknown (possibly nonlinear) measurement function and $\mathbf{v}_t$ is a corrupting n-dimensional measurement noise, drawn from an unknown stationary pdf $q(\mathbf{v})$ and independent of $\mathbf{y}_t$.

The description of θ_t constitutes a parametric manifold that controls the accessible measurements at hand. Our goal is to reveal the underlying process θ_t and its dynamics based on a sequence of measurements $\mathbf{z}_t$.

Let $p(\mathbf{z}; \theta)$ denote the pdf of the measured process $\mathbf{z}_t$ controlled by θ_t, which satisfies the following property.

Proposition 10.1 *The pdf of the measured process $\mathbf{z}_t$ is a linear transformation of the pdf of the clean observation component $\mathbf{y}_t$.*

Proof: The proof is obvious, relying on the independence of $\mathbf{y}_t$ and $\mathbf{v}_t$, the pdf of the measured process is given by

$$p(\mathbf{z}; \theta) = \int_{g(\mathbf{y},\mathbf{v})=\mathbf{z}} f(\mathbf{y}; \theta) q(\mathbf{v}) d\mathbf{y} d\mathbf{v}. \tag{10.3}$$

---

[1] x^i denotes access to the ith coordinate of a point $\mathbf{x}$.

We note that in the common case of additive measurement noise, that is, $g(\mathbf{y}, \mathbf{v}) = \mathbf{y} + \mathbf{v}$, only a single solution $\mathbf{v}(\mathbf{z}) = \mathbf{z} - \mathbf{y}$ exists. Thus, $p(\mathbf{z}; \theta)$ in (10.3) becomes a linear convolution

$$p(\mathbf{z}; \theta) = \int_{\mathbf{y}} f(\mathbf{y}; \theta) q(\mathbf{z} - \mathbf{y}) d\mathbf{y} = f(\mathbf{z}; \theta) * q(\mathbf{z}).$$

The dynamics of the underlying process is conveyed by the time-varying pdf of the measured process. Thus, this pdf may be very useful in revealing the desired underlying process and its dynamics. Unfortunately, the pdf is unknown and needs to be estimated. Assume we have access to a class of estimators of the pdf over discrete bins which can be viewed as linear transformations. Let $\mathbf{h}_t$ be such an estimator with m bins which is viewed as an m-dimensional process and is given by

$$p(\mathbf{z}; \theta_t) \overset{\mathcal{T}}{\mapsto} \mathbf{h}_t, \tag{10.4}$$

where $\mathcal{T}$ is a linear transformation of the density $p(\mathbf{z}; \theta)$ from the infinite sample space of $\mathbf{z}$ into a finite interval space of dimension m. By Proposition 10.1 and by (10.4) we get the following results: (i) The process $\mathbf{h}_t$ is a linear transformation of the pdf of the clean observation component $\mathbf{y}_t$. (ii) The process $\mathbf{h}_t$ can be described as a deterministic nonlinear map of the underlying process θ_t.

Histogram is an example for a linear estimator of the pdf. Let $\mathbf{h}_t$ be the empirical local histogram of the measured process $\mathbf{z}_t$ in a short-time window of length L at time t. Let $\mathcal{Z}$ be the sample space of $\mathbf{z}_t$ and let $\mathcal{Z} = \bigcup_{j=1}^{m} \mathcal{H}_j$ be a finite partition of $\mathcal{Z}$ into m disjoint histogram bins. Thus, the value of each histogram bin is given by

$$h_t^j = \frac{1}{L} \sum_{s=t-L+1}^{t} \mathbf{1}_{\mathcal{H}_j}(\mathbf{z}_s), \tag{10.5}$$

where $\mathbf{1}_{\mathcal{H}_j}(\mathbf{z}_t)$ is the indicator function of the bin $\mathcal{H}_j$. By assuming (unrealistically) that infinite number of samples are available and that their density in each histogram bin is uniform, (10.5) can be expressed as

$$h_t^j = \int_{\mathbf{z} \in \mathcal{H}_j} p(\mathbf{z}; \theta) d\mathbf{z}. \tag{10.6}$$

Thus, ideally the histograms are *linear transformations* of the pdf. In addition, if we shrink the bins of the histograms as we get more and more data, the histograms converge pointwise to the pdf

$$\mathbf{h}_t \xrightarrow[|\mathcal{H}_j| \to 0]{L_1 \to \infty} p(\mathbf{z}; \theta). \tag{10.7}$$

In practice, since the computation of high-dimensional histograms is challenging, we propose to preprocess high-dimensional data by applying random filters in order to reduce the dimensionality without corrupting the information.

10.2.2 Mahalanobis distance

We view $\mathbf{h}_t$ (the linear transformation of the local densities, for example, the local histograms) as feature vectors for each measurement $\mathbf{z}_t$. By (10.1), the process $\mathbf{h}_t$ satisfies the dynamics given by Itô's lemma

$$h_t^j = \sum_{i=1}^{d} \left(\frac{1}{2} \frac{\partial^2 h^j}{\partial \theta^i \partial \theta^i} + a^i \frac{\partial h^j}{\partial \theta^i} \right) dt \tag{10.8}$$

$$+ \sum_{i=1}^{d} \frac{\partial h^j}{\partial \theta^i} dw_t^i, \; j = 1, \ldots, m.$$

For simplicity of notation, we omit the time index t from the partial derivatives. According to (10.8), the (j,k)th element of the $m \times m$ covariance matrix $\mathbf{C}_t$ of $\mathbf{h}_t$ is given by

$$C_t^{jk} = \mathrm{Cov}(h_t^j, h_t^k) = \sum_{i=1}^{d} \frac{\partial h^j}{\partial \theta^i} \frac{\partial h^k}{\partial \theta^i}, \; j, k = 1, \ldots, m. \tag{10.9}$$

In matrix form, (10.9) can be rewritten as

$$\mathbf{C}_t = \mathbf{J}_t \mathbf{J}_t^T \tag{10.10}$$

where $\mathbf{J}_t$ is the $m \times d$ Jacobian matrix, whose (j,i)th element is defined by

$$J_t^{ji} = \frac{\partial h^j}{\partial \theta^i}, \; j = 1, \ldots, m, \; i = 1, \ldots, d.$$

Thus, the covariance matrix $\mathbf{C}_t$ is a semi-definite positive matrix of rank d.

We define a symmetric $\mathbf{C}$-dependent squared distance between pairs of measurements as

$$d_{\mathbf{C}}^2(\mathbf{z}_t, \mathbf{z}_s) = 2(\mathbf{h}_t - \mathbf{h}_s)^T (\mathbf{C}_t + \mathbf{C}_s)^{-1} (\mathbf{h}_t - \mathbf{h}_s). \tag{10.11}$$

Since usually the dimension d of the underlying process is smaller than the number of histogram bins m, the covariance matrix is singular and non-invertible. Thus, in practice we use the pseudo-inverse to compute the inverse matrix in (10.11).

The distance in (10.11) is known as the *Mahalanobis distance* with the property that it is invariant under linear transformations. Thus, by Proposition 10.1, it is invariant to the measurement noise and function (e.g., additive noise or multiplicative noise). We note however that the linear transformation employed by the measurement noise on the observable pdf (10.3) may degrade the available information. For example, an additive Gaussian noise employs a low-pass "blurring" filter on the clean observation component. In case the dependency on the underlying process is manifested in high-frequencies, the linear transformation employed by the noise significantly attenuates the connection between the measurements and the underlying process. Therefore, we expect to exhibit noise resilience up to a certain noise level as long as the observable pdfs can be regarded as non-degenerate functions of the underlying process. Above this level, we expect to experience a sudden drop in the performance.

In addition, by (10.10), the Mahalanobis distance in (10.11) approximates the Euclidean distance between samples of the underlying process. Let θ_t and θ_s be two samples of the underlying process. Then, the Euclidean distance between the samples is approximated to a second order by a local linearization of the nonlinear map of θ_t to $\mathbf{h}_t$, and is given by

$$\|\theta_t - \theta_s\|^2 = d_{\mathbf{C}}^2(\mathbf{z}_t, \mathbf{z}_s) + O(\|\mathbf{h}_t - \mathbf{h}_s\|^4). \tag{10.12}$$

For more details see Refs. [19] and [17].

Assuming there is an intrinsic map $i(\mathbf{h}_t) = \theta_t$, the approximation in (10.12) is equivalent to the inverse problem defined by the following nonlinear differential equation

$$\sum_{i=1}^{m} \frac{\partial \theta^j}{\partial h^i} \frac{\partial \theta^k}{\partial h^i} = \left[C_t^{-1}\right]^{jk}, \ j,k = 1,\ldots,d. \tag{10.13}$$

This equation which is nothing more than a discrete formulation of the definition of a Riemannian metric on the manifold is empirically solved through the eigenvectors of the corresponding discrete Laplace operator. Thus, the approximation in (10.12) recovers the intrinsic distances on the parametric manifold and is obtained empirically from the noisy measurements by "infinitesimally" inverting the measurement function.

For further illustration of the geometric notion, see Figure 10.3.

In order to compute the Mahalanobis distance (10.11), the local covariance matrices $\mathbf{C}_t$ have to be estimated. Given a sequence of measurements, we estimate the covariance in short time windows. By (10.1) and (10.8), the samples in the short window can be seen as small perturbations of a pivot sample created by the noise $\mathbf{w}_t$. Thus, we assume that the samples share similar local probability densities.[2]

We remark that the rank of the covariance matrix d is usually smaller than the covariance matrix dimension m. Thus, in order to compute the inverse matrix we use only the d principal components of the matrix. This operation "cleans" the matrix and filters out noise. In addition, when the empirical rank of the local covariance matrices

[2]We emphasize that the local covariance takes into account the dynamics of the histogram and not the distribution of the measurements, given by the histograms themselves.

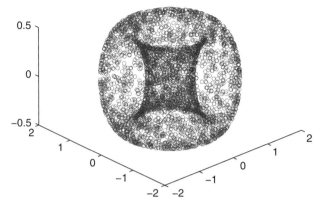

FIGURE 10.3 Consider a set of points on a two-dimensional torus in $\mathbb{R}^3$ ("the manifold") which are samples of a Brownian motion on the torus. The geometric interpretation of the intrinsic notion is the search for a canonical description of the set, which is independent of the coordinate system. For example, the points can be written in three cartesian coordinates or in the common parameterization of a torus using two angles; however, the intrinsic model (constructed based on the points) describing the torus should be the same. The Mahalanobis distance attaches to each point a Riemannian metric that corresponds to a probability measure that is driven by the underlying dynamics (the Brownian motion in this particular case), and therefore, it is invariant to the coordinate system.

of the histograms is lower than d, it indicates that the available information is insufficient. In addition, the estimations of the time-varying pdfs and the local covariance matrices define two time scales on the sequence of measurements. The fine time scale is defined by short-time windows of L measurements to estimate the temporal pdf. The coarse time scale is defined by the short-time window of pdf estimates. As a result, the approximation in (10.12) is valid as long as the statistics of the noise is locally fixed in the short-time windows of length L (i.e., slowly changing compared to the fast variations of the underlying process) and the fast variations of the underlying process can be detected in the difference between the pdfs in the windows used to estimate the local covariance.

10.3   ORGANIZATION AND HARMONIC ANALYSIS OF DATABASES/MATRICES

Let M be a matrix and denote its column set by X and its row set by Y. M can be viewed as a function on the product space, namely $M: X \times Y \to \mathbb{R}$. Our first step in processing M, regardless of the particular problem, is to simultaneously organize X and Y, or in other words, to construct a product geometry on $X \times Y$, in which proximity (in some appropriate metrics) implies predictability of matrix entries. Equivalently, we would like the function M to be "smooth" with respect to the product geometry in its domain. As we will see, smoothness, compressibility, and having low entropy are all interlinked in our organization.

To illustrate the basic concept underlying the simultaneous row-column organization, consider the case of a vector (namely, a matrix with one row). In this case, the only reasonable organization would be to bin the entries in decreasing order. This decreasing function is obviously smooth outside a small exceptional set (being of bounded variation). Our approach extends this simple construction—which is just a one-dimensional quantization tree—to a coupled quantization tree.

We now describe a mathematical framework for organization and analysis of matrices. Observe that a hierarchical vector quantization tree on the set of columns, X, as vectors in Euclidean space, induces a tree metric ρ_X for which the rows are Hölder smooth. A similar hierarchical organization on Y induces a similar tree metric ρ_Y on the columns with a similar smoothness property. A useful combination should enable us to predict or regress missing values. We claim that we can get an organization, which satisfies a *Mixed-Hölder* condition, enabling the estimation of one value in terms of three neighbors with a higher order error in the two metrics

$$|f(x_0, y_0) - f(x_0, y_1) - f(x_1, y_0) + f(x_1, y_1)| \leq C \cdot \rho_X(x_0, x_1)^\alpha \cdot \rho_Y(y_0, y_1)^\alpha.$$

In the Euclidean setting, this would be a relaxation of the bounded mixed derivative condition $|\partial^2 f / \partial x \partial y| \leq C$, which has been studied in the context of approximation in high dimensions [5].

Therefore, our main object of study becomes a real matrix equipped with a partition trees pair—one on each dimension—rather than a stand-alone real matrix. We study in detail the Mixed-Hölder function class on an abstract product set equipped with partition tree pair. The main tool is an orthogonal transform for the space of matrices, naturally induced by the pair of partition trees (or the tensor product of the corresponding martingale difference transforms). Specifically, taking the tensor product of the *Haar-like bases* induced on X and on Y by their respective partition trees.

The Mixed-Hölder arises naturally in several different ways. First, any matrix can be given Mixed-Hölder structure. Second, any bounded matrix decomposes into a sum of a Mixed-Hölder part and a part with small support (as for the one row example).

10.3.1 Haar bases

A hierarchical partition tree on a dataset X allows harmonic analysis of real-valued functions on X, as it induces special orthonormal bases called *Haar bases* [14].

A Haar basis is obtained from a partition tree as follows. Suppose that a node in the tree has n children, that is, that the set described by the node decomposes into n subsets in the next, more refined, level. Then this node contributes $n - 1$ functions to the basis. These functions are all supported on the set described by the node, are piecewise constant on its n subsets, are all mutually orthogonal, and are orthogonal to the constant function on the set.

Observe that just like the classical Haar functions, coefficients of an expansion in a Haar basis measure variability of the conditional expectations of the function in subnodes of a given node.

Tasks such as compression of functions on the data set, as well as subsampling, denoising, and learning of such functions, can be performed in Haar coefficient space using methods familiar from Euclidean harmonic analysis and signal processing [14].

Some results for the classical Haar basis on $[0, 1]$ extend to generalized Haar bases. Recall that the classical Haar functions are given by

$$h_\ell(x) = \left(|I|^{-\frac{1}{2}} \right) (\chi_- - \chi_+),$$

where χ_- is the indicator of the left half of I and χ_+ is the indicator of the right half of I.

Let us first note that the classical Haar basis on $[0, 1]$ is the Haar basis induced by the partition tree of dyadic subintervals of $[0, 1]$. This tree defines a natural dyadic distance $d(x, y)$ on $[0, 1]$, defined as the length of the smallest dyadic interval containing both points. Hölder classes in the metric d are characterized by the Haar coefficients $a_I = \int f(x) h_I(x) dx$:

$$|a_I| < c |I|^{\frac{1}{2} + \beta} \iff |f(x) - f(x')| < c \cdot d(x, x')^\beta.$$

This result holds for any Haar basis when d is the tree metric induced by the partition tree and $|I| = \#I / \#X$ is the normalized size of the folder I.

10.3.2   Coupled partition trees

Consider for example a matrix M whose columns are documents, whose coordinates are terms-frequency, and rows corresponding to terms occurrences within documents. The partition tree construction can be applied to the columns of M, where the affinity matrix A is obtained from local distances between the columns as vectors in Euclidean space. Each partition in the resulting partition tree can be interpreted as a division of the documents into contexts or subjects. However, it can also be applied to the set of rows of M, resulting in partitions of the terms into concepts, or word bags.

Coupling the construction of the two partition trees—on the columns and the rows—takes us away from the representation of the dataset as a point cloud in Euclidean space, toward representation of the dataset as a function on the product set $\{\text{rows}\} \times \{\text{columns}\}$. We now consider data matrix M and assume two partition trees—one on the column set of M and one on the row set of M—have already been constructed. Each tree induces a Haar basis and a tree metric as above.

The tensor product of the Haar bases is an orthonormal basis for the space of matrices of the same dimensions as M. We now consider an analysis of M in this basis.

Denote by $R = I \times J$ a "rectangle" of entries of M, where I is a folder in the column tree and J is a folder in the row tree. Denote by $|R| = |I||J|$ the volume of the rectangle R. Indexing Haar functions by their support folders, we write $h_I(x)$ for a

Haar function on the rows. This allows us to index basis functions in the tensor product basis by rectangles and write $h_R(x,y) = h_I(x)h_J(y)$.

Analysis and synthesis of the matrix M in the tensor Haar basis are simply

$$a_R = \int M(x,y)h_R(x,y)dxdy$$

$$M(x,y) = \sum_R a_R h_R(x,y).$$

The characterization of Hölder functions mentioned above extends to Mixed-Hölder matrices [6, 13]:

$$|a_R| < c|R|^{1/2+\beta} \Leftrightarrow |f(x,y) - f(x',y) - f(x,y') + f(x',y')| \leq cd(x,x')^\beta D(y,y')^\beta$$

where d and D are the tree metrics induced by the partition trees on the rows and columns, respectively. Observe that this condition implies the conventional two dimensional Hölder condition

$$|f(x,y) - f(x',y')| \leq d(x,x')^\beta + D(y,y')^\beta$$

Simplicity or sparsity of an expansion is quantified by an entropyâ such as

$$e_\alpha(f) = \left(\sum |a_R|^\alpha\right)^{1/\alpha}$$

for some $\alpha < 2$. We comment that this norm is just a Besov norm that easily generalizes Earth Mover's distances, when scaled correctly, adding flexibility to our construction below. The relation between this entropy, that is, the efficiency of the representation in tensor Haar basis, and the Mixed-Hölder condition, is given by the following two propositions [6, 13].

Proposition 10.2   *Assume $e_\alpha(f) = (\sum |a_R|^\alpha) \leq 1$. Then the number of coefficients needed to approximate the expansion to precision $\varepsilon^{1-\alpha/2}$ does not exceed $\varepsilon^{-\alpha}$ and we need to only consider large coefficients corresponding to Haar functions whose support is large. Specifically, we have*

$$\int \left| f - \sum_{|R|>\varepsilon, |a_R|>\varepsilon} a_R h_R(x) \right|^\alpha dx < \varepsilon^{1-\alpha/2}$$

The next proposition shows that $e_\alpha(f)$ estimates the rate at which f can be approximated by Hölder functions outside sets of small measure.

Proposition 10.3   *Let f be such that $e_\alpha \leq 1$. Then there is a decreasing sequence of sets E_ℓ such that $|E_\ell| \leq 2^{-\ell}$ and a decomposition of Calderon Zygmund type $f = g_\ell + b_\ell$. Here, b_ℓ is supported on E_ℓ and g_ℓ is Hölder $\beta = 1/\alpha - 1/2$*

with constant $2^{(\ell+1)/\alpha}$. *Equivalently, g_ℓ has Haar coefficients satisfying* $|a_R| \leq$ $2^{(\ell+1)/\alpha}|R|^{1/\alpha}$.

To connect the bounded entropy condition with classical Euclidean harmonic analysis, consider the tensor product of smooth wavelet basis on $[0, 1]$ with itself, yielding an orthonormal basis for $L_2\left([0, 1]^2\right)$. The class of functions, whose wavelet expansion has finite entropy, has been characterized in Ref. [11]. There, it is shown that this class is independent of the wavelet used and equals the class of functions having a Harmonic extension whose corresponding fractional derivative is integrable in the disk (or bidisk). The dual spaces are also characterized as Bloch spaces, which in our dyadic structure case are just functions with bounded Haar coefficients. Also observe that for $f: [0, 1]^2 \to \mathbb{R}\ f = \sum |R|^{1/2} a_R |R|^{-1/2} h_R$ is a special atomic decomposition of $(\partial/\partial x)^{1/2}(\partial/\partial y)^{1/2}f$, which is, therefore, in the Hardy space H^1 of the bi-disk. A similar result holds for the other entropies, implying a fractional derivative in the Hardy space.

Proposition 10.3 decomposes any matrix into a "good," or Mixed-Hölder part, and a "bad" part with small support. Mixed-Hölder matrices indeed deserve to be called "good" matrices, as they can be substantially subsampled. To see this, note that the number of samples needed to recover the functions to a given precision is of the order of the number of tensor Haar coefficients needed for that precision. For balanced partition trees, this is approximately the number of bifolders R, whose area exceeds the precision ε. This number is of the order of $1/\varepsilon \log(1/\varepsilon)$.

Propositions 10.2 and 10.3 imply that the entropy condition quantifies the compatibility between the pair of partition trees (on the rows and on the columns) and the matrix on which they are constructed. In other words, to construct useful trees we should seek to minimize the entropy in the induced tensor Haar basis.

For a given matrix M, finding a partition tree pair, which is a global minimum of the entropy, is computationally intractable and not sensible, as the matrix could be the superposition of different structures, corresponding to conflicting organizations. At best, we should attempt to peel off organized structured layers.

Let us now describe iterative procedures for building tree pairs that often perform well in practice. Observe that the simplest geometry that can be built to fit a given function is to postulate that level sets are points whose proximity is defined by the values of the function. This corresponds to rearranging the function's level sets by decreasing values of the function. In reality, we are always confronting a collection of functions, such as the coordinates of our cloud of points, in high dimensions (or the rows of a matrix viewed as a collection of functions on the columns, representing points in space) and are therefore forced to tune the geometry to render all coordinates as smooth as possible.

If instead of rearranging scalars, we organize the columns in a quantization tree, as in the construction of partition trees above, then tautologically each coordinate (row) is Hölder relative to the tree metric on the points, enabling us to organize the rescaled (or differentiated) Haar coefficients of the rows in a quantization tree, thereby guaranteeing that the matrix is Mixed-Hölder. The procedure can then be iterated.

We illustrate a simpler procedure on the example of a questionnaire matrix M, where $M_{i,j}$ is the response of person j to question i. In other words, columns correspond to persons and rows correspond to questions in the questionnaire. We start by building an affinity graph on the columns (persons) using correlation or any other affinity measure between columns. Bottom-up hierarchical clustering is then performed on the resulting graph, producing a partition tree on the columns. Folders in the different partitions correspond to demographic groups in the population. We append the demographic groups as a new stereotype (or meta-person) to the data set. The response of a demographic group, stereotype to question i is the average response to this question in the demographic group.

With this extended generalized population at hand, we proceed to build an affinity graph on the questions using correlation (or any other affinity measure) between the extended rows. Namely, the affinity between rows i_1 and i_2 is the correlation, say, along the answers of both real people and stereotypes (demographic groups). The process is now iterated: bottom-up hierarchical clustering is performed on the resulting graph of the questions (rows), producing a partition tree on the rows. Folders in this tree correspond to conceptual groups of questions. We append conceptual groups as new meta-questions to the data set. The response of person j to a conceptual group "meta-question" is their average response along that group of questions.

This procedure thus alternates between construction of partition trees on rows and on columns. Empirically, the entropy of the tree pair converges after a few iterations. Figures 10.4 and 10.5 show the resulting organization of persons (columns)

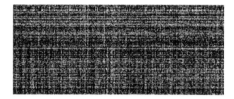

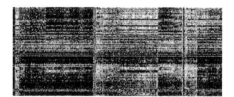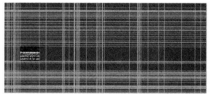

FIGURE 10.4   Organization of a questionnaire. A disorganized questionnaire, on the top left; the columns represent people and the row are binary questions. An organized questionnaire after mutual multiscale bilearning, on the bottom left. On the bottom right, the questionnaire is split on a two-scale grid. The consistency of responses of a demographic group (context) to a group of questions (concept) is shown by highlighted rectangle.

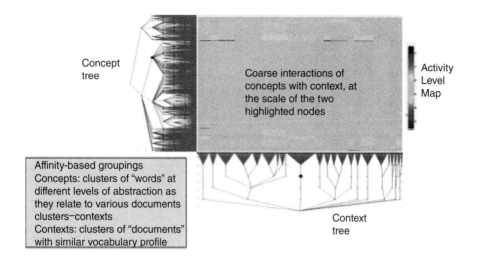

FIGURE 10.5 Concept tree and context tree. Mutual Organization/Tree Structures for context–concept duality. Although we use linguistic analogies, these trees were built on time series of observations of 500 object; the concepts are seenarios of times with similar responses among the population while the contexts are group of objects with similar temporal responses.

and questions (rows) in a personality questionnaire. The resulting partition tree pair—a partition tree of demographic groups, or contexts, on the persons—and a partition tree of conceptual groups on the questions is illustrated in Figure 10.5.

10.4 SUMMARY

To conclude, we see emerging a signal processing toolbox for digital data as a first step to analyze the geometry of large data sets in high-dimensional space and functions defined on such data sets. The number of problems and applications include, among many others, multidimensional document rankings extending the Google PageRank, information navigation, heterogeneous material modeling, multiscale complex structure organization, and so on.

The ideas described above are strongly related to nonlinear principal component analysis, kernel methods, spectral graph embedding, and many more, at the intersection of several branches of mathematics, computer science, and engineering. They are documented in literally hundreds of papers in various communities.

A simple description of many of these ideas and more is given through diffusion geometries. We refer the interested reader to the July 2006 issue of *Applied and Computational Harmonic Analysis* and references therein.

REFERENCES

1. W. K. Allard, G. Chen, and M. Maggioni, "Multi-scale geometric methods for data sets II: Geometric multi-resolution analysis," *Applied and Computational Harmonic Analysis*, 32(3), pp. 435–462 (2012).

2. B. Alpert, G. Beylkin, R. R. Coifman, and V. Rokhlin, "Wavelet-like bases for the fast solution of second-kind integral equations," *SIAM Journal on Scientific Computing*, 14(1), pp. 159–184 (1993).

3. M. Belkin and P. Niyogi, "Laplacian eigenmaps and spectral techniques for embedding and clustering." *Advances in Neural Information Processing Systems*, 14 pp. 585–591 (2001).

4. M. Belkin and P. Niyogi, "Laplacian eigenmaps for dimensionality reduction and data representation," *Neural Computing*, (13), pp. 1373–1397 (2003).

5. H. J. Bungartz and M. Griebel, "Sparse grids," *Acta Numerica*, 13, pp. 147–269 (2004).

6. R. R. Coifman and M. Gavish, "Harmonic analysis of digital data bases," *in: Wavelets and Multiscale Analysis (Cohen, J., Zayed, A., Eds.)*, Birkhäuser, Boston (2011) pp. 161–197.

7. R. R. Coifman and S. Lafon, "Geometric harmonics: a novel tool for multiscale out-of-sample extension of empirical functions," *Applied and Computational Harmonic Analysis*, 21(1), pp. 31–52 (2006).

8. R. R. Coifman, S. Lafon, A. Lee, M. Maggioni, B. Nadler, F. Warner, and S. Zucker, "Geometric diffusions as a tool for harmonic analysis and structure definition of data. Part I: Diffusion maps," *Proceedings of the National Academy of Sciences*, 102, pp. 7426–7431 (2005).

9. R. R. Coifman, S. Lafon, A. Lee, M. Maggioni, B. Nadler, F. Warner, and S. Zucker, "Geometric diffusions as a tool for harmonic analysis and structure definition of data. Part II: Multiscale methods," *Proceedings of the National Academy of Sciences*, 102, pp. 7432–7438 (2005).

10. R. R. Coifman and M. Maggioni, "Diffusion wavelets," *Applied and Computational Harmonic Analysis*, 21(1), pp. 54–95 (2006).

11. R. R. Coifman and R. Rochberg, "Another characterization of B.M.O," *Proceedings of the American Mathematical Society*, 79, pp. 249–254 (1980).

12. C. J. Dsilva, R. Talmon, N. Rabin, and R. R. Coifman and I. G. Kevrekidis, "Nonlinear intrinsic variables and state reconstruction in multiscale simulations," *The Journal of Chemical Physics,* 139(18), 184109 (2013).

13. M. Gavish and R. R. Coifman, "Sampling, denoising and compression of matrices, by coherent matrix organization," *Applied and Computational Harmonic Analysis*, 33(3), pp. 354–369 (2012).

14. M. Gavish, B. Nadler, and R. R. Coifman, "Multiscale wavelets on trees, graphs and high dimensional data: theory and applications to semi supervised learning," *Proceedings of the 27th International Conference on Machine Learning, ICML* Haifa, Israel, June 21–24 (2010).

15. L. Greengard and V. Rokhlin, "A fast algorithm for particle simulations," *Journal of Computational Physics*, 73, pp. 325–348 (1987).

16. C. E. Priebe, D. J. Marchette, Y. Park, E. J. Wegman, J. L. Solka, D. A. Socolinsky, D. Karakos, K. W. Church, R. Guglielmi, R. R. Coifman, D. Link, D. M. Healy, M. Q. Jacobs, and A. Tsao, "Iterative denoising for cross-corpus discovery," *Proceedings of COMPSTAT 2004*, Physica-Verlag/Springer, pp. 381–392 (2004).

17. D. Kushnir, A. Haddad, and R. Coifman, "Anisotropic diffusion on sub-manifolds with application to earth structure classification," *Applied and Computational Harmonic Analysis*, 32 (2), pp. 280–294 (2012).

18. A. Singer, "Angular synchronization by eigenvectors and semidefinite programming," *Applied and Computational Harmonic Analysis*, 30(1), pp. 20–36 (2011).

19. A. Singer and R. R. Coifman, "Non-linear independent component analysis with diffusion maps," *Applied and Computational Harmonic Analysis*, 25, pp. 226–239 (2008).

20. S. A. Smolyak, "Quadrature and interpolation formulas for tensor products of certain classes of functions," *Soviet Mathematics Doklady*, 4, pp. 240–243 (1963).

21. A. D. Szlam, M. Maggioni, and R. R. Coifman, "Regularization on graphs with function-adapted diffusion processes," *Journal of Machine Learning Research* 9, pp. 1711–1739 (2008).

22. R. Talmon, I. Cohen, S. Gannot, and R. R. Coifman, "Diffusion maps for signal processing: A deeper look at manifold-learning techniques based on kernels and graphs," *IEEE Signal Processing Magazine,* 30(4), pp. 75–86 (2013).

23. R. Talmon and R. R. Coifman, "Empirical intrinsic geometry for intrinsic modeling and nonlinear filtering," *Proceedings of the National Academy Sciences,* 110(31), pp. 12535–12540 (2013).

SECTION 5

MATHEMATICAL METHODS IN SOCIAL SCIENCES AND ARTS

11

SATISFACTION APPROVAL VOTING

STEVEN J. BRAMS[1] AND D. MARC KILGOUR[2]

[1]*Department of Politics, New York University, New York, NY, USA*
[2]*Department of Mathematics, Wilfrid Laurier University, Waterloo, Ontario, Canada*

11.1   INTRODUCTION

Approval voting (AV) is a voting system in which voters can vote for, or approve of, as many candidates as they like. Each approved candidate receives one vote, and the candidates with the most votes win.

This system is well suited to electing a single winner, which almost all the literature on AV since the 1970s has addressed (see, e.g., Brams and Fishburn [7, 8] and Brams [6, chaps. 1 and 2]). But for multiwinner elections, such as for seats on a council or in a legislature, AV's selection of the most popular candidates or parties can fail to reflect the diversity of interests in the electorate.

We certainly are not the first to address the problem of selecting multiple winners and will later reference relevant work in social choice theory. There has also been much interest in other fields, such as computer science and psychology, in developing methods for aggregating preferential and nonpreferential information, assessing the properties of these methods, and applying them to empirical data. See, for example, Refs. [11, 18, 19].

As a possible solution to the problem of electing multiple, representative candidates when voters use an approval ballot,[1] in which they can approve or not approve

---

[1] Merrill and Nagel [17] were the first to distinguish between approval balloting, in which voters can approve of one or more candidates, and approval voting (AV), a method for aggregating approval ballots. SAV, as we will argue, is a method of aggregation that tends to elect more representative candidates in.

Mathematical and Computational Modeling: With Applications in Natural and Social Sciences, Engineering, and the Arts, First Edition. Roderick Melnik.

of each candidate, we propose. Satisfaction approval voting (SAV) works as follows when the candidates are individuals. A voter's satisfaction score is the fraction of his or her approved candidates who are elected, whether the voter is relatively discriminating (i.e., approves of few candidates) or not (approves of many candidates). In particular, it offers a strategic choice to voters, who may bullet vote (i.e., exclusively for one candidate) or vote for several candidates, perhaps hoping to make a specific set of candidates victorious.

Among all the sets of candidates that might be elected, SAV chooses the set that maximizes the sum of all voters' satisfaction scores. As we will show, SAV may give very different outcomes from AV; SAV outcomes are not only more satisfying to voters but also tend to be more representative of the diversity of interests in an electorate.[2] Moreover, they are easy to calculate.

In Section 11.2, we apply SAV to the election of individual candidates (e.g., to a council) when there are no political parties. We show, in the extreme, that SAV and AV may elect disjoint subsets of candidates. When they differ, SAV winners will generally represent the electorate better—by at least partially satisfying more voters—than AV winners. While maximizing total voter satisfaction, however, SAV may not maximize the number of voters who approve of at least one winner—one measure of representativeness—though it is more likely to do so than AV.

This is shown empirically in Section 11.3, where SAV is applied to the 2003 Game Theory Society (GTS) election of 12 new council members from a list of 24 candidates (there were 161 voters). SAV would have elected 2 winners different from the 12 elected under AV and would have made the council more representative of the entire electorate. We emphasize, however, that GTS members might well have voted differently under SAV than under AV, so one cannot simply extrapolate a reconstructed outcome, using a different aggregation method, to predict the consequences of SAV.

In Section 11.4, we consider the conditions under which, in a 3-candidate election with 2 candidates to be elected, a voter's ballot might change the outcome, either by making or breaking a tie. In our decision-theoretic analysis of the 19 contingencies in which this is possible, approving of one's two best candidates induces a preferred outcome in about the same number of contingencies as bullet voting, even though a voter must split his or her vote when voting for 2 candidates. More general results on optimal voting strategies under SAV are also discussed.

In Section 11.5, we apply SAV to party-list systems, whereby voters can approve of as many parties as they like. Parties nominate their "quotas," which are based on their vote shares, rounded up; they are allocated seats to maximize total voter satisfaction, measured by the fractions of nominees from voters' approved parties that are elected. We show that maximizing total voter satisfaction leads to the

---

[2]Representing this diversity is not the issue when electing a single winner, such as a mayor, governor, or president. In such an election, the goal is to find a consensus choice, and we believe that AV is better suited than SAV to satisfy this goal. Scoring rules, in which voters rank candidates and scores are associated with the ranks, may also serve this end, but the optimal scoring rule for achieving particular standards of justice (utilitarianism, maximin, or maximax) is sensitive to the distribution of voter utilities [2].

proportional representation (PR) of parties, based on the Jefferson/d'Hondt method of apportionment, which favors large parties.

SAV tends to encourage multiple parties to share support, because they can win more seats by doing so. At the same time, supporters of a party diminish its individual support by approving of other parties, so there is a trade-off between helping a favorite party and helping a coalition of parties that may be able to win more seats in toto. Some voters may want to support only a favorite party, whereas others may want to support multiple parties that, they hope, will form a governing coalition. We argue that this freedom is likely to make parties more responsive to the wishes of their supporters with respect to (i) other parties with which they coalesce and (ii) the candidates they choose to nominate.[3]

In Section 11.6, we conclude that SAV may well induce parties to form coalitions, if not merge, before an election. This will afford voters the ability better to predict what policies the coalition will promote, if it forms the next government, and, therefore, to vote more knowledgeably.[4] In turn, it gives parties a strong incentive to take careful account of their supporters' preferences, including their preferences for coalitions with other parties.

11.2 SATISFACTION APPROVAL VOTING FOR INDIVIDUAL CANDIDATES

We begin by applying SAV to the election of individual candidates, such as to a council or legislature, in which there are no political parties. We assume in the subsequent analysis that there are at least two candidates to be elected and that more than this number run for office (to make the election competitive).

To define SAV formally, assume that there are $m > 2$ candidates, numbered $1, 2, \ldots, m$. The set of all candidates is $\{1, 2, \ldots, m\} = [m]$, and k candidates are to be elected, where $2 \le k < m$. Assume voter i approves of a subset of candidates $V_i \subseteq [m]$, where $V_i \neq \emptyset$. (Thus, a voter may approve of only 1 candidate, though more are to be elected.) For any subset of k candidates, S, voter i's satisfaction is $|V_i \cap S|/|V_i|$, or the fraction of his or her approved candidates that are elected.[5] SAV elects a subset of k candidates that maximizes

---

[3]The latter kind of responsiveness would be reinforced if voters, in addition to being able to approve of one or more parties, could also use SAV to choose a party's nominees.

[4]More speculatively, SAV may reduce a multiparty system to two competing coalitions of parties. The majority coalition winner would then depend, possibly, on a centrist party that can swing the balance in favor of one coalition or the other. Alternatively, a third moderate party (e.g., Kadima in Israel) might emerge that peels away supporters from the left and the right. In general, SAV is likely to make coalitions more fluid and responsive to popular sentiment.

[5]One interesting modification of this measure has been suggested [16]. When a voter approves of more candidates than are to be elected, change the denominator of the satisfaction measure from $|V_i|$ to $min\{|V_i|, k\}$. Thus, for example, if voter i approves of 3 candidates, but only $k = 2$ can be elected, i's satisfaction would be $2/2$ (rather than $2/3$) whenever any two of his or her approved candidates are elected. This modification ensures that a voter's influence on the election is not diluted if he or she approves of more candidates than are to be elected, but it does not preserve other properties of SAV.

$$s(S) = \sum_i \frac{|V_i \cap S|}{|V_i|} \tag{11.1}$$

which we interpret as the total satisfaction of voters for S. By convention, $s(\emptyset) = 0$.

To illustrate SAV, assume there are $m = 4$ candidates, a, b, c, and d, and 10 voters who approve of the following subsets[6]:

- 4 voters: ab
- 3 voters: c
- 3 voters: d

Assume $k = 2$ of the 4 candidates are to be elected. AV elects $\{a,b\}$ because a and b receive 4 votes each compared to 3 votes each that c and d receive. By contrast, SAV elects $\{c,d\}$ because the satisfaction scores of the six different two-winner subsets are as follows:

- $s(a,b) = 4(1) = 4$
- $s(a,c) = s(a,d) = s(b,c) = s(b,d) = 4(\frac{1}{2}) + 3(1) = 5$
- $s(c,d) = 3(1) + 3(1) = 6$.

Thus, the election of c and d gives 6 voters full satisfaction of 1, which corresponds to greater total satisfaction, 6, than achieved by the election of any other pair of candidates.[7]

More formally, candidate j's satisfaction score is $s(j) = \sum_i |V_i \cap j|/|V_i|$, whereas candidate j's approval score is $a(j) = \sum_i |V_i \cap j|$. Our first proposition shows that satisfaction scores make it easier to identify all winning subsets of candidates under SAV, that is, all subsets that maximize total satisfaction.

Proposition 11.1   *Under SAV, the k winners are any k candidates whose individual satisfaction scores are the highest.*

Proof: Because $V_i \cap S = \bigcup_{j \in S}(V_i \cap j)$, it follows from (11.1) that

$$s(S) = \sum_i \left(\frac{1}{|V_i|}\right) \sum_{j \in S} |V_i \cap j| = \sum_{j \in S} \sum_i \frac{|V_i \cap j|}{|V_i|} = \sum_{j \in S} s(j)$$

Thus, the satisfaction score of any subset S, $s(S)$, can be obtained by summing the satisfaction scores of the individual members of S. Now suppose that $s(j)$

---

[6]We use ab to indicate the strategy of approving of the subset $\{a,b\}$, but we use $\{a,b\}$ to indicate the outcome of a voting procedure. Later we drop the set-theoretic notation, but the distinction between voter strategies and election outcomes is useful for now.

[7]Arguably, candidates c and d benefit under SAV by getting bullet votes from their supporters. While their supporters do not share their approval with other candidates, their election gives representation to a majority of voters, whereas AV does not.

has been calculated for all candidates $j = 1, 2, \ldots, m$. Then, for any arrangement of the set of candidates $[m]$ so that the scores $s(j)$ are in non-increasing order, the first k candidates constitute a subset of candidates that maximizes total voter satisfaction. □

As an illustration of Proposition 11.1, consider the previous example, in which

$$s(a) = s(b) = 4\left(\frac{1}{2}\right) = 2 \quad s(c) = s(d) = 3(1) = 3.$$

Because c and d have higher satisfaction scores than any other candidates, the subset $\{c, d\}$ is the unique winning subset if $k = 2$ candidates are to be elected under SAV.

One consequence of Proposition 11.1 is a characterization of tied elections: There are two or more winning subsets if and only if the satisfaction scores of the kth and $(k+1)$st candidates are tied in their satisfaction score when the candidates are arranged in descending order, as described in the proof of Proposition 11.1. This follows from the fact that tied subsets must contain the k most satisfying candidates, but if those in the kth and the $(k+1)$st positions give the same satisfaction, a subset containing either would maximize total voter satisfaction. Ties among three or more sets of candidates are, of course, also possible.

It is worth noting that the satisfaction that a voter gains when an approved candidate is elected does not depend on how many of the voter's other approved candidates are elected, as some multiple-winner systems that use an approval ballot prescribe.[8] This renders candidates' satisfaction scores additive: The satisfaction from electing subsets of two or more candidates is the sum of the candidates' satisfaction scores. Additivity greatly facilitates the determination of SAV outcomes when there are multiple winners—simply choose the subset of individual candidates with the highest satisfaction scores.

The additivity of candidate satisfaction scores reflects SAV's equal treatment of voters: Each voter has one vote, which is divided evenly among all his or her approved candidates. Thus, if two candidates are vying for membership in the elected subset, then gaining the support of an additional voter always increases a candidate's score by $1/x$, where x is the number of candidates approved of by that voter.[9] This is a

---

[8]Two of these systems—proportional AV and sequential proportional AV—assume that a voter's satisfaction is marginally decreasing; the more of his or her approved candidates are elected, the less satisfaction the voter derives from having additional approved candidates elected. See http://www.nation-master.com/encyclopedia/Proportional-approval-voting and http://www.nationmaster.com/encyclopedia/Sequential-proportional-approval-voting for a description and examples of these two systems and Ref. [1] for an axiomatic treatment of systems in which the points given to a candidate are decreasing in the number of candidates of whom the voter approves, which they call "size approval voting." More generally, see Kilgour [15] and Kilgour and Marshall [16] for a comparison of several different approval-ballot voting systems that have been proposed for the election of multiple winners, all of which may give different outcomes.

[9]By contrast, under cumulative voting (CV), a voter can divide his or her votes—or, equivalently, a single vote—unequally, giving more weight to some candidates than others. However, equal and even cumulative voting (EaECV), which restricts voters to casting the same number of votes for all candidates whom they support, is equivalent to SAV, though its connection to voter satisfaction, as far as we know, has

consequence of the goal of maximizing total voter satisfaction, not an assumption about how approval votes are to be divided.

We next compare the different outcomes that AV and SAV can induce.

Proposition 11.2   *AV and SAV can elect disjoint subsets of candidates.*

Proof: This is demonstrated by the previous example: AV elects $\{a,b\}$ whereas SAV elects $\{c,d\}$.   □

For any subset S of the candidates, we say that S *represents* a voter i if and only if voter i approves of some candidate in S. We now ask how is the set of candidates who win under SAV or AV, that is, how many voters approve of at least one elected candidate.

SAV winners usually represent at least as many, and often more, voters than the set of AV winners, as illustrated by the previous example, in which SAV represents 6 voters and AV only 4 voters. SAV winners c and d appeal to distinctive voters, who are more numerous and so win under SAV, whereas AV winners a and b appeal to the same voters but, together, receive more approval and so win under AV.

But there are (perhaps unlikely) exceptions:

Proposition 11.3   *An AV outcome can be more representative than a SAV outcome.*

Proof: Assume there are $m = 5$ candidates and 13 voters, who vote as follows:

- 2 voters: a
- 5 voters: ab
- 6 voters: cde

If 2 candidates are to be elected, the AV outcome is $\{a,c\}$, $\{a,d\}$, or $\{a,e\}$ (7 approvals for a and 6 each for c, d, and e), whereas the SAV outcome is $\{a,b\}$, because

- $s(a) = 2(1) + 5(\frac{1}{2}) = 4\frac{1}{2}$
- $s(b) = 5(\frac{1}{2}) = 2\frac{1}{2}$
- $s(c) = s(d) = s(e) = 6(\frac{1}{3}) = 2.$

---

not previously been demonstrated. While CV and EaECV have been successfully used in some small cities in the United States to give representation to minorities on city councils, it seems less practicable in large elections, including those in countries with party-list systems in which voters vote for political parties (Section 11.5). See http://en.wikipedia.org/wiki/Cumulative_voting for additional information on cumulative voting.

Thus, whichever of the three AV outcomes is selected, the winning subset represents all 13 voters, whereas the winners under SAV represent only 7 voters.   □

The "problem" for SAV in the forgoing example would disappear if candidates c, d, and e were to combine forces and became one candidate (say, c), rendering $s(c) = 6(1) = 6$. Then the SAV and AV outcomes would both be $\{a,c\}$, which would give representation to all 13 voters. Indeed, as we will show when we apply SAV to party-list systems in Section 11.5, SAV encourages parties to coalesce to increase their combined seat share.

But first we consider another possible problem of both SAV and AV.

Proposition 11.4   *There can be subsets that represent more voters than either the SAV or the AV outcome.*

Proof: Assume there are $m = 5$ candidates and 12 voters, who vote as follows:

- 4 voters: *ab*
- 4 voters: *acd*
- 3 voters: *ade*
- 1 voter: *e*

If 2 candidates are to be elected, the AV outcome is $\{a,d\}$ (11 and 7 votes, respectively, for a and d), and the SAV outcome is also $\{a,d\}$, because

- $s(a) = 4(\frac{1}{2}) + 7(\frac{1}{3}) = 4\frac{1}{3}$
- $s(b) = 4(\frac{1}{2}) = 2$
- $s(c) = 4(\frac{1}{3}) = 1\frac{1}{3}$
- $s(d) = 7(\frac{1}{3}) = 2\frac{1}{3}$
- $s(a) = 3(\frac{1}{3}) + 1(1) = 2.$

While subset $\{a,d\}$ represents 11 of the 12 voters, subset $\{a,e\}$ represents all 12 voters.   □

Interestingly enough, the so-called (for representativeness) would select $\{a,e\}$. It works as follows. The candidate who represents the most voters—the AV winner—is selected first. Then the candidate who represents as many of the remaining (unrepresented) voters as possible is selected next, then the candidate who represents as many as possible of the voters not represented by the first two candidates is selected, and so on. The algorithm ends as soon as all voters are represented, or until the required number of candidates is selected. In the example used to prove Proposition 11.4, the greedy algorithm first chooses candidate a (11 votes) and then candidate e (1 vote).

Given a set of ballots, we say a is a subset of candidates with the properties that (i) every voter approves at least one candidate in the subset and (ii) there are no smaller subsets with property (i). In general, finding a minimal representative set is computationally difficult.[10] Although the greedy algorithm finds a minimal representative set in the previous example, it is no panacea.

Proposition 11.5   *SAV can find a minimal representative set when both AV and the greedy algorithm fail to do so.*

Proof: Assume there are $m = 3$ candidates and 17 voters, who vote as follows:

- 5 voters: ab
- 5 voters: ac
- 4 voters: b
- 3 voter: c

If 2 candidates are to be elected, the AV outcome is $\{a,b\}$ (a gets 10 votes and b 9), which is identical to the subset produced by the greedy algorithm.[11] On the other hand, the SAV outcome is $\{b,c\}$, because

- $s(a) = 5(\frac{1}{2}) + 5(\frac{1}{2}) = 5$
- $s(b) = 5(\frac{1}{2}) + 4(1) = 6\frac{1}{2}$
- $s(c) = 5(\frac{1}{2}) + 3(1) = 5\frac{1}{2}$

Not only does this outcome represent all 17 voters but it is also the minimal representative set.   □

The greedy algorithm fails to find the minimal representative set in the previous example because it elects the "wrong" candidate first—the AV winner, a. Curiously, a closely related example shows that none of these methods may find a minimal representative subset:

Proposition 11.6   *SAV, AV, and the greedy algorithm can all fail to find a unique minimal representative set.*

Proof: Assume there are $m = 3$ candidates and 9 voters, who vote as follows:

- 3 voters: ab
- 3 voters: ac

---

[10]Technically, the problem is NP hard (http://en.wikipedia.org/wiki/NP-hard) because it is equivalent to the hitting-set problem, which is a version of the vertex-covering problem (http://en.wikipedia.org/wiki/Vertex_cover) discussed in Ref. [14]. Under SAV, as we showed at the beginning of this section, the satisfaction-maximizing subset of, say, k candidates can be calculated efficiently, as it must contain only candidates with satisfaction scores among the k highest. Because of this feature, the procedure is practical for multiwinner elections with many candidates.

[11]Candidates a, b, and c receive, respectively, 10, 9, and 8 votes; the greedy algorithm first selects a (10 voters) and then b (4 more voters).

- 2 voters: b
- 1 voter: c

AV and the greedy algorithm give $\{a,b\}$, as in the previous example but so does SAV because

- $s(a) = 3(\frac{1}{2}) + 3(\frac{1}{2}) = 3$
- $s(b) = 3(\frac{1}{2}) + 2(1) = 3\frac{1}{2}$
- $s(c) = 3(\frac{1}{2}) + 1(1) = 2\frac{1}{2}$

As before, $\{b,c\}$ is the minimal representative set. □

Minimal representative sets help us assess and compare the outcomes of elections; the greedy algorithm contributes by finding an upper bound on the size of a minimal representative set, because it eventually finds a set that represents all voters, even if it is not minimal. But there is a practical problem with basing an election procedure on the minimal representative set: Only by chance will that set have k members. If it is either smaller or larger, it must be "adjusted."

But what adjustment is appropriate? For example, if the minimal representative set is too small, should one add candidates that give as many voters as possible a second representative, then a third, and so on? Or, after each voter has approved of at least one winner, should it, like SAV, maximize total voter satisfaction? It seems to us that maximizing total voter satisfaction from the start is a simple and desirable goal, even if it sometimes sacrifices some representativeness.

Another issue, addressed in the next proposition, is vulnerability to candidate cloning. AV is almost defenseless against cloning, whereas SAV exhibits some resistance.[12]

A clone of a candidate is a new candidate who is approved by exactly the supporters of the original candidate. We call a candidate, h, a minimal winning candidate (under AV or SAV) if the score of every other winning candidate is at least equal to the score of h; otherwise, h is a nonminimal winning candidate. We consider whether a clone of a winning candidate is certain to be elected; if so, a minimal winning candidate will be displaced.

We say that a winning candidate can clone successfully if its clone is certain to be elected. For any two candidates j and h, denote the set of voters who support both j and h by $V(j,h) = \{i : j \in V_i, h \in V_i\}$ and denote the set of voters who support j but not h by $V(j,-h) = \{i : j \in V_i, h \notin V_i\}$.

Proposition 11.7 *Under AV, any nonminimal winning candidate can clone successfully. Under SAV, a nonminimal winning candidate, j, cannot clone successfully if and only if, for every winning candidate $h \neq j$,*

---

[12]AV-related systems, like proportional AV and sequential proportional AV (see note 8), seem to share AV's vulnerability, but we do not pursue this question here.

$$\sum_{i \in V(j,-h)} \frac{1}{|V_i|+1} < \sum_{i \in V(h,-j)} \frac{1}{|V_i|} < \sum_{i \in V(j,-h)} \frac{1}{|V_i|}$$

Proof: Suppose that j, a nonminimal winning candidate under AV, clones. After cloning, the approval scores of all original candidates, including j, are unchanged, and the approval score of j's clone is the same as j's. Therefore, both j and its clone have approval scores that exceed that of the original minimal-winning candidate(s), and both j and the clone will belong to the winning set, to the exclusion of an original minimal-winning candidate.

Now suppose that j, a nonminimal winning candidate under SAV, clones. Clearly, j succeeds at cloning if and only if j and its clone displace some winning candidate h whose satisfaction score is necessarily less than $s(j)$. For such a candidate, h, we must have

$$s(j) = \sum_{i \in V(j,h)} \frac{1}{|V_i|} + \sum_{i \in V(j,-h)} \frac{1}{|V_i|} > s(h) = \sum_{i \in V(h,j)} \frac{1}{|V_i|} + \sum_{i \in V(h,-j)} \frac{1}{|V_i|}$$

or, in other words,

$$\sum_{i \in V(h,-j)} \frac{1}{|V_i|} < \sum_{i \in V(j,-h)} \frac{1}{|V_i|}$$

Let $s^n(j)$ and $s^n(h)$ be the satisfaction scores of j and h after cloning. If cloning fails to displace h, then it must be the case that

$$s^n(h) = \sum_{i \in V(j,h)} \frac{1}{|V_i|+1} + \sum_{i \in V(h,-j)} \frac{1}{|V_i|} > \sum_{i \in V(h,j)} \frac{1}{|V_i|+1} + \sum_{i \in V(j,-h)} \frac{1}{|V_i|+1} = s^n(j)$$

or, in other words,

$$\sum_{i \in V(j,-h)} \frac{1}{|V_i|+1} < \sum_{i \in V(h,-j)} \frac{1}{|V_i|},$$

which is easily seen to complete the proof. $\square$

Note that the second inequality of Proposition 11.7 is equivalent to $s(j) > s(h)$, which means that the original satisfaction score of h must be less than the original satisfaction score of j, so that the clone displaces a lower-ranked candidate.

To see that the condition of Proposition 11.7 has bite, consider an example with $m = 4$ candidates and 17 voters—who are to elect 2 candidates—and vote as follows:

- 6 voters: *ab*
- 6 voters: *ac*
- 5 voters: *d*

Under SAV, the scores are $s(a) = 6$, $s(b) = s(c) = 3$, and $s(d) = 5$, so the winning subset is $\{a,d\}$. If a clones, then both a and its clone have satisfaction scores of 4, whereas the score of d remains 5, so d is not displaced by a's clone and cloning is unsuccessful.

We conclude that, relative to AV, SAV discourages the formation of clones unless a candidate's support is sufficiently large that he or she can afford to transfer a substantial part of it to a clone and still win—in which case the clone as well as the original candidate would both seem deserving of election.

We turn next to a real election, in which AV was used to elect multiple winners, and assess the possible effects of SAV, had it been used. We are well aware that voters might have voted differently under SAV and take up this question in Section 11.4.

11.3 THE GAME THEORY SOCIETY ELECTION

In 2003, the Game Theory Society used AV for the first time to elect 12 new council members from a list of 24 candidates. (The council comprises 36 members, with 12 elected each year to serve 3-year terms.[13]) We give below the numbers of members who cast votes for from 1 to all 24 candidates (no voters voted for between 19 and 23 candidates):

Cast	1	2	3	4	5	6	7	8	9	10	11	12	13	14	15	16	17	18	24
Voters	3	2	3	10	8	6	13	12	21	14	9	25	10	7	6	5	3	3	1

Casting a total of 1574 votes, the 161 voters, who constituted 45% of the GTS membership, approved, on average, $1574/161 \approx 9.8$ candidates; the median number of candidates approved of, 10, is almost the same.[14]

The modal number of candidates approved of is 12 (by 25 voters), echoing the ballot instructions that 12 of the 24 candidates were to be elected. The approval of candidates ranged from a high of 110 votes (68.3% approval) to a low of 31 votes (19.3% approval). The average approval received by a candidate was 40.7%.

Because the election was conducted under AV, the elected candidates were the 12 most approved, who turned out to be all those who received at least 69 votes (42.9% approval). Do these AV winners best represent the electorate? With the caveat that the voters might well have approved of different candidates if SAV rather than AV had been used, we compare next how the outcome would have been different if SAV had been used to aggregate approval votes.

[13]The fact that there is exit from the council after 3 years makes the voting incentives different from a society in which (i) members, once elected, do not leave and (ii) members decide who is admitted [5].
[14]Under SAV, whose results we present next, the satisfaction scores of voters in the GTS election are almost uncorrelated with the numbers of candidates they approved of, so the number of candidates approved of does not affect, in general, a voter's satisfaction score—at least if he or she had voted the same as under AV (a big "if" that we investigate later).

Under SAV, 2 of the 12 AV winners would not have been elected.[15] Each set of winners is given below—ordered from most popular on the left to the least popular on the right, as measured by approval votes—with differences between those who were elected under AV and those who would have been elected under SAV underscored:

AV : 111111111 1 1 1000000000000
SAV : 111111111 0 1 0 1 1 0000000000

Observe that the AV winners who came in 10th (70 votes) and 12th (69 votes) would have been displaced under SAV by the candidates who came in 13th (66 votes) and 14th (62 votes), according to AV, and just missed out on being elected.

Recall that a voter is *represented* by a subset of candidates if he or she approves of at least one candidate in that subset. The elected subset under SAV represents all but 2 of the 161 voters, whereas the elected subset under AV failed to represent 5 of the 161 voters. But neither of these subsets is the best possible; the greedy algorithm gives a subset of 9 candidates that represents all 161 voters, which includes 5 of the AV winners and 6 SAV winners, including the 2 who would have won under SAV but not under AV.

It turns out, however, that this is not a minimal representative set of winners: There are more than a dozen subsets with 8 candidates, but none with 7 or fewer candidates, that represent all 161 voters, making 8 the size of a minimal representative set.[16] To reduce the number of such sets, it seemed reasonable to ask which one maximizes the minimum satisfaction of all 161 voters.

This criterion, however, was not discriminating enough to produce one subset that most helped the least-satisfied voter: There were 4 such subsets that gave the least-satisfied voter a satisfaction score of $1/8 = 0.125$, that is, that elected one of his or her approved candidates. To select the "best" among these, we used as a second criterion the one that maximizes total voter satisfaction, which gives

• 100111000000110001000001.

Observe that only 4 of the 8 most approved candidates are selected; moreover, the remaining 4 candidates include the least-approved candidate (24th on the list).

But ensuring that every voter approves of at least one winner comes at a cost. The total satisfaction that the aforementioned minimal representative set gives is 60.9, whereas the subset of 8 candidates that maximizes total voter satisfaction—without regard to giving every voter an approved representative—is

---

[15]Under the "minimax procedure" [6, 10], 4 of the 12 AV winners would not have been elected. These 4 include the 2 who would not have been elected under SAV; they would have been replaced by 2 who would have been elected under SAV. Thus, SAV partly duplicates the minimax outcome. It is remarkable that these two very different systems agree, to an extent, on which candidates to replace to make the outcome more representative.

[16]We are grateful to Richard F. Potthoff for writing an integer program that gave the results for the GTS election that we report on next.

• 111110011000100000000000.

Observe that 6 of the 8 most approved candidates are selected (the lowest candidate is 13th on the list). The total satisfaction of this subset is 74.3, which is a 22% increase over the above score of the most satisfying minimal representative set. We leave open the question whether such an increase in satisfaction is worth the disenfranchisement of a few voters.

In choosing a minimal representative set, the size of an elected voting body is allowed to be endogenous. In fact, it could be as small as one candidate if one candidate is approved of by everybody.

By contrast, if the size of the winning set is fixed and once a minimal representative set has been selected—if that is possible—then one can compute the larger-than-minimal representative set that maximizes total voter satisfaction. In the case of the GTS, because there is a minimal representative set with only 8 members, we know that a 12-member representative set is certainly feasible.

In making SAV and related calculations for the GTS election, we extrapolated from the AV ballots. We caution that our extrapolations depend on the assumption that GTS voters would not have voted differently under SAV than under AV. In particular, under SAV, would GTS voters have been willing to divide their one vote among multiple candidates if they thought that their favorite candidate needed their undivided vote to win?

11.4 VOTING FOR MULTIPLE CANDIDATES UNDER SAV: A DECISION-THEORETIC ANALYSIS

To try to answer the forgoing question, we begin by analyzing a much simpler situation—there are 3 candidates, with 2 to be elected. As shown in Table 11.1, there are exactly 19 contingencies in which a single voter's strategy can be decisive—that is, make a difference in which 2 of the 3 candidates are elected—by making or breaking a tie among the candidates. In these contingencies are the so-called states of nature.

In Table 11.1, the contingencies are shown as the numbers of votes that separate the three candidates.[17] For example, contingency 4 $(1, 1/2, 0)$ indicates that candidate a is ahead of candidate b by $1/2$ vote and that candidate b is ahead of candidate c by $1/2$

---

[17]Notice that the numbers of votes shown in a contingency are all within 1 of each other, enabling a voter's strategy to be decisive; these numbers need not sum to an integer, even though the total number of voters and votes sum to an integer. For example, contingency 4 can arise if there are 2 ab voters and 1 ac voter, giving satisfaction scores of $3/2, 1$, and $1/2$, respectively, to a, b, and c, which sum to 3. But this is equivalent to contingency 4 $(1, 1/2, 0)$, obtained by subtracting $1/2$ from each candidate's score, whose values do not sum to an integer. Contingencies of the form $(1, 1/2, 1/2)$, while feasible, are not included, because they are equivalent to contingencies of the form $(1/2, 0, 0)$—candidate a is $1/2$ vote ahead of candidates b and c.

TABLE 11.1 Strategies and outcomes for 19 contingencies in 3-candidate, 2-winner Elections in which one voter can be decisive

Strategy	1	2	3	4	5	6	7	8	9	10
	1,1,0	1,0,1	0,1,1	1,$\frac{1}{2}$,0	1,0,$\frac{1}{2}$	$\frac{1}{2}$,1,0	0,1,$\frac{1}{2}$	$\frac{1}{2}$,0,1	0,$\frac{1}{2}$,1	1,0,0
a	ab*	ac*	a−b|c*	ab*	ac	ab*	ab*	ac*	ac*	a−b|c
ab	ab*	ac*	bc	ab*	a−b|c	ab*	b−a|c	ac*	bc	ab*
b	ab*	a|b|c*	bc	ab*	ab*	ab*	bc	bc	bc	ab*
bc	ab*	ac*	bc	ab*	ac	b−a|c	bc	ac	bc	a−b|c
c	a|b|c	ab*	bc	ac	ac	bc	bc	c−a|b	bc	ac
ac	ab*	ab*	bc	a−b|c	ac	ab*	bc	ac	c−a|b	ac

Strategy	11	12	13	14	15	16	17	18	19
	0,1,0	0,0,1	$\frac{1}{2}$,$\frac{1}{2}$,0	$\frac{1}{2}$,0,$\frac{1}{2}$	0,$\frac{1}{2}$,$\frac{1}{2}$	$\frac{1}{2}$,0,0	0,$\frac{1}{2}$,0	0,0,$\frac{1}{2}$	0,0,0
a	ab*	ac*	ab*	ac	a−b|c*	a−b|c	ab*	ac*	a−b|c
ab	ab*	c−a|b	ab*	a−b|c*	b−a|c	ab*	ab*	a|b|c*	ab*
b	b−a|c	bc	ab*	b−a|c	bc	ab*	b−a|c	bc	b−a|c
bc	bc	bc	b−a|c	c−a|b	bc	a|b|c	bc	bc	bc
c	bc	c−a|b	c−a|b	ac	bc	ac	bc	c−a|b	c−a|b
ac	b−a|c	ac*	a−b|c	ac	c−a|b	ac	a/b/c	ac*	ac

The outcomes produced by a voter's strategies in the left columns of Table 11.1 are indicated (i) by the two candidates elected (e.g., ab), (ii) by a candidate followed by two candidates who tie for second place, separated by a vertical line (e.g., a − b|c), or (iii) by the candidates in a three-way tie (a|b|c). For the focal voter with preference a ≻ b ≻ c, outcomes marked with an asterisk indicate his or her best or tied-for-best outcomes for each contingency; underscores indicate a uniquely best outcome.

vote.[18] The outcomes produced by a voter's strategies in the left column of Table 11.1 are indicated (i) by the two candidates elected (e.g., ab), (ii) by a candidate followed by two candidates who tie for second place, indicated by a vertical line (e.g., $a - b|c$), or (iii) by all the candidates in a three-way tie ($a|b|c$).

A voter may choose any one of the six strategies by approving either one or two candidates. (Approving all three candidates, or none at all, would have no effect on the outcome, so we exclude them as strategies that can be decisive.[19]) To determine the optimal strategies of a voter, whom we call the *focal voter*, we posit that he or she has strict preference $a \succ b \succ c$.

We assume that the focal voter has preferences not only for individual candidates but also over sets of two or three candidates. In particular, given this voter's strict preference for individual candidates, we assume the following preference relations for pairs and triples of candidates:

$$ab \succ a - b|c \succ ac \approx b - a|c \approx a|b|c \succ c - a|b \succ bc,$$

where "$\approx$" indicates indifference, or a tie, between pairs of outcomes: One outcome in the pair is not strictly better than the other. Thus, the certain election of a and c (ac) is no better nor worse than either the certain election of b and the possible election of either a or c ($b - a|c$), or the possible election of any pair of a, b, or c ($a|b|c$).[20]

We have indicated with asterisks, for each contingency, outcomes that are the best or the tied-for-best for the focal voter; underscores indicate a *uniquely* best outcome. In contingency 4, for example, there are four starred ab outcomes, all of which give the focal voter's top two candidates. These outcomes are associated with the focal voter's first four strategies; by contrast, his or her other two strategies elect less preferred sets of candidates.

In contingency 7, outcome ab, associated with the focal voter's strategy a, is not only starred but also underscored, because it is a uniquely best outcome. A strategy that is associated with a uniquely best outcome is *weakly undominated*, because no other strategy can give at least as good an outcome for that contingency.

Observe from Table 11.1 that strategy a leads to a uniquely best outcome in 4 contingencies (3, 7, 9, and 15), strategy ab in 2 contingencies (14 and 19), and strategy b in 1 contingency (5), rendering all these strategies weakly undominated. It is

---

[18]We have not shown contingencies in which any candidate is guaranteed a win or a loss. The 19 contingencies in Table 11.1 represent all states in which the strategy of a voter can make each of the three candidates a winner or a loser, rendering them *3-candidate competitive contingencies*.

[19]If there were a minimum number of votes (e.g., a simple majority) that a candidate needs in order to win, then abstention or approving everybody could matter. But here we assume the two candidates with the most votes win, unless there is a tie, in which case we assume there is (unspecified) tie-breaking rule.

[20]Depending on the tie-breaking rule, the focal voter may have strict preferences over these outcomes, too. Because each allows for the possibility of any pair of winning candidates, we chose not to distinguish them. To be sure, $a - b|c$ (second best) and $c - a|b$ (second worst) also allow for the possibility of any pair of winning candidates, but the fact that the first involves the certain election of a, and the second the certain election of c, endows them with, respectively, a more-preferred and less-preferred status than the three outcomes among which the focal voter is indifferent.

not difficult to show that the focal voter's other three strategies, all of which involve approving of c, are weakly dominated:

1. a, ab, and b weakly dominate bc
2. a and ab weakly dominate c
3. a weakly dominates ac.

In no contingency does a weakly dominated strategy lead to a better outcome than a strategy that dominates it, and in at least one contingency, it leads to a strictly worse outcome.

Among the weakly undominated strategies, a leads to at least a tied-for-best outcome in 14 contingencies, ab in 13 contingencies (9 of the a and ab contingencies overlap), and b in 8 contingencies. In sum, it is pretty much a toss-up between weakly undominated strategies a and ab, with b a distant third-place finisher.

It is no fluke that the focal voter's three strategies that include voting for candidate c (c, ac, and bc) are all weakly dominated.

Proposition 11.8   *If there is more than one candidate, a strategy that includes approving of a least-preferred candidate is weakly dominated under SAV.*

Proof: Let W be a focal voter's strategy that includes approving of a least-preferred ("worst") candidate, w. Let $\overline{W}$ be the focal voter's strategy of duplicating W, except for approving of w, unless W involves voting only for w. In that case, let $\overline{W}$ be a strategy of voting for any candidate other than w.

Assume that that the focal voter chooses $\overline{W}$. Then $\overline{W}$ will elect the same candidates that W does except, possibly, for w. However, there will be at least one contingency in which $\overline{W}$ does not elect w with certainty (e.g., in a contingency in which w is assigned 0) and W does, but none in which the reverse is the case. Hence, $\overline{W}$ weakly dominates W.   □

In Table 11.1, voting for a second choice, candidate b, is a weakly undominated strategy, because it leads to a uniquely best outcome in contingency 5. This is not the case for AV, in which a weakly undominated strategy includes always approving of a most-preferred candidate—not just never approving of a least-preferred candidate [7].

Thus, SAV admits more weakly undominated strategies than AV. In some situations, it may be in the interest of a voter to approve of set of strictly less-preferred candidates and forsake a set of strictly more-preferred candidates. As a case in point, assume a focal voter strictly ranks 5 candidates as follows, $a \succ b \succ c \succ d \succ e$, and 2 candidates are to be elected. In contingency $(a,b,c,d,e) = (0,0,3/4,1,1)$, strategy ab elects candidates d and e, the focal voter's two worst choices, whereas strategy cd, comprising less-preferred candidates, elects candidates c and d, which is a strictly better outcome.

To conclude, our decision-theoretic analysis of the 3-candidate, 2-winner case demonstrates that voting for one's two most-preferred candidates leads to the same number of uniquely best and about the same number of at least tied-for-best out-comes, despite the fact that voters who vote for more than one candidate must split their votes evenly under SAV. We plan to investigate whether this finding carries over to elections in which there are more candidates and more winners, as well as the effect that the ratio of candidates to winners has.

Unlike AV, approving of just a second choice when there are 3 competitive can-didates is a weakly undominated strategy under SAV, though it is uniquely optimal in only one of the 19 contingencies.[21] More generally, while it is never optimal for a focal voter to select a strategy that includes approving of a worst candidate (not sur-prising), sometimes it is better to approve of strictly inferior candidates than strictly superior candidates (more surprising), though this seems relatively rare.

11.5   VOTING FOR POLITICAL PARTIES

In most , voters vote for political parties, which win seats in a parliament in propor-tion to the number of votes they receive. We now propose a SAV-based party voting systems in which voters would not be restricted to voting for one party but could vote for as many parties as they like. If a voter approves of x parties, each approved party's score would increase by $1/x$.

Unlike standard methods, some of which we will describe shortly, our SAV system does *not* award seats according to the *quota* to which a party is entitled. (A party's quota is a number of seats such that its proportion of the seats is exactly equal to the proportion of its supporters in the electorate. Note that a quota is typically not an integer.) Instead, parties are allocated seats to maximize total voter satisfac-tion, measured by the fractions of nominees from voters' approved parties that are elected.

We begin our discussion with an example, after which we formalize the application of SAV to party-list systems. Then we return to the example to illustrate the possible effects of voting for more than one party.

11.5.1   Bullet voting

Effectively, SAV requires that the number of candidates nominated by a party equal its *upper quota* (its quota rounded up). To illustrate, consider the following 3-party, 11-voter example, in which 3 seats are to be filled (we indicate parties by capital letters).

---

[21]To the degree that voters have relatively complete information on the standing of candidates (e.g., from polls), they can identify the most plausible contingencies and better formulate optimal strategies, taking into account the likely optimal strategies of voters with opposed preferences. In this situation, a game-theoretic model would be more appropriate than a decision-theoretic model for analyzing the consequences of different voting procedures. We plan to investigate such models in the future. For models of strategic behavior in proportional-representation systems—but not those that use an approval ballot—see Ref. [20].

- 5 voters support A
- 4 voters support B
- 2 voters support C.

Assume that the supporters of each party vote exclusively for it. Party i's quota, q_i, is its proportion of votes times 3, the number of seats to be apportioned:

- $q_A = (5/11)(3) \approx 1.364$
- $q_B = (4/11)(3) \approx 1.091$
- $q_C = (2/11)(3) \approx 0.545$.

Under SAV, each party is treated as if it had nominated a number of candidates equal to its upper quota, so A, B, and C have effectively nominated 2, 2, and 1 candidates, respectively—2 more than the number of candidates to be elected. We emphasize that the numbers of candidates nominated are not a choice that the parties make but follow from their quotas, based on the election returns.

SAV finds apportionments of seats to parties that (i) maximize total voter satisfaction and (ii) are *monotonic*: A party that receives more votes than another cannot receive fewer seats.

In our previous example, there are three monotonic apportionments to parties (A, B, C)—$(3,0,0)$, $(2,1,0)$ and $(1,1,1)$—giving satisfaction scores of

- $s(3,0,0) = 5(1) + 4(0) + 2(0) = 5$
- $s(2,1,0) = 5(1) + 4(\frac{1}{2}) + 2(0) = 7$
- $s(1,1,1) = 5(\frac{1}{2}) + 4(\frac{1}{2}) + 2(1) = 6\frac{1}{2}$.

Apportionment $(2,1,0)$ maximizes the satisfaction score, giving

- 5 A voters satisfaction of 1 for getting A's 2 nominees elected
- 4 B voters satisfaction of $1/2$ for getting 1 of B's 2 nominees elected
- 2 C voters satisfaction of 0, because C's nominee is not elected.

11.5.2 Formalization

In a election of k candidates from lists provided by parties $1, 2, \ldots, p$, suppose that party j has v_j supporters and that $\sum_{j=1}^{p} v_j = n$. Then party j's quota is $q_j = (v_j/n)k$. If q_j is an integer, party j is allocated exactly q_j seats.

We henceforth assume that all parties' quotas are nonintegral. Then party j receives either its lower quota, $_j = \lfloor q_j \rfloor$, or its upper quota, $u_j = \lceil q_j \rceil$. Of course, $u_j = l_j + 1$. In total, $r = k - \sum_{j=1}^{p} l_j$ parties receive their upper quota rather than their lower quota. By assumption, $r > 0$. The set of parties receiving upper quota, $S \subseteq [p] = \{1, 2, \ldots, p\}$, is chosen to maximize the total satisfaction of all voters, $s(S)$, subject to $|S| = r$.

Recall that when electing individual candidates, SAV chooses *candidates* that maximize total voter satisfaction. When allocating seats to parties, SAV finds *apportionments* of seats that maximize total voter satisfaction.

The apportionment in our example is *not* an apportionment according to the Hamilton method (also called "largest remainders"), which begins by giving each party the integer portion of its exact quota (1 seat to A and 1 seat to B). Then any remaining seats go to the parties with the largest remainders until the seats are exhausted, which means that C, with the largest remainder (0.545), gets the 3rd seat, yielding the apportionment $(1,1,1)$ to (A,B,C).

There are five so-called apportionment [4]. Among these, only the Jefferson/ d'Hondt method, which favors larger parties, gives the SAV apportionment of $(2, 1, 0)$ in our example.[22] This is no accident, as shown by the next proposition.

Proposition 11.9   *The SAV voting system for political parties gives the same apportionment as the method, but with an upper-quota restriction.[23] SAV apportionments also satisfy lower quota and thus satisfy quota.*

Proof: Each of party j's v_j voters gets satisfaction of 1 if party j is allocated its upper quota and satisfaction l_j/u_j if party j is allocated its lower quota. If the subset of parties receiving upper quota is $S \subseteq [p]$, then the total satisfaction over all voters is

$$s(S) = \sum_{j \in S} v_j + \sum_{j \notin S} v_j \left(\frac{l_j}{u_j} \right) = \sum_{j=1}^{p} v_j - \sum_{j \notin S} \frac{v_j}{u_j} \tag{11.2}$$

where the latter equality holds because $l_j/u_j = 1 - 1/u_j$. The SAV apportionment is, therefore, determined by choosing S such that $|S| = r$ and S maximizes $s(S)$, which by (11.2) can be achieved by choosing $S^c = [p] - S$ to minimize $\sum_{j \in S^c} v_j/u_j$. Clearly, this requirement is achieved when S contains the r largest values of v_j/u_j.

To compare with the Jefferson/d'Hondt apportionment, assume that all parties have already received l_j seats. The first party to receive u_j seats is, according to Jefferson/d'Hondt, the party, j, that maximizes $v_j/l_j + 1 = v_j/u_j$. After this party's allocation has been adjusted to equal its upper quota, remove it from the set of parties. The next party to receive u_j according to Jefferson/d'Hondt is the remaining party with the greatest value of v_j/u_j, and so on. Clearly, parties receive seats in decreasing order of their values of v_j/u_j. Because Jefferson/d'Hondt apportionments always satisfy

---

[22] The Jefferson/d'Hondt method allocates seats sequentially, giving the next seat to the party that maximizes $v/a + 1$, where v is its number of voters and a is its present apportionment. Thus, the 1st seat goes to A, because $5 > 4 > 2$ when $a = 0$. Now $a = 1$ for A and remains 0 for B and C. Because $4/1 > 5/2 > 2/1$, B gets the 2nd seat. Now $a = 1$ for A and B and remains 0 for C. Because $5/2 > 4/2 = 2/1$, A gets the 3rd seat, giving an apportionment of $(2,1,0)$ to (A,B,C). The divisor method that next-most-favors large parties is the Webster/Sainte-Laguë method, under which the party that maximizes $(v/a + 1/2)$ gets the next seat. After A and B get the first two seats, the 3rd seat goes to C, because $\frac{2}{1/2} > \frac{5}{3/2} > \frac{4}{3/2}$, so the Webster/Sainte-Laguë method gives an apportionment of $(1, 1, 1)$ to (A,B,C).

[23] There are objective functions with a min/max operator that Jefferson/d'Hondt also optimizes ([4], p. 105; [13]), but they are more difficult to justify in the context of seat assignments.

lower quota [4, pp. 91, 130], SAV apportionments satisfy both quota, that is, upper and lower).[24]   □

A consequence of this procedure is that SAV apportionments are certain to satisfy upper quota, unlike (unrestricted) Jefferson/d'Hondt apportionments. Effectively, parties cannot nominate candidates for, and therefore cannot receive, more seats than their quotas rounded up.[25]

Because SAV produces Jefferson/d'Hondt apportionments, except for the upper-quota restriction, SAV favors large parties. Nevertheless, small parties will not be wiped out, provided their quotas are at least 1, assuming that no threshold, or minimum vote to qualify for a seat, is imposed (in some countries, the threshold is 5% or more of the total vote).

11.5.3 Multiple-party voting

If a voter votes for multiple parties, his or her vote is equally divided among all his or her approved parties. To illustrate in our previous example, suppose parties B and C reach an agreement on policy issues, so that their $6(=4+2)$ supporters approve of both parties. Meanwhile, the 5 party A supporters continue to vote for A alone.

Now the vote totals of B and C are taken to equal $6(\frac{1}{2}) = 3$, making the quotas of the three parties the following:

1. $q_A = (5/11)3 \approx 1.364$
2. $q_B = (3/11)3 \approx 0.818$
3. $q_C = (3/11)3 \approx 0.818$

By the algorithm above, party seats are allocated in decreasing order of $\frac{v_j}{u_j}$. Because these ratios are $5/2 = 2.5$, $3/1 = 3.0$, and $3/1 = 3.0$ for parties A, B, and C respectively, it follows that the apportionment of seats is $(1,1,1)$. Compared with apportionment $(2,1,0)$ earlier with bullet voting, A loses a seat, B stays the same, and C gains a seat.

In general, parties that are too small to be represented at all cannot hurt themselves by approving of each other. However, the strategy may either help or hurt the

---

[24]The Jefferson/d'Hondt method with an upper-quota constraint is what Balinski and Young [4, p. 139] call Jefferson-Quota; SAV effectively provides this constraint. Balinski and Young [4, chap. 12] argue that because it is desirable that large parties be favored and coalitions encouraged in a parliament, the Jefferson/d'Hondt method should be used, but they do not impose the upper-quota constraint that is automatic under SAV. However, in earlier work [3], they—along with [21]—looked more favorably on such a constraint.

[25]This SAV-based system could be designed for either a closed-list or an open-list system of proportional representation. In a closed-list system, parties would propose an ordering of candidates prior to the election; the results of the election would tell them how far down the list they can go in nominating their upper quotas of candidates. (For a different approach to narrowing the field in elections, see Ref. [9].) By contrast, in an open-list system, voters could vote for individual candidates; the candidates' vote totals would then determine their positions on their party lists.

combined seat count of parties that achieve at least one seat on their own. In the previous example, B and C supporters together ensure themselves of a majority of 2 seats if they approve of each other's party, but they may nonetheless choose to go their separate ways.

One reason is that B does not individually benefit from supporting C; presumably, B's supporters would need to receive some collective benefit from supporting C to make it worth their while also to approve of C. Note that if only 2 of B's supporters also approve C, but both of C's supporters approve of B, the vote counts, $(5, 2 + 4/2, 4/2) = (5, 4, 2)$, would be exactly as they were originally, so the outcome of the election would be unchanged.

A possible way around this problem is for B and C to become one party, assuming that they are ideologically compatible, reducing the party system to just two parties. Because the combination of B and C has more supporters than A does, this combined party would win a majority of seats.

11.6 CONCLUSIONS

We have proposed a new voting system, SAV, for multiwinner elections. It uses an approval ballot, whereby voters can approve of as many candidates or parties as they like, but they do not win seats based on the number of approval votes they receive.

We first considered the use of SAV in elections in which there are no political parties, such as in electing members of a city council. SAV elects the set of candidates that maximizes the satisfaction of all voters, where a voter's satisfaction is the fraction of his or her approved candidates who are elected. This measure works equally well for voters who approve of few or of many candidates and, in this sense, can mirror a voter's personal tastes.

A candidate's satisfaction score is the sum of the satisfactions that his or her election would give to all voters. Thus, a voter who approves of a candidate contributes $1/x$ to the candidate's satisfaction score, where x is the total number of candidates of whom the voter approves. The winning set of candidates is the one with the highest individual satisfaction scores.

Among other findings, we showed that SAV and AV may elect disjoint sets of candidates. SAV tends to elect candidates that give more voters either partial or complete satisfaction—and thus representation—than does AV, but this is not universally true and is a question that deserves further investigation.

Additionally, SAV inhibits candidates from creating clones to increase their representation. But voting for a single candidate can be seen as risky for a voter, as the voter's satisfaction score will be either 0 or 1, so risk-averse voters may be inclined to approve of multiple candidates.

SAV may not elect a representative set of candidates—whereby every voter approves of at least one elected candidate—as we showed would have been the case in the 2003 election of the Game Theory Society Council. However, the SAV outcome would have been more representative than the AV outcome (given the approval ballots remained the same as in the AV election). Yet we also showed that a fully representative outcome could have been achieved with a smaller subset of candidates (8 instead of 12).

Because SAV divides a voter's vote evenly among the candidates he or she approves of, SAV may encourage more bullet voting than AV does. However, we found evidence that, in 3-candidate, 2-winner competitive elections, voters would find it almost equally attractive to approve of their two best choices as their single best choice. Unlike AV, they may vote for strictly less-preferred candidates if they think their more-preferred candidates cannot benefit from their help.

We think the most compelling application of SAV is to party-list systems. Each party would provide either an ordering of candidates or let the vote totals for individual candidates determine this ordering. Each party would then be considered to have nominated a number of candidates equal to its upper quota after the election. The candidates elected would be those that maximize total voter satisfaction among monotonic apportionments.

Because parties nominate, in general, more candidates than there are seats to be filled, not every voter can be completely satisfied. We showed that the apportionment of seats to parties under SAV gives the Jefferson/d'Hondt apportionment method with a quota constraint, which tends to favor larger parties while still ensuring that all parties receive at least their lower quotas.

To analyze the effects of voting for multiple parties, we compared a scenario in which voters bullet voted with a scenario in which they voted for multiple parties. Individually, parties are hurt when their supporters approve of other parties. Collectively, however, they may be able to increase their combined seat share by forming coalitions—whose supporters approve all parties in it—or even by merging. At a minimum, SAV may discourage parties from splitting up unless to do so would mean they would be able to recombine to form a new and larger party, as Kadima did in Israel.

Normatively speaking, we believe that better coordination by parties should be encouraged, because it would give voters a clearer idea of what to expect when they decide which parties to support—compared to the typical situation today, when voters can never be sure about what parties will join in a governing coalition and what its policies will be. Because this coordination makes it easier for voters to know what parties to approve of, and for party coalitions to form that reflect their supporters' interests, we believe that SAV is likely to lead to more informed voting and more responsive government in parliamentary systems.

11.7 SUMMARY

Under approval voting (AV), voters can approve of as many candidates or as many parties as they like. We propose a new system, SAV, that extends AV to multiwinner elections. However, the winners are not those who receive the most votes, as under AV, but those who maximize the sum of the satisfaction scores of all voters, where a voter's satisfaction score is the fraction of his or her approved candidates who are elected. SAV may give a different outcome from AV—in fact, SAV and AV outcomes may be disjoint—but SAV generally chooses candidates representing more diverse interests than does AV (this is demonstrated empirically in the case of a recent election of the Game Theory Society). A decision-theoretic analysis shows that all strategies under SAV, except approving of a least-preferred candidate, are undominated, so voters may rationally choose to approve of more than one candidate. In party-list systems, SAV apportions seats to parties according to the Jefferson/d'Hondt method with a quota constraint, which favors large parties and gives an incentive to smaller parties to coordinate their policies and forge alliances, even before an election, that reflect their supporters' coalitional preferences.

Acknowledgments

We thank Joseph N. Ornstein and Erica Marshall for valuable research assistance, and Richard F. Potthoff for help with computer calculations that we specifically acknowledge. We also appreciate the helpful comments of an anonymous reviewer. This chapter appears in Fara et al. [12]; we are grateful to Springer for giving us permission to republish it.

REFERENCES

1. Jorge Alcalde-Unzu and Marc Vorsatz, "Size Approval Voting," *Journal of Economic Theory*, **144** (3) 1187–1210 (2009).

2. Jose Apesteguia, Miguel A. Ballester, and Rosa Ferrer, "On the Justice of Decision Rules," *Review of Economic Studies*, **78** (1) 1–16 (2011).

3. Michel Balinski and H. Peyton Young, "Stability, Coalitions and Schisms in Proportional Representation Systems," *American Political Science Review*, **72** (3) 848–858 (1978).

4. Michel Balinski and H. Peyton Young, *Fair Representation: Meeting the Ideal of One Man, One Vote*, Yale University Press, New Haven, CT, 1982. Revised, Brookings Institution, Washington DC (2001).

5. Salvador Barberà, Michael Maschler, and Jonathan Shalev, "Voting for Voters: A Model of Electoral Evolution," *Games and Economic Behavior*, **37** (1) 40–78 (2001).

6. Steven J. Brams, *Mathematics and Democracy: Designing Better Voting and Fair-Division Procedures*, Princeton University Press, Princeton, NJ, (2008).

7. Steven J. Brams and Peter C. Fishburn, "Approval Voting," *American Political Science Review*, **72** (3) 831–847 (1978).

8. Steven J. Brams and Peter C. Fishburn, *Approval Voting*, Springer, New York, (1983). Reprinted, 2007.

9. Steven J. Brams and D. Marc Kilgour, "Narrowing the Field in Elections: The Next-Two Rule," *Journal of Theoretical Politics*, **24** (4) 507–525 (2012).

10. Steven J. Brams, D. Marc Kilgour, and M. Remzi Sanver, "A Minimax Procedure for Electing Committees," *Public Choice*, **132** (3–4) 401–420 (2007).

11. Edith Elkind, Jérôme Lang, and Abdallah Saffidine, "Choosing Optimal Sets of Alternatives Based on the Condorcet Criterion," *Twenty-Second International Joint Conference on Artificial Intelligence*, July, 19–22, 2011, Barcelona, Spain (2011).

12. Rudy Fara, Dennis Leech, and Maurice Salles (eds.), *Voting Power and Procedures: Essays in Honor of Dan Felsenthal and Moshe Machover*, Springer, Heidelberg, (2014).

13. Gonzalo Fernández de Córdoba and Alberto Penandés, "Institutionalizing Uncertainty: The Choice of Electoral Formulas," *Public Choice*, **141** (3–4) 391–403 (2009).

14. Richard M. Karp, "Reducibility among Combinatorial Problems." In R.E. Miller and J. W. Thatcher (eds.) *Complexity of Computer Calculations*, Plenum, New York, 85–103 (1972).

15. D. Marc Kilgour, "Using Approval Balloting in Multi-Winner Elections,". In Jean-François Laslier and M. Remzi Sanver (eds.) *Handbook on Approval Voting*, Springer, Berlin, 105–124 (2010).

16. D. Marc Kilgour and Erica Marshall, "Approval Balloting for Fixed-Size Committees." In Moshé Machover and Dan Felsenthal (eds.) *Electoral Systems: Paradoxes, Assumptions, and Procedures*, Springer, Berlin, 305–326 (2011).

17. Samuel Merrill, III and Jack H. Nagel, "The Effect of Approval Balloting on Strategic Voting under Alternative Decision Rules," *American Political Science Review*, **81** (2) 509–524 (1987).

18. Ariel D. Procaccia, Jeffrey S. Rosenschein, and Aviv Zohar, "On the Complexity of Achieving Proportional Representation," *Social Choice and Welfare*, **30** (3) 353–362 (2008).

19. Michel Regenwetter, Moon-Ho R. Ho, and Ilia Tsetlin, "Sophisticated Approval Voting, Ignorance Priors, and Plurality Heuristics: A Behavioral Social Choice Analysis in a Thurstonian Framework," *Psychological Review*, **114** (4) 994–1014 (2007).

20. Arkadii Slinko and Shaun White, "Proportional Representation and Strategic Voters," *Journal of Theoretical Politics*, **22** (3) 301–332 (2010).

21. Jonathan W. Still, "A Class of New Methods for Congressional Apportionment," *SIAM Journal on Applied Mathematics*, **37** (2) 401–418 (1979).

12

MODELING MUSICAL RHYTHM MUTATIONS WITH GEOMETRIC QUANTIZATION

GODFRIED T. TOUSSAINT

Department of Computer Science, New York University Abu Dhabi, Abu Dhabi, United Arab Emirates

12.1 INTRODUCTION

The field of evolutionary musicology is concerned with characterizing what music is, determining its origins and cross-cultural universals, evaluating its usefulness for human survival and evolution, and creating a phylogeny of types of music [5, 32]. For the purpose of mathematical modeling, music may be viewed as a multidimensional vector of music's fundamental features. Such features include rhythm, melody, harmony, timbre, tempo, volume, and pitch, among others. Most scholars consider rhythm to be the fundamental and oldest (in evolutionary terms) of these properties. It is perhaps for this reason that a phylogeny of music may sometimes be correctly constructed from rhythmic features alone, in spite of neglecting the non-rhythmic features of music [26, 31, 34]. In order to perform a phylogenetic analysis of a family of rhythms, a convenient approach makes use of a distance (dissimilarity) matrix calculated from all pairs of rhythms in the family [26]. A popular and successful distance measure between two rhythms modeled as sequences of symbols is the *edit* (also Levenshtein) distance, which measures the distance in terms of the minimum number of mutations needed to transform one sequence into the other [16, 19, 23, 30].

Mathematical and Computational Modeling: With Applications in Natural and Social Sciences, Engineering, and the Arts, First Edition. Roderick Melnik.
© 2015 John Wiley & Sons, Inc. Published 2015 by John Wiley & Sons, Inc.

Typically the mutations employed in the edit distance are simple local transformations consisting of insertions, deletions, and substitutions of symbols.

A mathematical analysis of musical rhythm presupposes what exactly a musical rhythm is and how it is represented. In the music literature, there are more than 50 ways to define musical rhythm (see chapter 1 of *The Geometry of Musical Rhythm*, pp. 1–4) [31]. None of these definitions are necessarily any more "correct" than any other. They just describe different attributes of musical rhythm when it is viewed through different lenses. Divergent analyses are served well by distinct definitions. For the mathematical approach to modeling transformations of one musical rhythm to another, it is convenient to represent rhythms as sequences of binary symbols (pulses) that indicate note onsets or rests, where each note or rest occupies one unit of time [18]. For example, the well-known *clave son* rhythm widely popular in disparate cultures spread all over the world [29] may be notated in this way by the sequence [x . . x . . x . . . x . x . . .], where the symbol "x" denotes the onset of a sound and "." a silent pulse.

Here we are concerned with transforming rhythms from one category of rhythms to another. Musical rhythms may be classified into a variety of categories depending on their properties. One category called *aksak* rhythms consists of patterns made up of the concatenations of the two-pulse (binary) and three-pulse (ternary) patterns [x .] and [x . .], respectively [4, 10]. In the music literature, these units are called *duple*-time and *triple*-time, respectively. The present study is concerned with two classes of rhythms here referred to as *binary* and *ternary*. Rhythms in a cycle of 16 pulses (also two, four, and eight) are called *binary*. Essentially, binary rhythms have an even number of pulses that is divisible by two without remainder and is not divisible by three. By contrast, rhythms made up of 12 pulses (also three and six) are called *ternary*. Ternary rhythms are divisible by three without remainder. The distinction between binary and ternary rhythms is useful because binary rhythms are considered to be simpler than ternary rhythms [13] and they permit a hierarchical division into two equal parts at every level of the hierarchy. This is not true of ternary rhythms [7, 17]. That ternary rhythms are generally considered to be more complex than binary rhythms is one possible reason for their relative perceptual instability and their resulting mutation, since complex rhythms tend to be perceptually mapped into simpler categories [8].

One of the most prominent West African ternary rhythms is the twelve-pulse, five-onset *fume-fume* (also standard pattern) with sequence [x . x . x . . x . x . .] [1]. The ethnomusicologist Rolando Pérez Fernández [20, 21] has postulated the hypothesis that ternary rhythms such as the *fume-fume* were mutated into binary rhythms, in this case the *clave son*, through a process of cultural blending, by which the ternary rhythms of the African slaves brought to the New World mixed with Spanish songs characterized by binary rhythms. He calls this process *binarization*. By contrast, some ethnomusicologists such as Gerhard Kubik have argued that rhythms either travel intact during cultural migrations or they do not travel at all [14, 15]. Here several binarization algorithms that result from geometric quantization schemes are examined using the transformation of the *fume-fume* into the *clave son* as a case study, in order to assess how well the geometric models fit the musicological rules described by Pérez Fernández [20], and which of these binarization schemes may have more relevance to psychological theories of perceived rhythm similarity. It is hoped that this study sheds some light on the likelihood of this mutation between these two rhythms and that it suggests psychological experiments to determine which binarization algorithm is the better predictor of human perceived similarity.

12.2 RHYTHM MUTATIONS

In a biological context, a mutation is a change in the DNA sequence (molecule) of an organism. On the other hand, in this study, a mutation is defined as a transformation of one binary sequence into another. Transformations may be classified into two broad categories: global and local. In a local transformation, a sounded pulse and a silent pulse (rest) that are adjacent to each other in the sequence may swap positions in the rhythmic cycle. For example, the 16-pulse, 5-onset clave son rhythmic pattern [x . . x . . x . . . x . x . . .] may be transformed into the more syncopated rumba rhythmic pattern [x . . x . . . x . . x . x . . .] by swapping the third sounded pulse with the silent pulse that follows it. This is the mathematical definition of a swap that results from modeling a rhythm as a sequence of bits (binary digits). By contrast, a musical definition of a swap would emphasize the shifting of a note onset. Other mutation operations often include the additional operations employed in the edit distance [23], including insertions, deletions, and substitutions of onsets and rests that may lengthen or shorten the duration of the rhythm. For a discussion of various other types of rhythmic mutations employed in music see Ref. [3].

12.2.1 Musicological rhythm mutations

Rolando Pérez Fernández [20] provides a list of 6- and 12-pulse rhythms along with their binarized counterparts. This contains two binarizations of the 12-pulse ternary rhythm [x . x . x . . x . x . .]. The first of these is the clave son with duration pattern [x . . x . . x . . x . x . . .], and the second is a variant of the Brazilian bossa-nova rhythm given by [x . . x . . x . . x . . x . . .], one of the most common rhythms used in electronic dance music [6]. Pérez Fernández constructs his list of binarized rhythms by first decomposing the ternary rhythms into the shorter ternary metric *feet* described in Greek antiquity, such as the *trochee* [x . x], the *iamb* [x x .], and the *tribrach* [x x x]. These feet are then matched to all their possible binary counterparts that appear in the practice of Latin-American music. Finally, these binary pieces are concatenated to produce the final binarized rhythms. Such an approach reflects an exhaustive combinatorial generation of the existing binary rhythms, and thus does not necessarily isolate the most probable matched pairs of binary-ternary mutations. Hence more than one binarization of the fume–fume may result, in spite of the fact that the clave son is considered to be its veridical binarization. Pérez Fernández does not offer any *a priori* algorithms for generating the most likely binarizations nor suggest perceptual models for explaining their occurrence.

12.2.2 Geometric rhythm mutations

A recent study of geometric algorithms for modeling four types of musical rhythm mutations [11] revealed that most of the binarizations of the ternary rhythms listed by Pérez Fernández [20] are accounted for by those obtained with one or another of these four geometric mutations. Figure 12.1 illustrates, using circular lattice notation, the four types of mutations for the case of three-pulse, three-onset ternary rhythms binarized to four-pulse binary rhythms with either three or two onsets. The binary rhythms have their onsets on the circular lattice consisting of the four equi-spaced points corresponding to the hours of noon, 3 P.M., 6 P.M., and 9 P.M., whereas the ternary rhythms have their onsets at the three equi-spaced lattice points corresponding to the hours of noon, 4 P.M., and 8 P.M. The arrows in Figure 12.1 indicate how

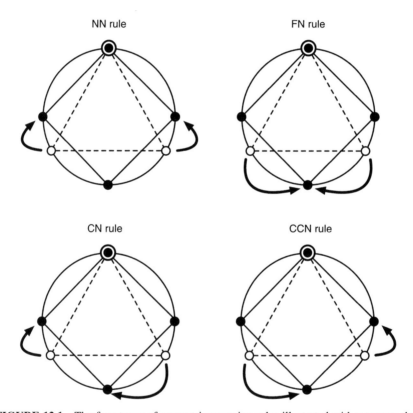

FIGURE 12.1 The four types of geometric snapping rules illustrated with a ternary three-pulse rhythm (dashed lines connecting white circles) binarized to a four-pulse binary rhythm (solid lines connecting black circles): the nearest neighbor rule (NN), furthest neighbor rule (FN), the clockwise neighbor rule (CN), and the counter-clockwise neighbor rule (CCN). Onsets that lie on locations where the ternary and binary lattice points overlap do not change position. A mutation may result in reducing the number of onsets in the cycle, as in the FN rule of the given example.

the onsets of the ternary rhythm move to the positions of the resulting binary rhythm. In the nearest neighbor rule (NN), ternary onsets move to their nearest binary lattice points. In the furthest neighbor rule (FN), ternary onsets move to their furthest adjacent binary lattice points. In the clockwise neighbor rule (CN), ternary onsets move to their nearest clockwise binary lattice points. In the counter-clockwise neighbor rule (CCN), ternary onsets move to their nearest counter-clockwise binary lattice points. (It is worth noting that this latter type of transformation has also been used by Jack Douthett to model dynamical voice leading in the pitch domain [9].) Thus the onsets of ternary rhythms positioned at "noon" do not move, since they are already in the locations of the binary rhythms. As an example, consider the three-pulse, three-onset ternary rhythm (x x x). With the NN, FN, CN, and CCN rules, this rhythm is mutated to (x x . x), (x . x .), (x . x x), and (x x x .), respectively. Although these four mutation rules have been selected for study purely on the basis of geometry and simplicity, they characterize most of the binarizations tabulated by Pérez Fernández [20] and may be justified to a greater or lesser extent by the Gestalt principles of perception, as discussed further in the following sections [25].

12.3 SIMILARITY-BASED RHYTHM MUTATIONS

One of the basic principles of Gestalt psychology of perception is that stimulus proximity causes perceptual grouping [25]. Therefore, one hypothesis is that the binarization process is determined by the nearest neighbor rule. The application of the 4 mutation rules to the 12-pulse, 5-onset *fume-fume* [x . x . x . . x . x . .] is shown in Figure 12.2. Note, however, that the nearest neighbor rule does not yield the clave son as its offspring, which is produced instead by the clockwise neighbor rule. Does this imply that the nearest-neighbor geometric binarization process fails as a perceptual model? Before discarding the NN-rule model altogether it should be pointed out that for the ternary rhythm it is common practice among the practitioners to begin the rhythmic cycle at different starting points [2, 27]. Furthermore, from the mathematical viewpoint, there exists an equivalence relation between the nearest integer of a number, and rounding that number down to the next integer. Indeed, rounding a real number x to its nearest integer is equivalent to rounding down the real number $x + 0.5$. This raises an interesting question: if the ternary fume–fume rhythm is continuously rotated through a full 360° revolution, thus taking on all the infinite possible positions, and at each moment in time the binarization is carried out using the nearest neighbor rule, how many different binarized rhythms are generated by this process, modulo rotations? Enumeration of all the possibilities reveals that there exist only three such different NN-rule binarizations of the fume–fume, and these have inter-onset intervals (3-2-4-3-4), (3-3-4-2-4), and (2-3-4-3-4). Furthermore, the clave son is included in this group of three, raising the additional question: on what mathematical grounds should the clave son be selected from among these three candidates, as the *bona fide* binarization of the fume–fume? One potential limitation of the geometric snapping rules that generate these binarizations is that they make use of *local* rules. Therefore,

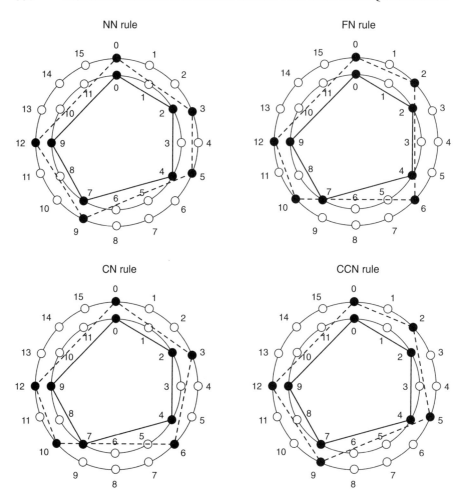

FIGURE 12.2 The four types of geometric snapping rules illustrated with the ternary fume-fume rhythm with inter-onset intervals (2-2-3-2-3). The binary 16-pulse circular lattice is on the outside, the ternary 12-pulse circular lattice on the inside. The fume-fume rhythm is the pentagon made up of solid lines. The binarizations are the pentagons with dashed lines. The nearest neighbor binarization rule (NN) does not yield the clave son, which instead results from the clockwise neighbor rule (CN). With all four snapping rules the first and last onsets in the cycle remain in their relative positions.

a natural alternative method for choosing the right binarization is to apply a global measure of similarity to select the one most similar to the fume–fume.

12.3.1 Global rhythm similarity measures

There are many ways to measure rhythm similarity [22, 28, 33]. One natural approach is to sum the discrepancies between all the corresponding pairs of onsets. For illustration, it is convenient to embed both the 16- and 12-pulse rhythms in the same lowest common multiple clock diagram as shown in Figure 12.3. However, the binary rhythm onsets must fall on every three marks starting at 0, and the ternary rhythm onsets must lie on every four marks. Figure 12.3 shows one possible rotation of the fume–fume rhythm (at marks 40, 0, 8, 20, and 28) along with its NN-rule binarization (at marks 39, 0, 9, 21, and 27). The distance of each onset of the fume–fume, on the lowest common multiple clock, from its corresponding clave son binary location is 0 for the onset at mark 0, and 1 for all the others, leading to a total distance of four units. On the other hand, the distance between the fume–fume and the two other binarizations (3-2-4-3-4) and (2-3-4-3-4) is only 3. Therefore, this distance measure fails to yield the clave son as the most similar binarization of the fume–fume.

The above measure of discrepancy is known as the *city-block* metric, *taxi-cab* distance, *Manhattan* metric, or the L_1 metric: it sums the absolute values of the individual differences between corresponding elements of two objects. Perhaps this is the wrong metric to distinguish the clave son from the other binarizations. There exist two other metrics often used to measure similarity: the *Euclidean* distance (or L_2 metric) and the *Maximum* metric (also *Sup* metric or L_∞ metric). The Euclidean metric takes the

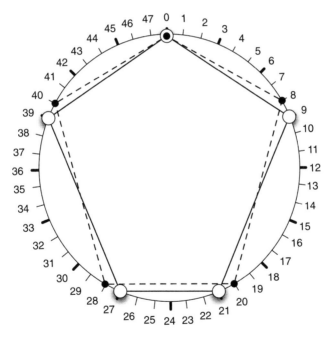

FIGURE 12.3 The fume–fume rhythm, on a clock diagram with a number of pulses equal to the lowest common multiple of 12 and 16, starting at pulse 40. The bold tick-marks (every three units: 0, 3, 6, 9, etc.) indicate the 16 locations for the onsets of the binary rhythms. The ternary rhythm onsets lie on every four units: 0, 4, 8, 12, 16, 20, etc. The fume–fume rhythm polygon is shown in dashed lines, and its nearest-neighbor binarization (the clave son) in solid lines. One corresponding pair of onsets overlap at pulse 0, and the other four are at distance one unit apart. Therefore, the distance between the two rhythms is 4 units.

square root of the sum of the squares of the individual differences between corresponding elements, and the L_∞ metric takes the maximum of these differences. In the case of musical temporal and pitch features, Tenney and Polansky argue that the city-block metric is to be preferred [25]. In the context of visual perception, Sinha and Russell found the city-block metric to be superior to the Euclidean metric [24]. In the present application, both the Manhattan metric and the Euclidean distance fail to produce the clave son. This leaves the L_∞ metric for consideration. In the context of geometric reconstruction problems, it was found that the L_∞ metric was more unstable than the Euclidean distance [12], and indeed, in the present application, all three binarizations are tied with distance 1 from the clave son.

12.4   CONCLUSION

The failure of the purely geometric models described above suggests that additional musicological knowledge may be required to single out the clave son as the proper binarization of the fume–fume from the perceptual viewpoint. A natural candidate source for this information is the underlying meter [35]. One possible hypothesis is that an onset is snapped to a neighboring position of greater metrical value. Referring to Figure 12.2, the strongest metrical locations for the binary rhythm occur at pulses 0, 4, 8, and 12. The onsets of the fume–fume at pulses 2 and 4 do not snap to the binary pulse 4. The next strongest metrical positions occur at pulses 2, 6, 10, and 14. Snapping the onset at pulse 4 of the fume–fume to pulse 6 of the binary cycle yields the correct position but the onset at pulse 2 of the fume–fume goes to the incorrect pulse 2 location on the binary cycle yielding the same binarization as the furthest-neighbor rule.

Another important musicological property of the fume–fume and the clave son is that they both start with two equal adjacent inter-onset intervals: two-pulse and three-pulse intervals, respectively, but neither interval corresponds to the regular beats [3-3-3-3] and [4-4-4-4] of their respective assumed meters. Of the three binarizations, only the clave son has this property, which could be used to break the tie obtained with the L_∞ metric. However, this leads to a rather complicated model of the perceptual process compared to the simpler clockwise nearest neighbor rule. The former also has the awkward feature that in order to compute the maximum of the five onset discrepancies, the listener has to wait until the entire rhythm is heard. By contrast, the simpler second model allows the listener to project the anticipated binary locations as the rhythm unfolds. Which of these two geometric approaches best models the underlying perceptual mechanism will have to be determined by psychological experimentation.

Acknowledgment

This research was supported by a grant from the Provost's Office of New York University Abu Dhabi, through the Faculty of Science, in Abu Dhabi, the United Arab Emirates. The author thanks the reviewers for providing helpful comments.

REFERENCES

1. K. Agawu, "Structural analysis or cultural analysis? Competing perspectives on the "standard pattern" of West African rhythm," *Journal of the American Musicological Society*, **59**, 1 (2006).

2. W. Anku, "Circles and time: A theory of structural organization of rhythm in African music," *Music Theory Online*, **6** (2000).

3. M. I. Arlin, "Metric mutation and modulation: The nineteenth-century speculations of F.-J. Fétis," *Journal of Music Theory*, **44**, 261 (2000).

4. S. Arom, "L'aksak: Principes et typologie," *Cahiers de Musiques Traditionnelles* **17**, 12 (2004).

5. S. Brown, B. Merker, and N. L. Wallin, "An introduction to evolutionary musicology," in *The Origins of Music*, Ed. N.L. Wallin, B. Merker, and S. Brown, MIT press, Cambridge, MA, pp. 3–24, 2000.

6. M. J. Butler, *Unlocking the Groove: Rhythm, Meter, and Musical Design in Electronic Dance Music*, Indiana University Press, Bloomington and Indianapolis, p. 93, 2006.

7. G. Cooper and L. B. Meyer, *The Rhythmic Structure of Music*, University of Chicago Press, Chicago, 1960.

8. P. Desain and H. Honing, "The formation of rhythmic categories and metric priming," *Perception*, **32**, 341 (2003).

9. J. Douthett, "Filter point-symmetry and dynamical voice-leading," in *Music Theory and Mathematics: Chords, Collections, and Transformations* (Eastman Studies in Music), Ed. J. Douthett, M. Hyde, and C. Smith, University of Rochester Press, Rochester, p. 72, 2008.

10. N. Fracile, "The "aksak" rhythm, a distinctive feature of the Balkan folklore," *Studia Musicologica Academiae Scientiarum Hungaricae*, **44**, 197 (2003).

11. F. Gómez, I. Khoury, J. Kienzle, E. McLeish, A. Melvin, R. Pérez-Fernández, D. Rappaport, and G. T. Toussaint, "Mathematical models for binarization and ternarization of musical rhythms," in *BRIDGES: Mathematical Connections in Art, Music, and Science*, Ed. R. Sarhangi and J. Barallo, Tarquin Publications, San Sebastian, p. 99, 2007.

12. R. I. Hartley and F. Schaffalitzky, "L-infinity minimization in geometric reconstruction problems," in *Proceedings of the IEEE Conference on Computer Vision and Pattern Recognition*, Washington, DC, June 27–July, 2, 2004.

13. C. F. Hasty, *Meter as Rhythm*, Oxford University Press, Oxford, 1997.

14. G. Kubik, "Analogies and differences in African-American musical cultures across the hemisphere: Interpretive models and research strategies," *Black Music Research Journal*, **18**, 203 (1998).

15. G. Kubik, *Africa and the Blues*, University of Mississippi Press, Jackson, 1999.

16. K. Lemström and A. Pienimäki, "On comparing edit distance and geometric frameworks in content-based retrieval of symbolically encoded polyphonic music," *Musicae Scientiae*, **4A**, 135 (2007).

17. F. Lerdahl and R. Jackendoff, *A Generative Theory of Tonal Music*, MIT Press, Cambridge, MA, 1983.

18. Y. Liu and G. T. Toussaint, "Mathematical notation, representation, and visualization of musical rhythm: A comparative perspective," *International Journal of Machine Learning and Computing*, **2**, 28 (2012).

19. M. Mongeau and D. Sankoff, "Comparison of musical sequences," *Computers and the Humanities*, **24**, 161 (1990).

20. R. Pérez-Fernández, *La Binarización de los Ritmos Ternarios Africanos en América Latina*, Casa de las Américas, Havana, 1986.

21. R. Pérez-Fernández, "El mito del carácter invariable de las lineas temporales," *Transcultural Music Review* **11**, 1–13 (2007).

22. L. Polansky, "Morphological metrics," *Journal of New Music Research*, **25**, 289 (1996).

23. O. Post and G. T. Toussaint, "The edit distance as a measure of perceived rhythmic similarity," *Empirical Musicology Review*, **6**, 164 (2011).

24. P. Sinha and R. Russell, "A perceptually-based comparison of image-similarity metrics," *Perception*, **40**, 1269 (2011).

25. J. Tenney and L. Polansky, "Temporal Gestalt perception in music," *Journal of Music Theory*, **24**, 205 (1980).

26. E. Thul and G. T. Toussaint, "A comparative phylogenetic analysis of African timelines and North Indian talas," in *Proceedings of the 11th BRIDGES: Mathematics, Music, Art, Architecture, and Culture*, Ed. R. Sarhangi and C. Sequin, Leeuwarden, the Netherlands, July 24–28. Tessellation Publishing, p. 187, 2008.

27. G. T. Toussaint, "Classification and phylogenetic analysis of African ternary rhythm timelines," in *Proceedings of BRIDGES: Mathematical Connections in Art, Music, and Science*, University of Granada, Granada, Spain, July 23–27, p. 25, 2003.

28. G. T. Toussaint, "A comparison of rhythmic dissimilarity measures," *FORMA*, **21**, 129, (2006).

29. G. T. Toussaint, "The rhythm that conquered the world: What makes a 'good' rhythm good," *Percussive Notes*, **2**, 52 (2011).

30. G. T. Toussaint, "The edit distance as a measure of rhythm complexity," in *Proceedings of the 2nd Stochastic Modeling Techniques and Data Analysis International Conference*, Chania, Crete, Greece, June 5–8, 2012.

31. G. T. Toussaint, *The Geometry of Musical Rhythm*, Chapman-Hall-CRC Press, Boca Raton, 2013.

32. G. T. Toussaint, "Phylogenetic tools for evolutionary musicology," in *Mathematical and Computational Musicology*, Ed. T. Klouche, National Institute for Music Research, Berlin, Springer, 2014.

33. G. T. Toussaint, M. Campbell, and N. Brown, "Computational models of symbolic rhythm similarity: Correlation with human judgments," *Analytical Approaches to World Music Journal*, **1** (2011).

34. M. L. West, *Ancient Greek Music*, Clarendon Press, Oxford, 1992.

35. W. L. Windsor, "Dynamic accents and the categorical perception of metre," *Psychology of Music*, **21**, 127 (1993).

INDEX

*Mathematical and Computational Modeling: With Applications in Natural and Social Sciences,
Engineering, and the Arts*, First Edition. Roderick Melnik.
© 2015 John Wiley & Sons, Inc. Published 2015 by John Wiley & Sons, Inc.

*SIEGEG—Topics in Complex Function Theory
     Volume 1—Elliptic Functions and Uniformization Theory
     Volume 2—Automorphic Functions and Abelian Integrals
     Volume 3—Abelian Functions and Modular Functions of Several Variables
SMITH and ROMANOWSKA—Post-Modern Algebra
ŠOLÍN—Partial Differential Equations and the Finite Element Method
STADE—Fourier Analysis
STAHL and STENSON—Introduction to Topology and Geometry, Second Edition
STAHL—Real Analysis, Second Edition
STAKGOLD—Green's Functions and Boundary Value Problems, Third Edition
STANOYEVITCH—Introduction to Numerical Ordinary and Partial Differential Equations
    Using MATLAB®
*STOKER—Differential Geometry
*STOKER—Nonlinear Vibrations in Mechanical and Electrical Systems
*STOKER—Water Waves: The Mathematical Theory with Applications
WATKINS—Fundamentals of Matrix Computations, Third Edition
WESSELING—An Introduction to Multigrid Methods
†WHITHAM—Linear and Nonlinear Waves
ZAUDERER—Partial Differential Equations of Applied Mathematics, Third Edition